REASON, REGULATION, AND REALISM

SUNY Series in
Philosophy and Biology

David Edward Shaner, Editor

C. A. Hooker

REASON, REGULATION, AND REALISM

Toward a Regulatory Systems Theory of Reason and Evolutionary Epistemology

STATE UNIVERSITY OF NEW YORK PRESS

Production by Ruth Fisher
Marketing by Bernadette LaManna

Published by
State University of New York Press, Albany

Printed in the United States of America

For information, address the State University of New York Press, State University Plaza, Albany, NY 12246

Library of Congress Cataloging-in-Publication Data

Hooker, C. A. (Clifford Alan), 1942–
Reason, regulation, and realism : toward a regulatory systems theory of reason and evolutionary epistemology / C.A. Hooker.
p. cm.—(SUNY series in philosophy and biology)
Includes bibliographical references and index.
ISBN 0-7914-2261-5 (alk. paper).—ISBN 0-7914-2262-3 (pbk. : alk. paper)
1. Knowledge, Theory of. 2. Naturalism. 3. system theory.
I. Title. II. Series.
BD181.H66 1995
121—dc20 94-5769
CIP

10 9 8 7 6 5 4 3 2 1

To Newcastle, Australia, for 10 good years: swimming in crystalline dawn-fired ocean on the Pacific rim, watching bowerbird mating dances in our bush garden, lunching among the wineries at the foot of the Broken Back range, walking in one of the oldest rainforests on earth, finding people who are loving and a community that is sane.

And to that salt-of-the-earth minority who quietly work out of love to raise the probability a smidgin above 0 that our grandchildren will enjoy such privilege.

CONTENTS

ACKNOWLEDGMENTS

The essays that form chapters 4–6 of this book were drafted during 1986–1989. During that time I benefited from discussions with Kai Hahlweg on evolutionary epistemology and with Norton Jacobi on Everything (from Buddhist trance states to limbic system function). These chapters were written into a first book draft during a research leave in 1990 during which I was the recipient of generous material support, scholarly criticism, and friendship, all from the department of philosophy, University of Western Ontario; I am especially grateful to Bob Butts, department chairman, for that. Chapters 1–3 were drafted in 1991 and the whole re- written over 1992/3. During that time I have benefited (and sometimes suffered) from critical feedback on the drafts and am especially grateful for detailed critiques of specific chapters to Hal Brown (1, 2, 5, 6), John Collier (2), Bill Herfel (2), Barry Hodges (co-writer, 3), Norton Jacobi (2, 6), Richard Kitchener (5), Ausonio Marras (5), David Naor (6), Bruce Penfold (2), and Ralph Robinson (2). During 1991–2 Bill, Norton, Bruce, Ralph, and I formed a wonderfully irreverent discussion group on these matters, joined by Hal and John whenever they were in town. The environment for all this activity is dominated by two key women in my life, each independent professionals, each irreplaceable, demanding, critical, funny, affectionate, and altogether bloody marvellous at providing realism, correction, sustenance, and support in magically right ways: Jean Hooker, my wife of 30 years, and Dorrit Nesmith, my secretary of 10 years. To all of this long list: thank you.

Research for this book has been conducted under Australian Research Council grants A18/(9)416–58, AC9031991, which are gratefully acknowledged from a small country in a big fierce world. I have appreciated the consistent and generous research support the University of Newcastle has provided over the years, in particular, for 1990 study leave support. I am grateful to the specific people who lie hidden beneath this institutional abstraction and who help to create an institutionalized life with a human face (or anyway offset the worst of the administrivia and bureaumania).

Chapter 1

Setting the Scene: Naturalism and the Prospects for Evolutionary Epistemology and Reason

Introduction

The business of philosophy, as I conceive it, is providing a general understanding of the nature of life and of how to live it. We humans go about understanding some facet or domain of life by creating theories about it, testing these out in experience, and reflecting on all this, we slowly elaborate both a conception of the domain in question and a conception of how we know (what we think we do know) about it. I shall describe this process as theorizing a domain or subject matter. Self-reflexively this is the naturalist theory of understanding itself, which I employ.

This is a rather radical naturalist conception of cognition, and of philosophy in particular—just how radical will be seen during the course of its defense in section 1.I below. But first the major aims of this book are briefly introduced below and placed in a larger research context. Subsequently, as remarked, an evolutionary naturalist realist philosophy is outlined and argued and, following on from that, a correlative approach to an evolutionary epistemology (section 1.II). These discussions are necessarily somewhat condensed and programmatic, taking their meaning and force from large bases of research and argument that cannot be explicitly reproduced here. Readers unfamiliar with these may like to read this chapter lightly now and return to it after having read the remainder of the book.

One of the immediate consequences of my naturalism is that philosophy should be deeply engaged with all areas of theoretical and practical understanding. For neither the problems nor the (best conjectural) solutions are given in advance, but only emerge from actual historical engagement. And that brings me to the first and most important theme of this book.

We are today, I believe, at the beginning of an important and fundamental revolution in the conceptual foundations of all the sciences, one with important consequences also for the professions: the shift from linear, reversible, and compositionally reducible mathematical models of dynamics to nonlinear, irreversible, and functionally irreducible complex dynamic systems models, especially for complex adaptive systems (which include all living systems). This claim is not my own; a substantial part of it has been spelled out by Jantsch 1980 and Prigogine and Stengers 1984, among others, complemented by Dyke 1988. As little as a decade ago its character and consequences were discussed only by a handful of the mathematically knowledgeable; today there is a veritable explosion of new literature pursuing these ideas in every field from irreversible thermodynamics and chaos theory in physics through engineering control theory and complex adaptive systems theory in computer science, self-organization and hierarchy theory in biology, and dynamical neural net theory and genetic algorithms in neurophysiology and cognitive psychology, to evolutionary economics and international relations, with increasing cross-application among these. It would be neither practical nor desirable to attempt to review these exciting and increasingly interrelated technical fields here; the interested reader is directed to the asterisked items in the references for an introduction. They provide what the pioneers of the general systems literature of the 1960s and 1970s (Banathy, von Bertalanffy,

Margalef, et al.) could not provide: the beginnings of detailed, powerful models of nonlinear self-organizing complex adaptive dynamic systems. (Cf. Margalef 1968 or Odum 1971 with Caplan/Essig 1983, Hannan/Freeman 1989, Peacocke 1983, Weber et al. 1988.) I am a long way from having understood and integrated the flood of systems literature of the past two decades, and with some works that are seminal and popular I nonetheless await the outcome of further examination (see the sequence Brooks/Wiley 1986, Morowitz 1986, Collier 1986, and Hooker 1984 on Prigogine 1980). In this book I am only concerned with trying to develop the right kind of philosophical theory; a detailed exploration of the accompanying science is beyond its scope. (That hard work belongs to a future book and to you, dear reader.)

For all its value and conceptual importance, the ideas deriving from this revolution have as yet scarcely touched philosophy. For the most part, philosophers still model rational agents (explicitly or by tacit presumption) in terms of simple logical structure, and the whole of science likewise. Individual rationality and rational scientific method alike, for example, are standardly seen as reducible to logical inference of some kind; the psychology and sociology (economics and politics) of decision making are at best formally irrelevant, merely part of the "implementation" of the formal program; at worst are the sources of causally interfering factors reducing rationality. But those who think in these terms are effectively locked into the dominant analytic philosophy paradigm—philosophy as logical analysis—inherited from the positivists and logicists early this century and expressed in the artificial intelligence (AI) model of cognition from the 1970s, the simple formal logical programming machine. There are various reasons why this is so, some of them deriving from evidence and the attraction of simplicity, some deriving from presumptions about the nature and status of normative principles, and no doubt some deriving from the attractiveness to a discipline of having clear ownership of a domain. (This latter is actually based on illusion, for today much of the creative work in logical theory comes from mathematics and computer science, not philosophy—further grist to my naturalist mill.)

Whatever the reasons, accepting this dichotomy has had unfortunate consequences. Theoretically it either falsely places psychosocial scientific explanation in a category (causality) wholly distinct from that of philosophical epistemology (logic) with the practical consequence that the two kinds of practitioners often regard themselves as offering competing rather than complementary accounts, or it reinforces the constraint to formal symbol manipulation (logic/language of thought) among cognitive psychologists and philosophers, dividing them from other practitioners. But scientists acting rationally are thereby acting socially, and cognition is far more than narrow formal symbol manipulation. One upshot is that at present many are falsely led from a critique of narrowly formalist logical/AI theory to the conclusion that rational epistemology is bankrupt. The theoretical result has been rampant relativisms, a confusion of context sensitivity with context incommensurability, and a profusion of supposed social replacement models ("conversations," "alliances," "stakeholders," etc.); this is accompanied by the practical danger that what we have of reasoned public life will be replaced by demagoguery and narrowly self-interested force.

Instead of leaping for an easy or politically correct relativism, I suggest that we must search for a better theoretical framework providing a richer class of models of action. Specifically, *the working hypothesis for this book is that the currently most fruitful and viable framework is that of dynamic nonlinear self-organizing complex adaptive systems*. The general idea is to reconstruct philosophical ideas, in particular rationality and epistemology, in terms of characteristic adaptive processes in such systems. In this conception, roughly, people are represented as strategic adaptive systems pursuing complex goals, and social processes are understood in terms of the self-organizing dynamics of many-person systems. To this end I propose that we adopt a strategic or decision theoretic conception of cognitive agency whose basic component is the epistemic utility increasing strategic decision in response to a problem posed in a particular decision context. This allows the explicit introduction of problem context to cognitive theory and so an explicit role for social structure, in particular a central role for the institutionalized social structure of science in scientific rationality. And it imports the decision theoretic framework of social context-dependent strategic interaction among rational agents as a basis for a dynamics for science. The complex interactions within these systems include both belief and goal formation and re-formation and structuring and re-structuring of roles/processes. All these processes occur within and between all system levels from subindividual to whole-society and now to whole-species, and they derive from interactions both within and between all system levels.

In this way science can be modeled as a dynamic self-organizing complex adaptive process embedded in wider social and biological complex adaptive systems, and the specifically cognitive is to be recharacterized as (roughly) the information-extracting and organizing aspect of these dynamics, not as either a separate "level" or part of it. Rationality can then be re-theorized in terms of the (self-) organizational properties of such systems; it includes, for example, the capacities to systematically collect and evaluate environmental information of various kinds and to systematically adapt context-specific goals as empirical understanding accumulates. Epistemology can be characterized in the same way, for example, in terms of the satisfaction of various systematically constructed and reconstructed inquiry procedures and the resulting social construction of statements and procedures invariant across various systematic contexts. In this way, I aim to employ our best current scientific models of living processes to provide a principled reconstruction of our conception of science. On the one hand, this conception is to understand science as a specialized case of biological dynamics generally. On the other hand, it is to be one that recognizes the essentially social character of science and provides for a cognitive sociology of science—though without recognizing sociological study as privileged or foundational vis-à-vis other disciplines, and while retaining normative epistemology. *The main aim of this book is to argue an initial case of this kind for the philosophical concepts of reason and epistemology and to show how the reconceptualization might fruitfully be developed.*

Some philosophical work of this kind has already begun. Hooker 1981b placed the concept of reduction within a complex systems framework and distinguished between compositional and functional reduction. That distinction, crucial to discussions of reduction in biology and the social sciences, still does not seem to be very widely understood. (In philosophical biology, for example, intermediate or interfield theories, such as operon models of aspects of genetic regulation, are urged against reduction; they certainly do show the inadequacy of a simple compositional/type reduction, but from a regulatory perspective they simply correspond, for example, to modular functional analysis in electrical circuit theory or cybernetics, all of which are strictly compatible with ontological deflation.) Cummins 1983 provides a thorough analysis of the philosophical concept of functional explanation along these same lines, extending systems functional analysis to cognitive psychology, artificial intelligence, and the relation of both to neurophysiology. (See Hahlweg/Hooker 1989a Part IV for a summary of Cummins's work and its marriage with the theory of systems reduction; this account is briefly sketched in chapter 2, section 2.III.1 below.) Over the past two decades, Paul Churchland has developed naturalized analyses of cognitive concepts (such as that of cognitive agent as epistemic engine), which, although not explicitly set in a systems metaphysics, are particularly appropriate for integrating therein (see Churchland 1979, 1989), and Pat Churchland has similarly produced a major reevaluation of the fundamental concepts appropriate to understanding and integrating a neurophysiological theory of mind (Churchland 1986). And Giere has introduced simple decision theoretic models of scientific decisions and related them to an account of theories themselves as models to yield a much richer conception of scientific method (Giere 1979, 1988), but again without much attention to embedding this account within dynamic systems. However, recently the philosophical specialization known as evolutionary epistemology has proven a fruitful locus for the introduction of functional systems ideas to epistemology (see especially Campbell 1974, 1986, 1987, and generally Hahlweg/Hooker 1989b). Indeed, in the form I shall give it in section 1.II below, its task is precisely to understand science as a dynamic process. So, *a second major aim of this book is to provide a re-working of evolutionary epistemology in the new systems framework.*

These researches are only a few of many that could be mentioned; they are simply examples where I have knowledge derived from personal involvement. My own attempts to begin thinking in these terms were first stimulated by reading Piaget in the late 1960s. But there was then a dearth of good scientific models (it was at the dawn of the discovery of chaos and before the seminal works by Morowitz 1968 and Odum 1971 appeared), and Piaget was under attack from all sides (partly for good reasons, but often, it turns out, for bad reasons; see chapter 5). So I turned instead to an elaboration and defense of the kind of general philosophical naturalist realism that I believed (correctly) would be required as a defensible framework for the rethinking of basic ideas when it became more fruitful to do so (see Hooker 1987, chapters 2--5, originally published 1974–1976). The limitations of empiricism were clear, and I was preoccupied with generalizing its critique to a matching analysis of Popper, Feyerabend, and others. An extension of that naturalist program, provides *the third major aim for this book: to show how an adequate philosophical conception of naturalizing epistemology and reason can be developed.*

By the early 1980s the required systems revolution had recognizably begun (see further below). A paper like Hooker 1987, chapter 7 ("Understanding and Control"), became writable. The wheel has now come full circle, and the conceptual tools now exist to understand Piaget and to integrate his work into the larger philosophical and scientific enterprise (chapter 5 and Hooker 1994d). The common problems of empiricism and Popper have also come into correlative focus as centering on the formalist logical models they employ, and it is now possible to provide a pointed critique of

Popper's evolutionary epistemology as well as his general philosophy and that of Feyerabend in a precisely complementary manner (Hooker 1991a). Indeed, it becomes possible to extract from Popper the beginnings of a very different theory of science from his official one, one suited to integration into the dynamic systems framework (chapter 3). In this way the philosophical rethinking presented in this book is self-reflexively a model of the developmental systems processes described therein. One can only hope that the result here also is widening theoretical adequacy and cognitive autonomy.

An enormous amount of work, mathematical, and scientific, but especially philosophical, still needs to be done on the fundamental regulatory systems concepts, for example, on the notions of functional versus compositional entropy and their relation to Shannon/Weaver information and to semantic information. Indeed, these and other fundamental concepts, such as self-organization, are as yet relatively poorly understood (not surprising in a young research field) and in urgent need of conceptual analysis. So the reader is not going to find a detailed theory of cognitive dynamics in this book. At this time we have neither the theorems nor even maturely developed appropriate conceptions of reason and cognition. Yet we are in a better position than our mentors, who nonetheless succeeded in providing us with key ideas: that cognitive dynamics is the key problem and is driven by knowledge begetting problems that disturb the present situation and so lead to new knowledge, that the connection of internal truth criteria to external success is complex but central for an adequate account of epistemic progress, that progress is to be linked to competence in a wider range of environments, and that it is also linked to self-organization and autonomy since it corresponds to a capacity to preserve internal operations invariant across a wider range of environments. Some or all of these themes are already in Popper and Rescher and especially Piaget (see quotes PQ1, 2, 19, 20, 22, 31 from chapter 5), and they form the backbone of the systems evolutionary epistemology espoused in Hahlweg/Hooker 1989a. They also have a long history in the scientific literature (cf. Margalef 1968, Buss 1987, Smith 1992, and their references). We do them most justice by trying to reunderstand their significance in the light of contemporary advances in knowledge.

Science is a complex system; because of that complexity we cannot reasonably expect detailed but general theorems about cognitive systems dynamics even when the subject has matured, if by that is meant being able to deduce from general principles precise behavior in individual contexts. Until recently established science had developed satisfactory detailed models only for systems with either just one or two components or with large numbers of components interacting randomly so that local detailed fluctuations could be ignored, whereas we are dealing with systems of many components all interacting nonrandomly in complex ways. While we have recently gained some insight into these latter systems, it has basically been insight into their general dynamical features; to obtain detailed models requires specialization to the contextual details case by case. Consider this simpler parallel: No one doubts that quantum mechanics applies to bridges, yet no rational engineer tries to write down Schrodinger's equation for a bridge as a basis for its engineering design. Rather, engineers continue to rely on simpler approximate models while using quantum theory selectively to shape general concepts (e.g., metal fatigue and surface corrosion), to predict relevant kinds of structural features (e.g., the role of impurities in tensile strength), and in this way reshape and place refined limits on the applicability of the simpler models. We should expect to deploy complex adaptive systems models in a similar way. Traditional logic/AI models of science were apparently detailed, but only at the cost of wiping out nearly all relevant contextual and goal structure, reducing all scientists to ciphers for an abstract rational mind. Complex adaptive systems models will be less sweeping, but only because they recognize a greater wealth of detail as relevant.

The research presented in this book is part of a larger philosophical program of re- conceptualization, which can be conveniently summarized in this way, where S-ORS refers to self-organizing regulatory systems:

LOGICAL CATEGORY	MAIN TASK
1. General Philosophical Doctrine	General Theory of the Nature of Reason
2. Specific Philosophical Doctrine	Reason as S-ORS
3. Psychosocial Implementation	Cognitive Psychology of Rational Processes; Theory of Rational Institutional Designs and Processes
4. S-ORS Models	Specific Models of Cognitive, Institutional Processes in S-ORS Terms
5. S-ORS Theory	General Theory of S-ORSs

Hooker 1991a focused first on level 1 theory, outlining the argument for the failure of formal reason in philosophy of science, and then turned to sketch a systems alternative at level 2. The latter is taken up in this book, which is focused at levels 2 and 3. The former level 1 project is to be taken up in Hooker 1994c and at length in Brown/Hooker 1994, which also develops a richer general philosophical account of reason with which to replace the formalist version. Other work at levels 3–5 is also underway on biologically and cognitively implementable control and cognitive modeling (Hooker et al. 1992a, b; Hooker 1994e), on cybernetic modeling vis-à-vis S-ORS models (cf. variously Churchland 1986; Churchland 1989; Cunningham 1972; Lloyd 1989; Powers 1973; Thagard 1988 with Jantsch 1981: Jantsch/Waddington 1976: Pines 1988: Stear 1987: Yates 1987) and on foundational concepts of S-ORS, but further discussion will be left for another occasion. Let us be clear, the hardest work is yet to come. When readers finish this book, I would like to have persuaded them that a naturalized regulatory systems approach is the most promising route into the future and that it has the support of much of the best of past theorists as well. But then it is necessary to create and investigate detailed working models of biology, cognition, social dynamics, AI, and engineering control, and so on, so that the promise of the ideas is put to the test. That exacting task, though well under way in the sciences, is largely yet to come for philosophers.

The basic naturalist realist philosophical and metaphilosophical framework within which this book is formulated is set out in the next section below, followed by a brief sketch of the advocated approach to evolutionary epistemology. The former is necessary because naturalist realism has radical consequences for the formulation of the philosophical enterprise itself and readers need to be clear about these before proceeding. (Self-reflexively, it is itself a fallible theory, but one for which the analyses of this book provide heartening support and promise of future fruitful engagement with science. In another context, the views developed here can be used to defend the account against relativism; see Hahlweg/Hooker 1988.) The latter is undertaken because I believe that the currently dominant models of evolutionary epistemology, though they do move traditional epistemology one step in the right direction, are not themselves appropriate for proper integration within a dynamic systems framework and it is necessary to set out my own account for the reader.

Following this, chapter 2 provides a first sketch of a dynamic systems account of science. This is a first attempt at theory construction, an attempt to develop enough concepts for the description of science as a dynamic system to motivate and direct further developing the approach. There is an enormous amount of work yet to be done, both by way of incorporating detailed systems models into the analysis on the one side and by co-opting useful philosophical analyses into the model on the other, and finally by working out more detailed historical applications to test the ideas. But I believe that there is already presented here a sufficiently substantive framework to be able to limn the outlines of a new paradigm for philosophy of science that can be presented as a principled alternative to the standard logicist/AI paradigm. In particular, after laying out some essential preliminaries, concepts of objectivity and scientific progress are developed in terms of the unfolding of regulatory systems organization. It is not, note, an attempt to develop an alternative normative epistemology in which the old logical rules (e.g., of induction) are replaced by some equally simple formal rules connecting the new concepts. Rather, normative epistemology is to be redeveloped within this dynamic account by relating norms to the fundamental capacities of the system processes to extract and systematize information. As noted earlier, epistemic norms can be characterized in terms of the satisfaction of various systematically constructed and reconstructed inquiry procedures together with the resulting social construction of statements and procedures invariant across various systematic contexts. From various angles most of the remainder of the book after chapter 2 is devoted to exploring this approach. What emerges is a conception of science as an abstraction from—an aspect or dimension of, but *not* a part of—a highly complex, highly interactive dynamic system of nested subsystems ranging all the way from internal neural organization to social institutions. As noted an important consequence of this is the capacity to provide a principled cognitive sociology of science, as epistemic system design; this provides the proper theoretical ground for those who wish to join the critics of traditional logicist/AI theories of science and yet do not wish to join the strong program sociological, Marxist, deconstructionist, Feyerabendian, and other relativisms in their flight from normative epistemology.

From this perspective it is satisfying to be able to take three evolutionary epistemologies regarded as friendly to rationalism, those of Piaget, Popper, and Rescher, and show how they provide fundamental insights for naturalized theories of evolutionary epistemology and reason. (Similarly, I argue in chapter 6 that Putnam's attempted rejection of naturalized reason only succeeds in providing desiderata for an adequate regulatory systems naturalization.) In Popper's case a thorough reconstruction from minority fragments is required, but required internally in any case by the failure of his theory on its own terms. Rescher's theory requires serious renovation but not reconstruction. Piaget requires only stripping and polishing to produce a vigorous naturalist regulatory systems framework. It is their amenability to naturalist reconstruction of the regulatory systems kind that accounts for their order of treatment. These three studies stand in their own right, and as well they contribute to developing a regulatory systems philosophy of science.

The point of chapter 3 is to take an important and well known philosophy (not to mention one from which I gained all my own initial conceptions of critical reason), namely Popper's, and use it to demonstrate (1) how metaphilosophical attachment to the logicist/AI model can defeat even the best intentions of an evolutionary epistemology and (2) how within someone as honest as Popper is in facing the functional reality of science one can still find and extract a more appropriate philosophical framework. A similar lesson applies exactly to Rescher, who is nearly unique among advocates of evolutionary epistemology in pressing the original formal analogy at least as far as incorporating one order of methodological dynamics. As with Popper, we may set aside the failures caused by attachment to the logicist/AI model and generalize the insights contained in his analysis to form an essential component of an adequate epistemology in a dynamic systems framework—see chapter 4. Indeed, there is a natural progression from the concluding analysis of Popper's alternative account of the control of decisions to Rescher's account, for the latter begins to furnish the control structure that Popper's rudimentary account left undeveloped.

And that brings us naturally to Piaget, where the Popper-Rescher account may be embedded in a still wider complementary framework. Piaget devoted his life to creating an integrated regulatory systems framework that would provide a unified account of knowledge from biological self-organization to the most esoteric reaches of science and mathematics. We shall be prepared by the preceding chapters to reread Piaget afresh in these terms and to appreciate the systematicity and fruitfulness of his conception of genetic epistemology. It incorporates an appropriate evolutionary epistemology within it as a special case and does so in a way that provides for a unified naturalist biologically based account. And as with Popper, we shall discover in Piaget a much neglected aspect, in this case a philosophically neglected account of the nature and formation of regulative ideals and their norms (now read literally as regulatory systems functions). Piaget's stripped and polished account, enriched by those of Rescher elaborated and the reread Popper, provides the foundations for a new paradigm of epistemology in a dynamic systems framework.

Epistemology is applied rationality (namely, rationality applied to information acquisition, whether sensory, revelatory, rationalist, or whatever). The notion of reason in the dynamic systems framework will have been running as an undercurrent throughout these chapters, so it is only appropriate that the book conclude with a study of the naturalization of reason and a discussion of its retheorizing in the new framework. It turns out that with naturalization appropriately understood, the allegedly strongest arguments against naturalizing reason instead simply provide so many desiderata for doing so. And the analyses of the preceding chapters provide a set of clues to how one might appropriately retheorize the nature of reason in terms of regulatory organization. This brings the initial tasks of philosophical rethinking, and with it the book, to a rounded close. The reader is reminded that, just as we are at the very beginning of a revolution in the foundations of the sciences, so we must be also at the very beginning of a correlative philosophical revolution. This book claims no more than to make a modest contribution to that beginning.

Terminology. Before proceeding it may prove helpful to insert some brief remarks about terminology: *Cognition* is normally understood to refer to the action and faculty of thinking, including perception and conception but excluding all else (e.g., feeling and volition). I don't want to be committed to these divisions; I believe that whole system evaluation (cognitive, moral, and aesthetic) is as essential to the knowing process as to all others. So first I need a term of wider reference encompassing the sum total of ways in which adaptive systems adapt, including marshaling feeling, volition, and evaluation (cognitive, moral, and aesthetic); for this, and despite its sometime usage in a narrowly cognitive sense, I shall use the term *intelligence*. Then I shall understand *cognition* in its broadest descriptive sense to refer to the thinking aspect or dimension of being intelligent, to the action and faculty of thinking, including perception and conception. Intelligence and cognition theories are descriptive, their construction is governed only by the norms for science in general. I shall take cognition to include public thought expressed in written, verbal, and non-linguistic forms (e.g., a laboratory exemplar) and so include science within its scope. Many sciences now contribute to a theory of intelligent systems; *psychology* I take to be the science of the actual intelligent systems historically available to us—living organisms, perhaps later to include robots—

whether or not their behavior is specifically cognitive. Psychology includes the psychology of intelligent behavior, and this in turn includes cognitive psychology.

By contrast with these terms, in its ordinary usage *knowledge* is a "success word," what is known must be true, and it comes weighed down with formal logical associations, the detritus of past attempts to guarantee that what in practice we rely on in science and common sense really is true. Naturalists who use the term have constantly to qualify it to mean just our currently most reasonable, but fallible, assertions—about whatever topic, including knowledge. *Epistemic* action is intelligent action when directed toward knowing and understanding. Its theory, epistemology, includes normative cognitive science (as much cognitive science tacitly is) and its structure as well as its construction is governed by the epistemic norms for knowing and understanding (i.e., by those for science generally). For naturalism epistemology is also a fallible descriptive science (see below).

The basic terms of systems science, such as *system*, *feedback*, *function*, and *stability* are more or less well established and will simply be used herein. Briefly, a system S is stable in parameter P, for example, if the system response to perturbations in the value of P always returns P to its original value and maintains it there. A refined quantitative analysis would introduce the range of perturbations over which stability holds, the limit to within which the system returns to its original value, the time it takes to do so, and the range of transient states it occupies while doing so. But I forego any further analysis here. Other terms, however, remain either vague or ambiguous or controversial, and it is useful to indicate how I use them. By a *regulatory system* I mean a system so equipped as to stabilize those parameters and processes necessary to its continued existence. These constitute respectively its homeostases (e.g., body temperature) and homeorheses (e.g., red blood cell production, immune system antibody adaptation, and cognitive development).[1] Many such systems are highly dynamic in the sense that, rather than maintain a particular form invariant, their interactions are nonlinear and they go on changing in historically idiosyncratic ways. If the system dynamics promotes certain processes (i.e., increases the value of relevant parameters, such as stored environmental information or response strategies) then I shall call the systems *self-organizing*.

Systems are *adaptive* if their interaction with their environment is of a kind that the dynamical sequence of system states is a nontrivial systematic function of the state of the environment. Adaptive processes include both adaptation and adaptability, that is, (respectively) alteration of specific traits so as to increase fitness in their environment and alteration of these alteration processes (chapter 2). The kinds of adaptive processes that chiefly interest us for a study of intelligence are those occurring in self-organizing regulatory systems characterized by homeorheses. In what follows I shall sometimes refer to them as self-organizing complex adaptive systems and often more broadly as self-organizing regulatory systems. Systems of this kind, among others, typically show a temporal development or ontogenesis that is characterized by the selective extraction of ordered energy (negentropy) from their environment to produce simultaneous system expansion and internal differentiation. The growth and development of any living organism is an example. I shall refer to this overall combination of expansion and differentiation as *superfoliation*. A superfoliating adaptive system may show increases in either or both adaptation and adaptability.

Finally, a note on the concepts of regulatory order and system level. These are distinct. A *system level* is a relatively functionally isolatable component of total system function grounded in causal processes relatively causally buffered or screened from causal influences entering from elsewhere, for example, kidney or immune system function vis-à-vis the statistical behavior of their individual cells (see Collier 1988b). Typically, levels can be associated with a particular system scale (e.g., cellular or phenotypic), and then this gives them a quasi-spatial sense, but this is not always the case; for example, the immune system runs across the cellular and organ scales. By contrast *regulatory order* is concerned with the sequence of conditionalization among regulatory outputs. If an output of regulator B is causally determined by its input and an output of regulator A, but not vice versa, then A is a higher order regulator than B. Thus a homeostatic regulative output occurring is conditional on the prior operation of the corresponding homeorhetic regulative output occurring. It is possible to causally instantiate regulatory order so that level structure materially reflects order structure; this is, for example, how a business office is often arranged, with the chain of command literally reflected in the height above ground of offices. But it is not necessary to do this, complex regulatory order may be modeled, for example, by a tuned neural net that is all at the one level, and an effective secretary may function across many levels of an institution. (Cf. the discussion in Hooker 1994e.) The latter may prove to be the commoner situation, and all kinds of intermediate designs are possible. I shall try to respect the distinction between orders and levels, which is constantly blurred by the casual use of *level* to refer indiscriminately to regulatory orders proper, metalinguistic relations, degrees of generality, and so on. But where others use it ubiquitously, I retain this usage for it, trying to confine it to contexts where it does little harm. (This is the policy pursued with Popper in chapter 3.) I retain *level* also for referring to levels of physical quantities and the like, just as I retain the term *order* in "thermodynamic order" and "mathematical order"; these should cause no confusion in context.

1.I. Evolutionary Naturalist Realism: A Philosophical Outline

The guiding spirit of all my philosophical work has been naturalism. Roughly, this means regarding all aspects of life as part of a single natural world, and this includes philosophy itself. This idea is developed in Hooker 1987. It will be discussed only briefly below. Naturalism, pursued systematically, has rather radical consequences (at least by contrast with much contemporary philosophizing), for example, philosophy is to be treated as fallible theorizing, like science; formal logic is abandoned as the paradigm of rationality, and evolutionary epistemology may be treated as only a first crude approximation to a more system(at)ic theory of knowledge. The purpose of this part is to set out these and other consequences of naturalism systematically, for they form the framework for all that follows.

But first a brief illustration of the general orientation of the position. A standard challenge to realism is that the world itself cannot be inspected independently of human conceptually interpreted sense perception, so how can those notions that lie at the heart of realism, correspondence truth, laws of nature, scientific progress, especially in method, and so on be determined or even coherently specified? Surely we have to confine ourselves to our perceptual experience and whatever can be defined from within its resources. The initial premise is granted, but before we leap to the proffered conclusion, let us pause to reflect that science itself understands us very differently. Current scientific theory presents us as one evolved species among many inhabiting a world we did not make and only very imperfectly understand, finding our way about through the use of highly fallible sensory and motor capacities orchestrated by equally highly fallible theories we construct and are constantly forced to reconstruct as our experience extends across ever wider environmental conditions. From this point of view nothing is certain; cognition is as problematic as the world. So those who counsel that we are to assume instead that our sensory experience is our foundation face an embarrassing dilemma: Either science is hopelessly wrong, or their position is internally inconsistent (science constructed on sensory foundations contradicts the existence of those foundations). By contrast, naturalist realism presents a self-consistent account, scientific and philosophic, and this speaks for it. Further, it leads to a proper analysis of the senses as dynamic adaptive systems, which shows how and why the notion of foundational sensory data fails (cf. Hooker 1978; see the similar response to arguments for relativism in Hahlweg/Hooker 1988). Self-reflexively, this framework is itself a fallible theory, but one for which scientific and philosophical analyses of this kind provide heartening support and promise of future fruitful engagement with science. In this way naturalist realism presents a coherent intellectual account of us as fallible, (partially) self-organizing adaptive systems. Now to develop the general principles of this position in a little more detail. The overall argument is summarized in Diagram 1.1.

From a naturalist viewpoint, human knowledge itself is a natural phenomenon, a complex of individual and species capacities with particular origins and related distinctive characters, to be studied as any other natural phenomenon—with the added complication that such theories must be reflexively consistent. It follows that epistemology (indeed, philosophy at large) should form a coherent unity with science, a single self-consistent conception of us and our cosmos.

From this, realism follows. For science presents a well-confirmed conception of an independent external world and of ourselves as a relatively recently evolved species in it, learning through causal interactions with it. This is the basic metaphysics for realism. It is of course possible to consistently doubt and dispute any particular part of the current scientific image, and that image is clearly only partial and then only approximately accurate. Naturalism requires a critical realism, not a naive one. Nonetheless, one who doubts only on reasonable grounds should acknowledge that at the present time there is no serious reason to question the general lineaments of the scientific image.

According to the scientific image, cognition involves the active construction of some kind of map or model in the head, beginning with elementary sensory-motor coordinations and elaborating on these. However the correct account of meaning goes, precisely, the basic referents of these constructions are situations in our external world causally mediated to us. This is the basic metaphysics for the correspondence theory of truth. To put the matter argumentatively: Either one adopts a coherence or a correspondence theory of truth, but the content of science rationally accepted under coherence truth yields correspondence truth metaphysics, so correspondence truth.

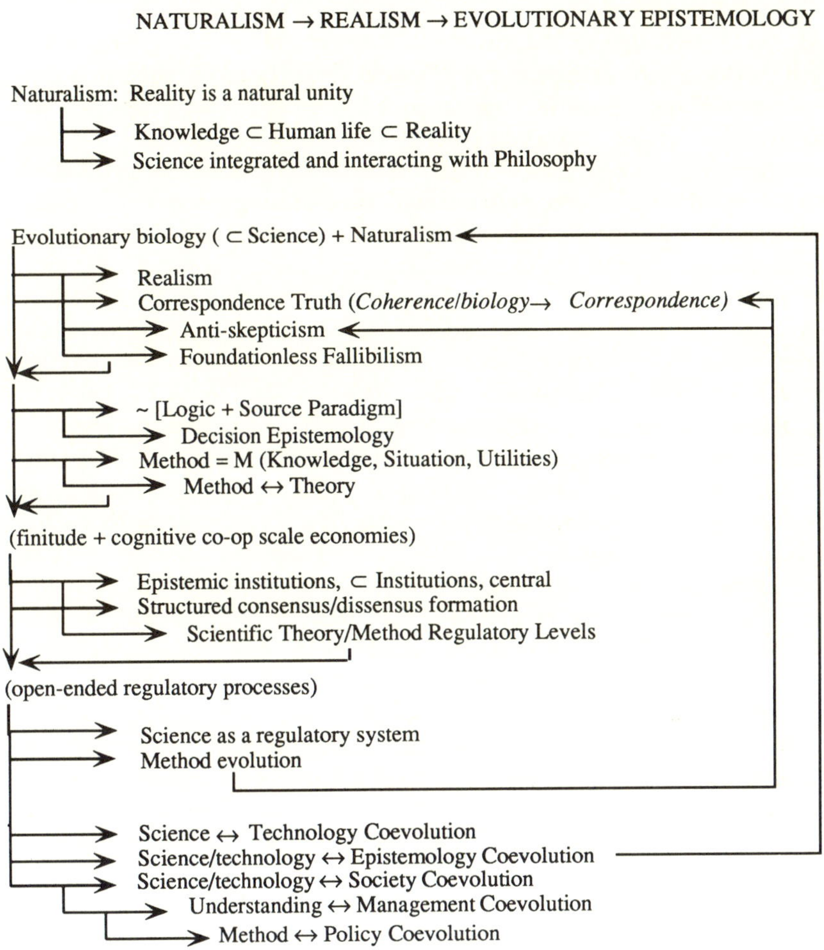

Diagram 1.1 Evolutionary Naturalist Realism
Schematic flow of argument from naturalism to its philosophical consequences. Descending arrows carry the argument forward; ascending arrows represent internal coherence and mutual reinforcement.

Another argument for correspondence truth comes from the open-endedness of the evolving cognitive process. The whole evolutionary process is an open-ended ascent of regulatory control (see below). As an extension of this process, cognition is, and should be, open-ended also because we are born in evolutionary ignorance; at our species birth and at each of our individual births, we arrive lacking explicit knowledge not only of the world, but also of every aspect of what knowing itself is. And as fast as our knowledge has expanded, our knowledge of the extent of our ignorance has grown even faster. But all coherence and pragmatist definitions of truth close off open-endedness. They do this because the truth criteria themselves, whatever they are, cannot in turn be examined for adequacy, cannot be opened up to learning. Coherence and pragmatist definitions of truth are attractive precisely because they offer epistemically accessible truth criteria. But it is just this feature that cuts off their open-endedness. Suppose, for example, one entertains the notion that for all propositions p, p is true $=_{df}$ p is entailed by science S arrived at through idealized use of methodology M. (It will have to be an idealized use in order to cover those parts/aspects of the universe that we would otherwise miss because they were too big, too small, too fast, too slow, too early, too late, etc.) But M is our construct, we can ask whether it is an adequate tool to achieve truth, for creatures like us, in our world. And this is something about which we can learn, and typically have learned, as we gain historical experience with methods and simultaneously improve our knowledge of ourselves and our world. So not only is our cognitive system open-ended, but it ought reasonably to be kept open-ended for creatures like us. But if M is accepted as definitive of truth, it cannot be critically assessed in this manner.[2] The argument generalizes: All definitions of truth employing epistemically accessible truth criteria are inadequate.

Correspondence truth, by contrast, is theorized as faithfulness to a reality that transcends, because it is "external" to, our cognitive or belief systems. Precisely for this reason, correspondence truth transcends the epistemic, so it can act as an ideal for an open-ended cognitive process.[3] One can always insist on epistemically accessible truth conditions, say by arguing that any others are meaningless (cf. empiricists) or confused (cf. idealists); these arguments aren't convincing, displaying instead just the kind of anthropocentrism from which naturalism is the best protection.

For if anything exists not logically or conceptually bound to cognition, then it is an open theoretical issue what exactly exists and how it is best known. Hence for naturalism all is fallible theory, from the lowliest factual claim to the highest metaphilosophical principle. This position is consistently supported by the scientific image. According to the evolutionary viewpoint, our entire cognitive capacity evolved from ignorance, a lack of knowledge not just of our latest theoretical concepts but even of the nature of conceiving itself. Our senses provide epistemically limited, partially biased, and moderately unreliable information channels, and we start in ignorance not only of our world, but also of the nature and reliability of our own perceptual access to it. The point generalizes to every aspect of cognition. These circumstances underwrite systematic fallibilism, applying at all orders of substantive conjecture, from perception through science to philosophy and metaphilosophy. Indeed, they underwrite a systematic foundationless fallibilism. There are no foundations because there are no guaranteed information channels that could act in this capacity. Not perception, certainly, and not reason either, since however a priori it may seem to the individual, it is a posteriori for the species (and that even over historical times).[4]

As remarked, our historical experience supports this self-conception. It has taken us two millennia to construct our present notions of geometry, for example, and these constructions have at every stage of their development arisen through interaction with our stumbling practical attempts to understand our world, from the mensuration of fields in ancient Egypt to the recent difficulties of reconciling classical mechanics and electromagnetism.

Nowhere is this experience of fallible historical development more vivid than in the development of our understanding of reason itself. There have been explosions in both logic and decision theory this century, resulting in a variety of distinct systems of both; contrast classical logic, for example, to intuitionist, n-valued, relevance, super-valuation, fuzzy, and quantum logics. These latter logical systems have all burst upon the scene this century, spoiling the air of necessity created by the preceding 2,500 years of logical theorizing in the Aristotelian mode. Their arrival followed the coupling of logic to algebra and geometry and the proliferation of mathematics consequent upon the discovery of non-Euclidean geometries and other nonstandard mathematical systems. A parallel story could be told for decision theory. The explosion of research in decision theory and logic has also resulted in the uncovering of numerous "paradoxes" (Arrow, Newcombe, Lowenheim-Skolem, etc.) and of startling theorems, such as the radical distinctness of quantum logic. There is increasingly intimate interaction of both these theories not only with mathematics, but also with economics, engineering, physics, and other scientific theories (cf. Hooker 1987). Again, conceptions of scientific method have changed substantially, especially across the last three centuries (Blake et al. 1960; Oldroyd 1989). And there is now intimate interaction between theory, technology, and method (cf. Hooker 1987 and below). All of this provides vivid examples of our learning historically about reason.

What then is a naturalist to say about the general nature and status of norms? Normative claims break up into two components, one functional and one about status. The functional component divides up the subject matter (inference, methodology, etc.) into the normatively permitted and the normatively excluded. Without this function, one does not have a normative claim at all. The status component claims that the normative principle should be obeyed for certain relevant reasons, for example, that it is a truth of reason. It has been characteristic of Western philosophy to try to defend normative claims by giving them some such privileged status. But from the point of view of naturalist realism, all claims have the status of fallible conjectures, there is no "higher" status with which to privilege normative claims. That leaves the question of status open. The status of philosophy as theory, together with our historical experience of their fallibility, supports the conjecture that there is no more to a claim for normative status than a claim of theoretical adequacy for describing some goal-directed activity. In this case, only their function remains as something distinctively normative about a claim.

This normative function is fulfilled by any adequate theory, for a theory is used to critically assess both methods and data within its domain. The primary methods of science are theoretical methods; that is, they are informed by our theories of the nature of the domain under investigation (one does count stable particles but not turbulent vortices, since theory tells us that the quantity of the former but not the latter is conserved, etc.), the nature of the instruments used to investigate that domain, the causal or statistical character of the relationships involved, and so on.[5] Theories have the role of partitioning methods and data into the acceptable and unacceptable. The more adequate the theory—however adequacy is assessed (explanatory power, empirical accuracy, etc.)—the more forceful is its critical, partitioning function. Thus for a naturalist claim to have a functional normative role for some domain is for it to be part of a fallible theory concerning that domain, and the rationally acceptable normative force it is taken to have is proportional to its adequacy.

In the old terminology, there is therefore a symmetry between normative and descriptive role and force. The more descriptively adequate a theory, the more it carries normative force, and the descriptive presuppositions of prescriptions that regularly lead to successful action thereby accrue respect. Conversely, the more serious the failure of descriptive (respectively prescriptive) adequacy, the more serious the weakening of normative (respectively descriptive) force. We simply don't take seriously methodological prescriptions based on seriously inaccurate or ad hoc conceptions of our situation.

Now we can return to the question of status. Although naturalism rules out any privileged status for normative theories, it does not rule out all differences between description and prescription. Naturalism should retain this traditional general skeleton of normative discourse: Normative theories are theories about what ought to be done and why, and the latter relate the prescribed action to the pursuit of relevant goals, which are derived from ideals, the traditional ones being complete-truth, full-rationality, perfect-goodness, and perfect-beauty. The importance of ideals to naturalism we have already seen in the case of truth: They hold the possibility of critique and progress permanently open and so are essential to an account of fallible historical progress. But of course a properly naturalist account of them will have to be given. A firm beginning toward meeting this requirement is provided in chapter 6, section 6.II.2, but it will also involve some substantial criticism of the traditional account in order to get free of formalist and antinaturalist assumptions buried in it.[6] It must be added that since ideals cannot be achieved in practice, indeed are not directly epistemically accessible, our operative goals must be formulated in terms of proxies for them that are epistemically accessible. These will be multiple and partial, and often conflicting, such is our situation. Proxies for truth, for example, form the epistemic utilities guiding methodological decision making (see below) and include empirical adequacy, security, explanatory and predictive powers, and so on; these are well known to be partially conflicting properties that may compete, context by context, for methodological influence.

The upshot of accepting this much of the traditional framework is that we can distinguish the normative aspect or content of theories by their reference to these ideals as goals. To use a theory T (functionally) normatively in pursuit of ideal I in some domain D is to accept that T is our best current guide to I in D (as measured by some function of I's proxies) and to apply T to distinguish what other acts of acceptance are also warranted. Thus an accepted scientific theory may be used to distinguish what data in its domain ought to be accepted, and to what extent, in the pursuit of truth, and an accepted theory of scientific method may be used to distinguish what procedures in its domain ought to be accepted, and to what extent, in the pursuit of rational assessment—and so also what theories ought then to be accepted. And so on. But note that we are not dealing with a linear hierarchy here, since some methods and theories will give rise to data that can be used to criticize other theories, or even themselves in appropriate circumstances; there is interaction between all theories, including those theories that are data statements.

The gulf between normative and descriptive is thus removed—they are two sides of the same coin—though nevertheless the normative function is retained. Just as no theory is reducible to its data domain or can be replaced by its empirical predictions (positivism is dead), so no normative theory is reducible to or replaceable by its descriptive subject matter. Thus philosophical theories retain their normative status, consistently with naturalism. And this position is reflexively self-consistent, for this present account is itself a fallible theory of the normative, functioning normatively, and warrantedly so, since it explains reasonable normative attribution.

Let us pass now from a discussion of norms generally in a naturalist framework to naturalized normative epistemology in particular. Our ideal of knowledge has been certainty achieved by making only guaranteed truth-preserving inferences from a guaranteed reliable source of basic information. The claimed source might be divinity, or reason, or the senses, but in each case it was the involuntariness, the "necessity," of its information that underwrote its status as incorrigible, the provider of guaranteed truth. Correspondingly, the inference mechanism should be normative, truth-tracking, and contentless (this latter to ensure that it cannot be contaminated by empirical error) and all of this somehow self-transparently so. Logic has typically been held to be the only candidate.[7] But an account of rationality is a philosophic theory, so it retains its normative status, but it is taken as a learnable, correctible, fallible theory. So the standard epistemic paradigm of privileged source of information plus logic as structuring principle must go. We have already noted that there are no privileged sources and there is no privileged logic. So there is reason to abandon both parts of the traditional epistemic paradigm.

But it is crucial to develop an appropriate naturalist perspective on how to replace this epistemically failed epistemic paradigm. Otherwise we might waste resources trying to replace it with another of essentially the same kind, for example, using Bayesian inference to try to construct a formal system to replace logic as the structure of method. And while there is pretty clearly no hope of repairing the fallibility of perception, one can look for extended logical systems that will allow elimination of data as well as of theory to cover this complication. But from the perspective of cognitive theory in a dynamic systems framework, logic is the wrong kind of basic structure, because it can't deliver the right kind of connection to the dynamics of cognitive processes. I will develop this point shortly, but first I shall pause to note some other difficulties with logic as the basic structuring principle for cognitive theory. (For these criticisms, see Brown 1988.)

First, there is the matter of misplaced foundationalism. Logic leads, via a regress of analyses (every argument must have premises there unargued), to foundations, that is, to self- authenticating terminators of analysis—or, if there is no termination, to a vicious regress. But there are no terminating foundations, so no satisfactory analysis of rational acceptance can be provided. We have already noted this, but there is also foundationalism of a second kind required, for we are equally in need of a foundational set of rules prescribing argument validity that are in some way self-justifyingly transparent to the rational intellect. But in fact there has been rational conflict over these very rules. Within the domain of logic itself, for example, the rule of subalternation was once widely accepted as a valid rule in logical reasoning and is now just as widely rejected by contemporary logical systems. Again, there is a diversity of approach to the law of excluded middle across classical, intuitionist, n-valued, relevance, super-valuation, and various quantum logics. These circumstances rather strongly suggest that there is no more hope of discovering foundational rules than there is of discovering foundational evidence.

Second, there is the correlative matter of misplaced formal structure. Reason as the application of formal rules is really committed to analyzing every case of rational behavior as rule following, even if it has to be hidden, subconscious rule following. This would be so, for example, in cases of arguments in science from new data to new theoretical conjectures, the choice of strategy and specific moves by a chess grand master, and so on. It must also apply when choosing among competing systems of rules, for example, when a logician is designing an elegant and powerful proof procedure. Indeed, it must even apply to the creation of systems of rules themselves; again the logician will serve as example. Unless, of course, all of these activities are to be dismissed as nonrational. But whether we dismiss all these activities as nonrational, or construe them as hidden, we enter an imposed division into the theory of mind without any fundamental rationale itself and without any fundamental supporting evidence from neuropsychology or cognitive psychology.

Now I shall turn to the central matter of dynamics. The formal inductive logic model has an associated dynamics given by the sequence of empirical data fed into the induction machine and the consequent sequence of its outputs. But even setting aside the problems with formal induction (see chapter 3), this gives no sense of the organization of perceptual direction (learning where to search for salient raw data), the self-investigation of perceptual limitations and errors (how to refine raw data into data), the dynamics of interacting choices between theory-based criticism of data and the use of data to criticize theory and between both and category/concept formation, the consequent reorganization of method in interaction with theory, and so on. In short, it gives no sense of the self-organizing dynamic capacities characterizing the real process of science. Here I briefly offer some reasons for accepting that we can only introduce those dynamics by dropping logical structure as basic; it can be recovered as a partially useful tool, determined context by context (cf. Cherniak 1986, Hooker 1994c).

(1) The representation of rationality as encapsulated in a set of formal rules requires that there be a universal and unchanging methodology for science. This methodology is expressed by the collection of formal rules for moving from evidence (and perhaps existing theory) to theory. But it is now widely recognized that such a model of science is wholly implausible.[8] Indeed, it is one of the chief characteristics of science that its method evolves along with its content, and this is fundamental to nonshallow learning systems (cf. chapters 5, 6). Essentially the only alternative available to this model of rationality is to postulate the a posteriori discovery of a priori rules. But this entails an accumulative model of scientific method (matching the accumulative empiricist model of scientific theory), and there is no evidence for this. Furthermore, formal rules by themselves simply do not suffice even to characterize the basic range of choices that must be made by scientists (see below).

(2) Formal rules place the emphasis in understanding rational decision making on proposition-proposition relations in abstraction from the cognitive agents involved and their circumstances. This immediately generates two unacceptable results: First, it fails to take seriously how difficult it often is to discover relations between propositions with the result that we would be committed to surprising denials of rationality in retrospect. If we consider just consistency alone as a necessary condition of rationality then, for example, after many years work one has to conclude that modern quantum field theory in its standard formulation is not rational and neither is modern naive set theory. Presumably by derivation, all of the activities that are involved in the construction and exploration of para-inconsistent logics, logical antimonies, and like formal systems are equally not rational. But this is a grossly uninsightful account of rational activity. Indeed, it ignores the irrationality of inflexibly imposing consistency on finite agents with finite reasoning capacities (see note 6). Second, Science is modeled as an abstract activity in which the diversity of scientists in ability, institutional role, historical location, and so on is essentially ignored. This leads to the divorce between the history of science and the theory of science as a rational activity—the notorious schizophrenia of "internal" and "external" science theory. And that leads in turn to the irrelevance

of institutional structure to a theory of the rationality of science, whereas science is a collective cognitive enterprise in which such structures are essential (see below). And as already noted, it leads ultimately to a dualistic representation of logical and causal descriptions of complex systems (Popper's Worlds 2 versus 3 is the glaring example) that confuses and confounds understanding cognition, and evolutionary epistemology in particular.[9]

(3) Logic provides acceptance criteria that are independent of consequences. All of the weight of substantive evaluation falls not on the acceptance criteria, but on the antecedent accepted premises. This is because logical principles have a certain kind of closedness; the principles of inference that make up acceptance strategies correspond to statements that are theorems of logic (i.e., analytic), and all analytic statements are strongly logically equivalent and in this sense free of subject matter. In this way, logic partitions epistemology into analytic form and substantive content (cf. the sharp analytic/synthetic distinction, see chapter 3 below). This removes acceptance strategies themselves (though not their results) from being improved as a consequence of learning from the outcomes they advise, that is, from naturalistic shaping via selection pressures. (This complements point 1 above.) By contrast, a shift to an alternative paradigm of acceptance as an act of practical commitment in principle opens the way for these strategic acts to be improved via selecting feedback from their consequences. This approach leads to a decision theoretic epistemology where consequences play a key role and strategies can more clearly be made subject to selective development. Moreover, there is no requirement for foundations.

(4) Risk taking finds no natural place within logical structure, and because of the same closedness noted above. The inference principles themselves, being analytic, are not supposed even to be candidates for riskiness. And these principles, both deductive and classical inductive, deliver statements directly as candidates for acceptance, so there is no natural purchase for a notion of the riskiness of the latter arising from their manner of derivation. Any riskiness imputed must take the form of a judgment imposed externally to the structure, affixed to the initial premises for some reason.[10] Even this possibility is undercut by empiricism, which confines premises to observation statements and treats these as incorrigible, hence as risk free. The classical ideal of science is designed to exclude risk completely, but for finite agents evolving from evolutionary ignorance risk stands at the center of rational procedure.

First, nothing is certain or guaranteed, including methods, much less observations and theoretical conjectures; there is risk in relying on any of them. From an evolutionary perspective, what is risky is cognition generally, since even if it were specifically selected for (and this is not guaranteed), it will be selected on the basis of being good enough for survival in those (past) circumstances and not on the basis of guaranteed truth tracking. And in any case, as already noted, we begin in ignorance of it all, its epistemic strengths and weaknesses, and even ignorance of what an epistemic feature is; we need to take the risk of learning about any of it. Second, agents with finite resources have constantly to make trade-off decisions between what some ideal rational method might enjoin and the value that following it can deliver; the result is constant, rational resort to risky heuristic methods. Thorough perceptual search and complete exploration of deductive consequences might seem attractive methods (cf. the total evidence principle and deductive closure as rational principles) but only to a god with unlimited resources. To a human facing a charging lion or a scientist facing a balance sheet and the limitations of laboratory life, they make little sense; in both cases, the law of diminishing returns from pursuing them sets in savagely. What are required instead are risky but intelligent (canny) judgments about how far to pursue these ideals, and in which directions, in the practical context to hand, so that the best possible decisions are made (note 6).

This situation is nicely illustrated by Popper's argument, stemming from the antagonism between security of truth and informativeness, that inductive support is not the support that one should rationally seek in science, the point being that there are at least two legitimate epistemic values driving science and that pursuing either, and even choosing the balance between them, is risky. But we shall see in chapter 3 that Popper seems not to have taken this lesson to heart, for his own method is a narrowly formalist one that causes a great deal of difficulty for him. Indeed, Popperian theory is riddled with ineliminable appeals to strategic human judgments and corresponding decisions: to choose what to observe and by what means, and how to process and assess the results; to choose theories to test, the tests to be performed and where to lay the error when theory and experimental result clash and when sufficient testing has been done for the moment; even the decision to cooperate in pursuing the rational life. The structure of decisions that face scientists is summarized in the decision tree of diagram 1.2. None of these decisions is dictated by formal logical considerations. If one operates with a metaphilosophical dichotomy in which decisions are either rational because logically dictated or nonrational and causal/social, then all these decisions must be relegated to the latter category.

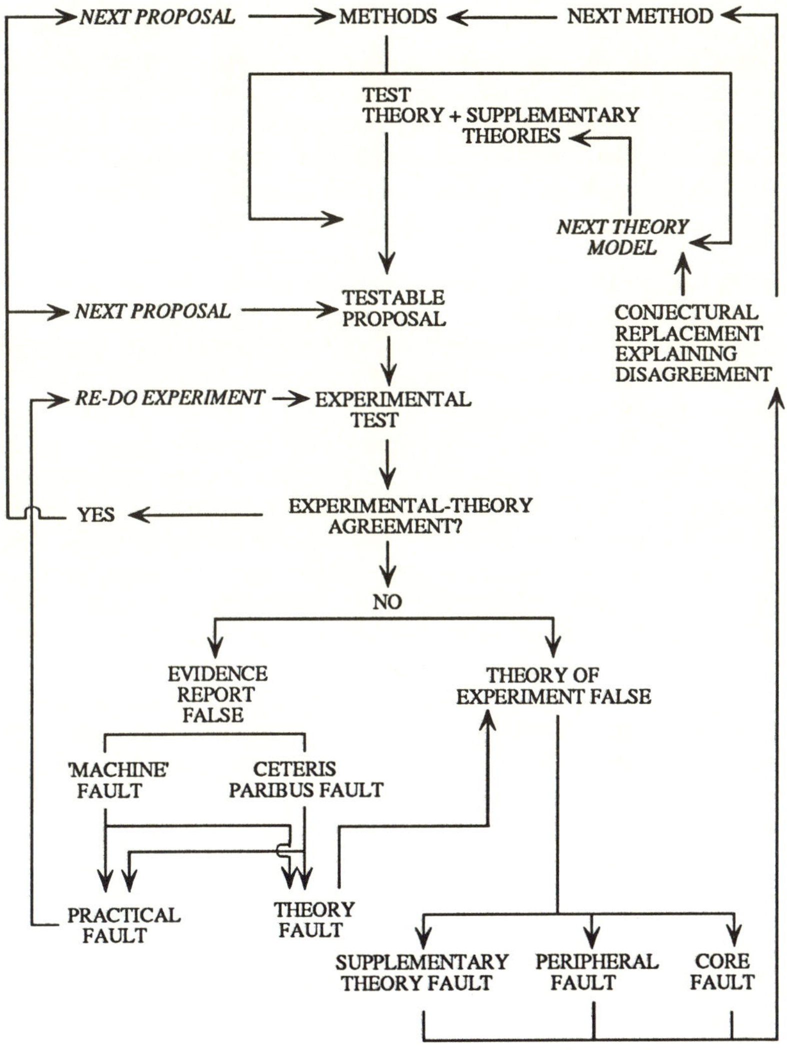

Diagram 1.2 Scientific Decisions
Simplified decision tree structure for scientific testing. Arrows indicate sequence of decisions facing scientists with tree branches indicating loci of key scientific choices determined by judgment.

After Popper, through Kuhn, Feyerabend, and all the others, the appeal to decisions by the scientific community widens rapidly—so rapidly that all of these latter have been accused of abandoning reason. Why? Only because of the tacit assumption that what cannot be reduced to logical method is nonrational. But this consequence is instead better taken as a reductio of this conception of rationality.

If risk is to be rationally managed, especially if it is to be reduced, these decisions require some kind of coordinated intersubjective procedure for their control; that is, they require an institutionalized procedure. This introduces a certain kind of social structure—epistemic institutions—as essential to science.[11] But logical models exclude institutional structure as irrelevant. Many theorists, having noticed the hidden dependency of scientific procedure on community decision making, have leapt to the conclusion that science cannot be a rational process or is at most a very limited rational process with much of its psychosocial substance arational. (This latter is Popper's conclusion.) Many have leapt to the conclusion that science is to be understood not on the basis of reason at all, but on the basis of those social forces that causally shape it through its social institutions.[12] We can now see more clearly what was asserted earlier, that these inferences all have as a suppressed premise the claim that logic is the epistemic paradigm. (If it is judged unlikely that a purely formal account of scientific reason can be given, and if it is assumed that no other account of reason is available, then science is partially or completely characterized by nonrational, merely causally specified, sociological processes; it is my conjecture that most theorists make this inference unaware of the suppressed premise.) But once we switch to theorizing science as the practical pursuit of risky acceptance strategies, the institutionalized coordination of decisions for risk control and the efficient use of finite resources generally becomes central to rational procedure. Attention now shifts to risk-taking strategies (including reliance on perceptual and other judgments) and to the division and distribution of cognitive risk and labor, concerns that stand at the heart of the evolution of intelligent behavior. Methods are conjectural, risk-taking, epistemic utility optimizing/satisficing, resource distributing strategies. (And judgment, risk-taking, and utility optimizing/satisficing are three concepts that do "grade off down" the evolutionary sequence of complexity.)

From this perspective, methods are regulating recipes for developing theories. Theories are understood as specifications of possibility structures for classes of contexts that provide a framework for, and so regulate, the development of more specific theoretical methods for more particularized contexts. Data are the most highly contextually confined theories, whose production is caused by interaction with the world and regulated by specific theory. Metaphysics and pure mathematics are also possibility frameworks, principally for developing methods. The focus in the theory of science shifts to the dynamics of regulated development of methods along with theories and data and as well incorporating the correlative dynamics of the development of epistemic institutions (see chapter 2). Thus we shift from the partial, limited dynamics allowed by the logic-based account to the full dynamics canvassed here.

This discussion of why logic-based models preclude developing adequate dynamic models for cognitive processes has also teased out many features that should be incorporated in any such model. A shift to an alternative paradigm of acceptance as an act of practical commitment leads to a decision theoretic formulation of rationality and epistemology which in principle statisfies these requirements (Hooker 1987, chapter 8). A sub-system, individual or institutional group, is modeled as possessing a set of epistemic values and rational method is the choice of most efficient means to pursue maximal value increase given the resources and accepted knowledge to hand. This permits methods to be tested against their consequences (criticism 3, cf. chapter 4) and thereby epistemic values also since the acceptance strategies they specify in turn correspond to patterns of method choice. This provides a framework for method dynamics (criticism 1). Since on this model rational decisions are strongly context-dependent, context specifying available resources and constraints, and even priority among epistemic values (think of practical versus theoretical contexts), the range of institutionally specified contexts finds natural expression within this approach (criticism 2). Risk-taking strategies also find a natural place within a decision theoretic model and can take advantage of a rich theoretical literature (criticism 4). To these strengths another can be added: the decision-theoretic framework provides the opportunity to unify the treatment of science methodology and public policies, including science policy, as various cases of institutionalized strategic action (see below). This is not only cognitively valuable but of practical importance in contemporary circumstances. Finally, there is no requirement for foundations, and the account fits well with naturalism. (As noted, the central capacities introduced here are ones which do 'grade off down' the evolutionary sequence of complexity and can be subject to selection.) There is neither pro-linguistic nor anthropomorphic bias in this conception of rationality and no specific barrier to reflective evaluation of both sources of knowledge and logical structures, though the latter can be recaptured as effective practices in appropriate cicrumstances (e.g., whenever formal truth preservation is the dominant value).

Is it possible to provide more detailed models of these decision processes? Only very partially at the present time. Here I shall confine myself to Holland et al. 1986, *Induction*, as illustration, chosen because it provides an example of a generalized model which at least seeks to replace static formal rules with a more dynamic process. In place of static formal logical structures, Holland et al. are led to introduce a generalized computational model equipped with hierarchies of rules each characterized by default conditions and permitting mutual reinforcement of competition among rules according to quantitative weighting processes along with processes for the introduction and deletion of rules. This is a somewhat more complex process that could be had with all-or-nothing, 1–0, logical rules. Although a conception of coherence emerges, for example, it is partial, probabilisticly structured and dynamic; again the simpler logical notion of logical consistency is at best an approximate special case. Holland et al. are quite explicit about the inadequacy of the traditional formal models—not only narrowly logical, but even set- theoretic models—insisting that they be replaced by procedural models of this kind.

The domain of nondeductive reasoning processes is very rich; in order to do justice to the empirical evidence, Holland et al. are forced to recognize many disparate processes that are only relatively loosely interrelated through the general model, for example, category formation, generalization, analogy, consilience, and unification. We have passed here well beyond a narrow conception of induction into the rich domain of all those processes by which complex adaptive systems improve their information-based adaptive responses. And they are able to model in their account responses to many of the criticisms just leveled against logic-based models. If, for example, we follow the naturalist regulatory conception of method above and construe hierarchically higher rules in Holland et al.'s model as encapsulating method, then their general model directly captures the requirement that method itself be tested against its consequences (criticism 3), since rules receive a weighting that reflects their support of successful action. And given the rule entry and deletion process, it follows that they have resources to build some conception of method dynamics as (higher) rule dynamics (criticism 1). Though, perhaps revealingly, they don't discuss these or related issues explicitly, their model already shows clearly how these dynamic features emerge from more empirically sensitive models than logic.

This is clearly an important step in the direction of replacing the logic-based model of induction with a dynamic process model. Insofar, I am happy to claim it as support for my general aims. But from my point of view, it does not yet go far enough. It is not germane to enter upon a detailed analysis here, and I shall confine myself to three general remarks. First, the model has been developed in most detail for individuals, and it is less easily and obviously applicable to science. Part of the reason for this lies in the complexity of scientific argumentation, for example, concerning the indirectness of theoretical concept formation, involving complexes of multidisciplinary theoretical models and technological practices, and concerning the compound closure structure of unification and the often competing strands within unification and even within our conception of rational representation.[13] These features are all deeply concerned with the characterization of scientific method, and it is unclear whether the rule structure of Holland et al.'s model will suffice for this. Another part of the reason lies in the omission of any explicit role for social structure, indeed for any context-dependence other than available information. The processes of scientific theory change are supposed to be formally identical to belief change in individuals. There is a potential basis for recognizing social structure in their representation of induction as a parallel distributed process, and socially institutionalized science is just that; but this is not developed, presumably because of the constraint just noted. There is, for example, no recognition that differing institutional orders and groups of scientists will quite reasonably have differing, and perhaps even competing, goals and that there will be no across- all-orders resolution of these but only resolution (if at all) in terms of the distribution of risk and effort across the entire institution.

Indeed, second, there is no discussion of risk at all. This is indicative of the fact that decision theoretic, strategic conceptions are formally absent from the characterization of decisions. The elements for introducing risk and strategy are present in the introduction of probabilities and of competition among rules, but they are not exploited in this way, because the notions of cost and benefit are absent. Again, a basis for these notions is implicit in the feedback loop from action via environment to the agent but is not explicitly constructed. This has got to do with the conception of rules in the model, which in certain respects retain a close similarity to the all-or-nothing rules of the standard system; for example, failure of expectations is treated very like Popperian falsification rather than a risky distributed search among diverse alternatives (diagram 1.2). Third, even for individuals, the models of category formation are restricted to rule-based versions. This is likely to prove restrictive, whereas those offered by neural network theory or by non-standard control theory (see, respectively, Churchland 1989 and Hooker et al. 1992a, b) are likely to prove more powerful in an adaptive systems context. Furthermore, it is hard to see that the rules provided can capture those processes of concept formation central to cognitive development and to science, for example, the distinction among generalization, completion, and constructive completion of conceptual schemes (see chapter 5). In sum, the model is still restricted by the basic features it still shares in common with logic/AI models.

My efforts in this book are largely complementary to those of the kind Holland et al. represent. I am interested in developing a general dynamic systems framework and in characterizing epistemology and rationality within it, rather than focusing on computational models for these processes and their interaction with individual psychology. (Indeed, I regard even the issue of whether the brain can be computationally modeled as open at this time; see, for example, the discussion in Penrose 1989.) As I see it, the work of Holland et al. provides an important and insightful example of a "half-way house" in the replacement of narrowly logical/AI models of induction with complex adaptive systems processes. Insofar as what they have accomplished can already support a conception of cognition as a complex dynamic process of adaptation, or can be modified to support it, I embrace it; insofar as it cannot, I shall try to develop a framework within which further work of their kind might take place.

The approach I advocate is nicely illustrated, and the conception of institution systematized, by seeing how technology fits within it. Cognition evolved because adaptive control of behavior was reinforced. Central to that control is anticipative intervention to disturb the environment in order to learn by its response and, subsequently, to modify the environment so as to reduce risk. From this, flows the central role of technology in the scientific process. Our earliest technologies were provided by our bodies. Subsequently, they evolved in delicate and increasingly intimate interactions with both practice and theory. Technologies are essentially amplifiers. With respect to science they (1) extend its information base through instrumentation (e.g., electron microscopes), (2) extend accessible methodologies (e.g., through numerical approximation techniques and automated data processing), (3) generate new concepts (e.g., computer models of cognition, (iv) extend epistemic institutions, e.g. through supporting global communications), and (5) provide the basic resource base for scientific activity, from economic surplus generally to rare earth metals and other specific resources. Conversely, the development of technology is directly impacted by science through (1) new theoretical conceptions, (2) new theory-driven designs, and (3) performance evaluation and redesign learning (fluid mechanics gained more from the development of reliable airplanes than vice versa). This coevolution of method, theory, and technology is of the essence of science and a vivid demonstration of its open-endedness. Its dynamics form a set of generic positive feedback/feedforward loops, which can be represented as in Diagram 1.3.

The mutual interaction of method, theory, and technology is nicely illustrated by the evolving role of the senses in science. On the one side, there has been an increasingly refined critique of natural sensory perception for its limitations (e.g., limited discriminations), biases (e.g., tracking delays) and imperfections (e.g., illusions). Technology and theory were essential here; for example, the camera and optics for detecting perspectival bias, infrared and x-ray technologies and backing theory for the use of the non-visible electromagnetic spectrum. On the other side there is the development of extensions to, and substitutions for, the senses, for example, telescopes, microscopes, micrometers, and x-ray photography. These allow us to confine use of our senses to those narrow circumstances where they work best (e.g., identifying and counting appropriately shaped and colored human-sized objects).

So we need to think of science in these respects as a dynamic system, transforming its own instrumental "body" as it evolves/develops. Through technology, science also transforms its social and natural environment (e.g., respectively, motor vehicles, world agricultural gene trade) and the natural and policy processes that yield the dynamics of that environment (e.g., respectively, the greenhouse effect, economic modelling for policy determination). And through the development of various science studies and science policy studies, which together form the so-called science of science, it is also transforming its own institutional design and social relations (e.g., respectively, the trend to larger specialized research groups and governmental science advice processes), including its own social evaluation. (This latter is marked by increasing ambivalence as felt powerless-

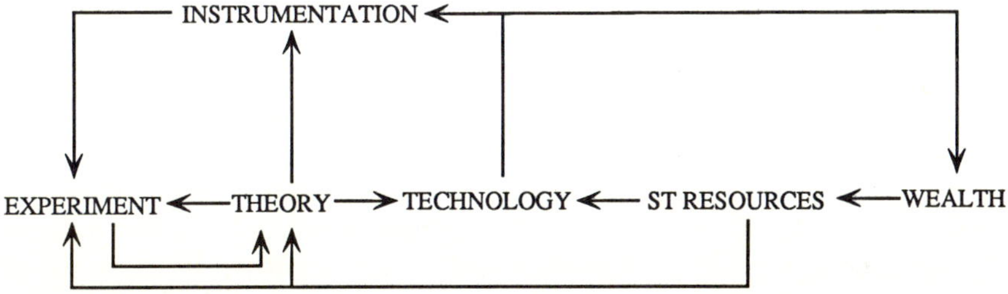

Diagram 1.3 The Science-Technology Machine
Arrows indicate positive feedback connections to produce mutually reinforcing loops that stimulate scientific, technological, and economic development.

ness increases; for example, reliance on, but distrust of, doctors or nuclear engineers.) This sophisticated and thorough transformative capacity is a crucial part of understanding science as an increasingly autonomous, dynamic cognitive system.[14] But this perspective is suppressed in the logic-centered conception of method and epistemology.

This instrumental transformation of science is itself part of a wider self-transformation of Western culture and organization, the "Change Machine," in which science-technology generates conceptual and technological changes that are fed into a wider economic and socio-political ecology of institutions designed to take advantage of them and that in turn feed back to reinforce the development of science-technology. The economic market is the central institution in Western society and it rewards and amplifies (salable) change at the momentary technological production frontier, for example both the better mousetrap and the new mouse gene technology. Science-technology is our most effective generator of such changes. Like evolution, the market has no global direction to it, it exploits change opportunistically. Parliament is a minimarket in votes; it rewards and amplifies (salable) change at the momentary political production frontier, for example both the better export stimulus and new taxes on manufacturers. As in the market, there is no intrinsic global directedness constraining the process, nothing to damp down the impact of momentarily advantageous change. Even Western religion has collaborated in getting out of the way of change. The Protestant's connection to God is individual and private, not through a public social form (as it was for the feudal Catholic); the church is wherever protestants gather, so Protestant Christianity is compatible with change in a way that communal faiths are not. The result is a highly dynamic, currently relatively unstable system that has been increasing in scale, intensity, and pace since the Renaissance and is now transforming the planet.

This is the wider proper setting for a theory of epistemic institutions. These will not be separate self-contained institutions in a collection of disparate institutions, but partially autonomous subsystems, together forming a cognitive institutional dimension or aspect of the whole institutional ecology of Western society. From the point of view of epistemic theory, relativist sociologies of knowledge and the like present useful information but fail to make a principled design distinction between epistemic and other social features of institutions. They thereby fail to do justice to the distinctive cognitive features of either science or technology (cf. Hooker 1987, sections 7.3, 8.4.2). But understanding these matters has an urgent practical dimension as well.

For this contemporary crystallization of human organization around cognition-driven institutions is the major contemporary transformation in planetary historical, ecological, and evolutionary conditions. If we are not to undermine our own survival (and that of many other species as well) with a resulting chaos, these developments must come to constitute the emergence of a new order of stable planetary regulatory structure. Designing such an integrated system is now our most urgent task. A small but important part of this task is to develop an adequate dynamic systems conception of science. It is to this that the present work is designed to make an initial contribution.

The philosophical theoretical conception of science that has emerged from this brief examination is a very different one from that offered by classical and contemporary philosophy. In place of the risk-free, logic centered, static conception of science as a uniquely determined structure, there now stands a conception of science as a dynamic process of strategic risk managing, resource allocating, and investigatory decision processes aimed at maximizing epistemic value (or at least satisficingly increasing it). The cognitive result is a complex of thoroughly fallible intertwined methods and theoretical frameworks growing increasingly broad in scope and autonomous in operation through the transformation of its instruments and the extension and refinement of its methods. These processes are realized in an intricate complex of social institutions and complex institutionalized roles filled by persons embodying specialized skills and information, the whole increasing in scale and internal differentiation and simultaneously showing increasing internal autonomy and external involvement and dependency.

In short, it is a system that is lifelike in its sense of organization and development. I propose modeling science as a component or aspect of the overall dynamics of evolution/development characterizing life on this planet, to understand the cognitive as an approximate abstraction from this overall process in much the way that population dynamics models of ecologies are approximate abstractions from the richer web of thermodynamic exchanges and economics is an approximate abstraction of priceable exchange processes from the richer web of biosocial relations. This is the proper context for considering an evolutionary epistemology, which must be capable of illuminating cognitive dynamics by exploiting the life- likeness metaphor. It is also the proper context for defending that epistemology against relativism while retaining fallibilism (see Hahlweg/ Hooker 1988). But following this overall approach will even lead us away from traditional forms of evolutionary epistemology (see below). Everyone familiar with conventional philosophy of science, even conventional realism, will recognize the radicalness of this rethinking of philosophy and metaphilosophy, science, and social practice. Yet it is only within such a radical evolutionary naturalist setting, I believe, that a thoroughgoing epistemology—not to mention appropriate social and science policy—can be adequately theorized.

1.II. Evolutionary Epistemology in a Naturalist Setting

Evolutionary epistemology is a doctrine that asserts some interesting relationship between the biological process of evolution and the development of science (and perhaps of cognition more generally). Notice that this is a relation between two processes operating across populations, nothing has been said about individual ontogenesis, including cognitive development, or about its relation to them. A relation of some kind is unavoidable once phenotypic viability constraints and behavioral capacities are acknowledged to play important roles in both genetic and scientific processes. Something will be said about this in chapter 2, and in chapter 5 we shall find Piaget arguing for an analogy between evolution and ontogenesis. But this relation is a further issue and, while no simple analogy among stages or other product states of these processes is likely to hold, it is appropriate and fruitful to develop a unified dynamical framework within which to represent them as interacting processes. Here a beginning is made on that project.

There are two somewhat different kinds of relationship one might try to develop between the biological process of evolution and the development of science, roughly formal analogy and direct embedding. The former is focused on arguing for, or exploiting, features of the evolutionary and theory succession processes which are held to be formally identical, for example, both are typically held to be variation, selection, and retention processes. The latter is focused on arguing for, or exploiting, a conception of cognition as a literal component or dimension of the evolutionary process, with science its current cutting edge so to speak. The bulk of the philosophical literature on the topic is either explicitly committed to, or presupposes, the formal analogy approach. In contradistinction to that, the naturalist doctrines developed above push toward a direct embedding account—and of a particular sort, namely to first model evolution as the dynamics of complex adaptive systems and then model science as an approximately extractable cognitive dimension of it.

Within the evolutionary epistemology literature, the dominant analogy approach takes its inspiration from Campbell 1974 while the direct embedding approach derives from Lorenz (1977) and Piaget (1971, 1974/80, 1976/78). Lorenz regards the whole of evolution as concerned with the encoding of environmental features in phenotypic structure—for example, the horse hoof encodes the ground structure of the steppes—and so a generalized learning process. Within that, he wants to understand the evolution of human cognition, especially perception and concept formation/categorization, with the aim of showing how the Kantian principles may be justified as a priori-cum-genetic for the individual while a posteriori for the species. Work in Lorenz's tradition is ably supported elsewhere (Plotkin 1982; Wuketits 1983). It is of limited interest here because it can at best only explain commonsense cognitive function (only that has had time to be selected for); its rigidly universal and unchanging principles are of dubious scope and perforce cannot explain either individual specialization in learning or the dynamics of scientific change so dominant today (cf. Hahlweg/Hooker 1989a, part I). For reasons already briefly described, and despite some reason to assume the contrary, I shall argue that the case with Piaget is exactly the opposite in both respects (see chapter 5 below).

The formal analogy and embedding approaches have fundamentally different aims. The formal analogy approach aims to increase the exactitude and scope of the formal analogy between the evolutionary and theory-succession processes. Criticisms of the exactitude or extent of the analogy become direct criticisms of the position (Thagard 1988). But there need be no ontological interest in the intellectual process vis-a-vis biological reality, because two quite disparate processes might share a formal analogy (e.g., heat flow and fluid flow), and often this approach simply leaves two unrelated processes running in formal tandem when the embodiment of minds provides an obvious reason to connect them. The embedding approach, by contrast, aims to understand cognition by exhibiting it as a natural phenomenon, embedded in the natural world and arising naturally within it. The focus is on establishing continuity between cognitive and more basic biological processes by understanding cognition as a complex of individual and group properties and processes that emerge increasingly clearly as neuro-biological complexity increases. The connection with evolution is made through the fact that neuro-biological complexity, however that difficult notion is finally to be defined, has increased over evolutionary time and through the strategic guess that that increase is not accidental but is fundamental to (at least some part of) the evolutionary process because of the dynamics of complex adaptive systems.

From this distinction of aims, it follows, first, that criticisms of the exactitude or extent of the analogy between the evolutionary and theory-succession processes do *not* in themselves become direct criticisms of the embedding position. It is not at all surprising that a version (implementation) of some process P in a simple system should be different in some respects from the implementation of P in some more complex system (think, for example, of information transmission as between telegraphed morse code and computerized video systems). Rather, the failure of the analogy would have to be of a kind that showed that the naturalist embedding aim was in difficulty. There are plenty of arguments of this kind, but they don't derive from the evolutionary epistemology literature; they are, rather, the usual antimaterialist arguments found in philosophy of mind. I do not find them convincing, not because I am severely materialist, but precisely because the traditional antimaterialist arguments presuppose too crude a conception of complex systems; however, this is not the place to argue the matter. There are also plenty of limitations to the evolution/theory succession analogy, but none, I think, that strike against an embedding thesis.[15]

Second, it also follows that the embedding approach is not logically wedded to any particular account of the evolutionary process; specifically, there is no conceptual constraint to current orthodox (neo-)Darwinism. This is the contrary of many of the formal analogy accounts where the orthodox position is simply assumed to form the biological side of the analogy. What the naturalist embedding position should be committed to is the scientifically most acceptable account of that process, whatever it be. Until recently this has been taken to be the orthodox neo-Darwinian account, and because of the broad range of evidence that supports it (see the nice summary in Ruse 1986) there is little doubt that it must continue to be an important part of the story. But the theory has its own theoretical and evidential difficulties (see Gould 1989, Ruse 1982, 1983); there is increasingly widespread support for supplementing the account with the introduction of phenotypic capacities, which can change it in significant respects (see below and chapter 2); and new theories centered around the notion of self-organizing processes have recently arisen to challenge at least its exclusivity and possibly its dominance (see Depew and Weber 1985, Kauffman 1993, Weber et al. 1988; Wesson 1993). The naturalist account will have to be adapted to the outcome of this debate. In what follows, I try not to take sides too heavily, though I avoid obvious forms of Lamarckism and support introduction of a theory of phenotypic capacities. (Note that while I do also try to develop conceptions of reason and public knowledge as organized systems properties, this in itself is neutral to much of this debate, for example, neutral between orthodox selectionist accounts of the emergence of these properties and self- organizing systems accounts.)

Promising formal analogies can be utilized within an embedding approach as heuristic guides to theory construction. A sensible initial strategy is to convert the analogy to an identity, that is, to claim that some cognitive process is literally the same (in the relevant respects) as some biological process. If we are lucky, a mature theory of self-organizing complex adaptive system dynamics will support the identity; but even if not, it will have been a usefully simple starting point. This is the approach taken here, so it makes sense to review briefly the formal analogies on offer. The obvious ordering principle is in terms of the kind of analogy proposed; here just three positions are distinguished.[16]

The *selective analogical evolutionary epistemology of science* (SA-EES) postulates a minimal selective or partial analogy, namely a common process of variation, selection, and retention (hereafter, VSR process). Biologically, this is the fundamental Darwinian process; cognitively, it is held to be the basic process of trial-and-error learning. Many prominent contemporary evolutionary epistemologists occupy this cautious position (e.g., Campbell 1974, 1977, 1986; Rescher 1977; Popper 1979).

Note the importance of insisting that only a highly selective analogy is being claimed between the two processes. Campbell, for example, does not wish to be faced with the problem of finding precise cognitive analogs for the genotype and phenotype or for the failure of transmission of information from cognitive phenotype to genotype. Even so, there are various disputes among proponents. Some (e.g., Campbell 1974) hold that the process of variation is equally blind in the scientific case as it is held to be in biology, whilst others (e.g., Rescher 1977) insist that it is consciously and rationally directed toward the most likely successful outcomes and therefore not blind. There is a similar split over the characterization of the selection process itself, some wanting to view it as a process that is essentially opportunistic and pragmatic, and perhaps purely causal, as the biological process itself is held to be, whilst others wish to emphasize the use of logic and insist that the process is rationally directed toward the conscious pursuit of truth. There will be further discussion of these issues in subsequent chapters; here it is noted again that while the role of VSR processes in biological evolution is assured, the nature and importance of that role is currently in question, and similarly for its role in cognitive processes.

The *extended analogical evolutionary epistemology of science* (EA-EES) is SA-EES augmented by the introduction of a genotype/phenotype structure to cognitive processes so as to extend the analogy between biological and cognitive evolution. Toulmin 1972 takes the units of variation to be individual concepts or ideas, the genotype being the particular constellation of ideas that are combined to constitute a scientific theory. Though Toulmin is less interested in the theory of the phenotype and offers it only sporadic and varying (inconsistent?) mention, the phenotype corresponds roughly to the theory itself. In sharp contrast to this, lies the position of Hahlweg 1983, for whom the genotype corresponds to the entire realm of language, while the phenotypes are actual scientists with their epistemic commitments and practices. There are many versions of this analogy in the literature, adding to the foregoing array such analogical genotypes as meme and mnemotype, rules and routines, and analogical phenotypes such as artifacts and institutional actions. There is in addition a range of choice as to how much additional biological structure is also imported into the cognitive case, for example, whether to add also cross-lineage borrowing, narrow selectionism, etc.

From a unified embedding position, this kind of extension has no intrinsic attractiveness; it needs specific evidential justification if it is to be adopted. For there are now two parallel genotypes and two corresponding phenotypes, one set biological and the other set intellectual. What is the significance of stating a relationship between them, especially if their utterly different status be emphasized? Popper places the biological entities in his Worlds 1 and 2 (respectively, material and psychological) and the intellectual entities in his World 3 (abstract), connecting World 3 to Worlds 1 and 2 by a "Principle of Transference," which simply baldly asserts that what is true logically in World 3 is true causally in Worlds 1 and 2. But this *labels* the problem rather than solves it (cf. Hooker 1981a and chapter 3 below). There is certainly the following argument: Trial-and-error learning is fundamental to cognitive process, and this is a VSR process, so a cognitive genotype and phenotype are required. But this argument is a nonsequitur. Its validity requires at least an additional premise to the effect that one cannot have VSR processes without the equivalent of genes, that is, basic functional units simultaneously of V, S, and R, instantiated causally as reproduction regulators in individuals. This is surely an efficient arrangement in certain respects, but it is not obvious that it is the only one (even biologically, cf. the variety within Hull's more general replicator/interactor structure, Hull 1988a, b, let alone shifting to developmental systems as the focus of selection—see Gray 1992). From a unified perspective, there is only one genetic process—that causally realized by DNA and the system of biosynthetic pathways. Cognitive processes are among those made possible within the complex systems that genes regulate and form an essential part of that regulatory structure, but why should the cognitive regulation exactly copy the genetic process?

The *phenotype extended analogical evolutionary epistemology of science* (PEA-EES) is the basic VSR analogy of SA-EES (not EA-EES) extended by adding a theory of phenotype capacities as essential to a characterization of both biological and cognitive evolutionary processes. Essentially, PEA-EES recognizes that biological selection acts on the phenoype, through the complex interactive ecology it and other phenotypes jointly create, not on the genotype directly, and, conversely, that variations cannot become eligible for environmental selection unless they have first led to a viable embryogenesis producing a viable phenotype (cf. Bonner 1982; Goodwin et al. 1983, 1989; Raff/Kaufman 1983). In the same way, a detailed theory of the development and functioning of the cognitive phenotype (human knowers), including its group or social behavior, is essential to understanding collective cognitive evolution, specifically science. Most recently, Hahlweg has developed a detailed theory of evolutionary epistemology of this kind (Hahlweg/Hooker 1989a, part II).

PEA-EES (taken generally, not as one of its many specific versions) is, I suggest, intrinsically compelling in a way that EA-EES is not. It is hard to see how any genetic regulatory process could escape having the two features cited above. Though in special circumstances, the phenotype might take on a degenerately simple "transparency" (e.g., when instantaneous panmixis and other idealizations are assumed so as to yield standard population genetics—cf. chapter 2), it would still be conceptually wrong to delete phenotypic capacities from the theory precisely because one can do so only in idealized cases. On the other hand, those forms of PEA-EES that also incorporate the additional structure of EA-EES carry with them the difficulties with constructing and justifying two parallel genotypes and phenotypes. In what follows, a generalized version of PEA-EES will be adopted as a proper part of a general regulatory systems conception of life, and specifically so as to situate the important adaptation/adaptability distinction (more generally, the vertical/horizontal regulatory superfoliation distinction, see chapter 2, section 2.II.2 below).

This brings us to the *embedded evolutionary epistemology of science* (E-EES). Viewed cybernetically, evolution presents a multilayered development of regulatory systems, from chemical auto-regulation, through various orders and levels of cellular and multicellular regulation to ecological regulations (local, regional, and planetary). Regulatory complexity may increase "horizontally," by increasing refinement within an existing regulatory structure, or it may increase "vertically" by adding new regulatory orders (regulatory ascent). From this point of, view the highly organized human cognition known as science represents an extension of both horizontal and vertical regulatory complexity across many orders and levels: Individuals carry much more complex cognitive regulatory maps, and the science-technology complex has driven both the planetary transformation of ecologies and the creation of planetary institutions of many kinds. This is the current cutting edge of the self-organizing adaptive processes that have been developing (no doubt irregularly and riskily) over evolutionary time. From this naturalist perspective, any biological/cognitive parallelism is eliminated. The aim instead is to understand the whole process, with cognition a naturally embedded, if distinctive, part of it.

Chapter 2

Toward a Regulatory Systems Conception of Science

Introduction: Five Regulatory Systems Ideas and Their Use

The purpose of the present chapter is to set out the regulating framework of ideas about regulatory systems that will support the reconceptualization of science as a distinctive dynamic process in a regulatory system, a specialization of self-organizing biological processes more generally. Five basic ideas about complex organized systems are introduced in this chapter.[1] Part 2.I lays out a framework for developing an account of complex regulatory systems using two of them, while Parts 2.II and 2.III introduce the other three ideas and explore some of their consequences for a theory of science. These ideas will then guide the studies in epistemology presented subsequently. They are, in the present historical circumstances (compounded by my personal limitations), tentative. They are also presented here in highly abbreviated form because of space limitations. Nonetheless, they are intended to be first steps toward creating a philosophical theory of science, and of cognition generally, in the terms characteristic of regulatory systems dynamics.

2.I. A Framework for Theorizing Complex Regulatory Systems.

Part 2.I provides a general framework for theorizing complex regulatory systems that shows how the various different approaches to them, while often treated separately and even regarded as mutual competitors, actually fit together in a complementary fashion to form a whole. It is a completely general account, applicable to all manner of such systems, but in pursuit of the goals just enunciated, it is developed using the general language of biology, and the discussion focuses on the development of a conception of science as dynamic process. The analysis will also allow the removal of some widespread assumptions and misconceptions that stand in the way of a naturalist, regulatory systems analysis.

2.I.1. The Distinction between Functional(/Informational) and Causal Systems Descriptions (Idea 1).

Complex systems, whether biological or cognitive, have proven elusively and confusingly difficult to theorize. There are several factors at work here, for example, the existence of multiple and conflicting versions of such basic concepts as organization and information, and theoretical uncertainty about principled relations between dynamic, thermodynamic, cybernetic, and informational characterizations of systems. One primary factor underlying these unclarities is the failure to apply a clear distinction between functional and causal characterizations of physical systems and their processes. Since this distinction has been discussed for genetic and cognitive systems in Hooker 1981b, Part III and presented for biology more widely in Hahlweg/Hooker 1989a, Part III, I shall be brief here.

A simple example of the distinction is provided by the thermostat, a device for regulating temperature. When one purchases a thermostat, say for the office hot water urn, one may know only that it stabilizes the temperature of the water. This is a purely functional characterization. It says that a thermostat-controlled urn given inputs of cold water and electricity will output a stable supply of hot water. Of course it (notoriously) does this only within certain limits. More detailed functional descriptions are therefore desirable and possible. For a particular thermostat and urn volume, we can specify a time-dependent input/output map so that, given a particular time distribution of urn use (output volume) and cold water and electricity input rates, the characteristic time distribution for urn output water temperature is given. For the class of thermostated urn systems, there is an obvious and important sense in which these functional characteristics are their most important characteristics. It is this knowledge to which explanatory appeal will be properly made, for example, when explaining why a demand pattern that exhibits large, sudden increases around 10:30 A.M. and 2:30 P.M. will produce subsequent 30 minute periods of tepid urn water. To use this functional representation to explain and predict a variety of urn temperature phenomena, it is not necessary to understand anything about the specific causal processes by which the thermostat works or by which the water or energy is switched on to the system. It is not, for example, necessary to understand whether the thermostat is essentially governed by a bi-metal strip, or a coiled spring or a chamber of gas. In this case, the very same function may be causally instantiated in any number of different ways, and the response analysis would continue to hold good.

Yet should the thermostat begin to malfunction, it will become crucial to understand its causal structure in order to understand how to correct it. Is the contact corroded, or merely oxidized? Is it the electrical connection or the bimetal strip that is broken? And so on. More importantly and more subtly, it is also necessary to understand its causal structure in order to understand the possible malfunctions and their likely consequences. Perhaps the bi-metal strip may fatigue and leave the urn permanently switched on, risking the creation of an electrical fire, while spring fatigue would instead fail to switch the urn on, leaving only disappointed customers. More generally, the causal analysis of the water heating process is what underwrites our capacity to set down general functional relations for the urn systems. (Crudely: Rate of water temperature increase is determined by total resistive heat energy delivered electrically and the current volume, temperature, and specific heat of the urn water. The formula is only complicated, but not essentially changed, by adding intermittent cup- sized emptying and continuous, depth-determined filling with cold water.) In short, knowing how a functional relation is causally instantiated is crucial to understanding its general form and its limits. But each can be known independently of the other.

The thermostated urn can be represented as making a simple decision: Is the water I hold colder than (some desired temperature) T? If no, switch off electricity; if yes, switch on electricity. Call this Program P. This decision tree can be easily elaborated by adding automatic switch-off protection when the urn is drained. The required electrical circuitry is thoroughly physical and causally specifiable, but the functional result is perspicuously represented as follows: Is the water of depth D or less? If yes, switch off electricity; if no, go to program P. In this way, the functional relation characterizing the urn is converted into an informational form. The thermostat-controlled urn system is represented as a virtual information processing system. Conversely, every information processing system, virtual or real, specifies a functional input/output relation. Thus information processing characterizations of systems have the same complex relation to causal characterizations as do functional ones.

Indeed, perhaps the most arresting examples of informational versus causal representations are provided by calculators and computers. Most of us understand only the broad outlines of the relevant informational representation, for example, that the desk computer before us implements Wordstar word processing or the programming language Turbo Pascal. Relatively few of us understand anything of the detailed engineering causal representation of how the particular devices actually carry out these functions. Again the same lessons apply: Often it suffices for explanation to have recourse only to some portion of the information specification of a calculator or computer, for example, to understand why it printed an error message rather than the desired screen configuration. But if one wants to understand the possible range of dynamics accessible to these systems, in particular if one wants to write new kinds of programs or understand their behavior during malfunctions and under non-standard conditions generally, then it is unavoidable and crucial that one have recourse to their causal engineering specifications.

Both notions are applied in theories of biology and science. What is passed on in heredity is the information of how to construct a new phenotype. This information is expressed in the organization of protein assembly and like processes, that is, as a set of cellular functions. It is the change in the informational content of the genome that ultimately accounts for novelty in evolution. On the other hand, evolution is clearly a causal process in the same vein as any other physical process. The survival and reproduction, or death, of an individual phenotype and hence also of populations of phenotypes is a causal process. Likewise, genetic mutation is always realized as a physical process (however exactly it comes about). Modern Mendelian or transmission genetics offers a functional account of evolution while molecular biology is now offering us the first partial specification of the complex causal account that subserves it.

Transmission genetics specifies the conversion of bundles of phenotypic features as inputs, understood to represent a population or group of phenotypes (i.e., roughly, the usual individuals), into output phenotypic feature bundles (a next generation) via genes, assumed to be simultaneously the unit of biological production for hereditary characteristics, of mutation and of combination,that is, as determining a stochastic transfer function. So transmission genetics is a functional theory about the processes by which phenotypic characteristics are transmitted from generation to generation in populations of organisms. Genotypes are defined functionally within the theory in terms of stable input/output phenotypic feature bundles. Thus they may equally be thought of as some kind of regulatory control structure whose exercise results, given characteristic initial fertilization conditions and subsequent environmental inputs, in a characteristic ontogenesis. This is the appropriate functional model for modern operon analysis. The population gene pool exhibits a temporal evolution as the causal results of reproduction and demise through selection combine with genetic regulatory shift (cross-over, inversion, and the like) and are reflected in the composition of the population. All this is functionally reflected in transmission genetics just by the specification of time dependent probabilities for gene expression, but more complex models are also available, for example, in terms of genetic algorithms (Holland 1992). The general conversion of functional into informational representations then licenses talk of genotypes as storing the information for constructing their phenotypes and of the genome as the storage of environmental information within the population.

The causal representation is concerned with the detailed causal structure of individuals and of both inter- and intrapopulational interactions among individuals and with their inanimate environment. The conditions under which a particular predator population is in equilibrium with a prey population may depend, for example, on the satiation time after feeding, and this time may in turn be a causal function of environmental features (e.g., ambient temperature) as well as biochemical features of the predator phenotypes. In terms of the simplest models, it is the difference between the Hardy-Weinberg and Lotka-Volterra equations. In contrast to transmission genetics, molecular genetics is concerned with the chemophysical mechanisms by which such genetic transmission occurs. Instead of a hypothetical partially ordered set of unit causes (genes) and an unanalyzed causal relation between these and phenotypic characteristics, molecular genetics has a set of complex molecules (DNA), an extremely complex set of biochemical processes by which specific proteins and other complex molecules are synthesized from these (the biosynthetic pathways), and finally a further relation between ordered sets of proteins and ordered sets of macroscopically observable phenotypic characteristics.

Some special features of functional descriptions are worth noting for later reference. A simple instance borrowed from Dewan 1976 and introduced in Hooker 1981b, Part III, is that of power systems engineers who describe a number of interconnected electrical generators as having a virtual governor with certain properties. But whereas each generator has a real governor, there is no object that is the virtual governor (hence the "virtual"); rather, the system output behaves as if the system were a single generator with a real governor. In short, governor talk is functional talk, and talk of virtual governors works because that representation is input/output equivalent to a detailed systems functional representation. But only the latter has a causal realization. So it is with many predicates of transmission genetics also (e.g., "——is dominant"). This holds the key to understanding the reduction of transmission genetic theories to molecular genetic theory. Genes are not things, but complexes of functional mechanisms.[2]

A similar case is that of claiming, roughly: The minds of humans are identical with human nervous systems. The basic framework of psychological and commonsense theories is functional; stimulus-response theory and computational cognitivism model behavior as a function of stimulus and mental state, dispositional analysis introduces similar functions. The objective is to give an account of how at least this functional framework relates to, and may be reduced to, neutral theory. Mental states stand to neural states primarily, though not wholly, in the relation in which states of virtual governors stand to systems-engineering states. Notice three consequences of this thesis: (1) Sentences containing expressions referring to mental states and/or properties are in general to be analyzed syncategorematically, their truth conditions requiring only appropriate internal causal processes and output (behavioral) characteristics. (2) Functional mental states and properties 'disappear' at the neural systems level in favor of component states and properties, structure, and causal processes (mechanisms). (3) Mental states, like virtual governors, by and large cannot be localized more closely than the entire system, here the person. What makes dualism and materialism both possible theories of mind is roughly that they are approximately input/out isomorphic.[3]

This framework offers the possibility of a unified theory of human knowledge as a natural product of natural processes, a direct extension of the inter-related phylogenetic, ecological, and ontogenetic life processes. This view is in contrast with acceptance of some kind of division between two different evolutionary accounts, that of our organs of cognition and that of the content of cognition (Bradie 1986). The distinction remains even within stronger formulations of evolutionary epistemology (see chapter 1 above). To a thoroughgoing naturalist's eye there is something wrong. There is still an artificial dichotomy. Knowledge remains represented as some kind of non-natural thing that can at best be shown to share common formal features with natural things but in general inhabits a world of its own with laws of its own. There is nothing in the diverse evidence for the gradual development of increasing physiological and neurological complexity to suggest anything but a single natural process. The correlative development of behavioral complexity, including social organization and communication, shows a similar continuity. Language in anything like its contemporary form will be noted as a very recent arrival, preceded by other, natural structures that are at least similar in richnees: song, mime/dance, and game playing.

Our currently best account is something like this: Living systems are dissipative, far-from-equilibrium structures stabilized by an energy flow and open to environmental interaction. Adaptive self-organizing processes in these systems incorporate order (negative entropy) available from the environment to produce function-supporting molecular structures which express it as increasing systemic regulatory structure. This corresponds functionally to the incorporation of environmental information since information is environmental order incorporated as meta-stable structures whose state transitions are a function of input. (Meta- stability represents uncertainty and the stimulus-dependent state transition, which initiates system response, represents reduction in uncertainty.[4]) In these terms the point of regulatory organization is to realize a sufficiently rich structure of meta-stable states. This is a quite general framework within which both evolution and ontogenesis may be represented as particular adaptive self-organizing processes. With respect to evolution, selection ensures that phenotypic structure maps relevant environmental features, but indirectly through the constraints imposed by natural laws and the regulatory systems context and only up to reproductive equivalence. Ontogeneses simultaneously express stored genetic order/information and that incorporated from the current local environment in those respects in which the ontogenetic process is sensitive to environmental factors, for example, the direction of light in relation to plant growth. (But though both phylogenesis and ontogenesis can be represented in the same dynamical framework and, as noted, there will be important interactions between them, the use of the framework per se does not imply any specific relation between them, that will be a matter of empirical detail and can be expected to vary with species and ecological environment.)

This common dynamical framework also makes available a unified treatment of cognition as an extension of biological regulatory self-organization. Consider, first, neural ontogenesis and individual cognitive development. The central nervous system continues to change throughout life, both in causal structure and, correlatively, functionally. There is no discontinuity at which the brain stops developing and starts thinking, nor is there any bifurcation after which the information that the brain stores accumulates independently of structural change. The characteristics of brain function and dynamics may change in important ways between early development and across adults (e.g., among builders, violinists, and mathematicians). Indeed, there is no sharp contrast between the causal structure of any organ and the information it contains, rather, the information is embedded as causally active meta- stable structure. From this point of view, the whole life-history of the brain is its ontogenesis and this includes all its cognitive history. As Piaget puts it, cognizing is an extension of embryogenesis, a natural process of causal alteration of meta-stable structure and correlatively changing functional activity. The processes of meta-stable structural development and those of information incorporation are one and the same.

Cognitive development, as a special case of information incorporation, is instantiated as a subset of these structural developments.[5] Similarly, every evolutionary change represents the incorporation of thermodynamic order and, when incorporated in appropriate metastable structures, is the incorporation of information—with the historical development of information within the human species, especially within its scientific institutions, as a specialized extension.[6]

Within this unifying framework there may be important differences of process, both between phylogenesis and ontogenesis and between the broadly biological and specific cognitive extensions of each, but these will be expressed as differences of dynamics. How deep the differences run and what the relationships are is then an empirical question. (This question has recently been reopened and is currently unsettled.) In any event, the naturalist evolutionary epistemology advocated herein offers a naturalist retheorizing of these processes within a unified metaphysics of regulatory systems.

2.1.2. The Distinction between Populations and Individuals: Not Collections to Members but Regulatory Systems to Subsystems (Idea 2).

The preceding discussion used the much more common distinction between individuals and populations. Though common, the distinction is not as simple as it may appear; it is necessary to discuss it briefly here in order to proceed with the development of the regulatory systems framework to be used in this book.

Darwin pioneered a sharp distinction between the two: Individuals develop according to genetically determined rules, but populations evolve through natural selection acting differentially upon phenotypic variation (Lewontin, 1982; Sober, 1984). But in physics, the distinction between the one and the many is not so clear cut. The "autonomy" of molecules varies sharply between gases, liquids, and solids. Gases exhibit maximal molecule-molecule independence, molecules influencing each other only during brief local interactions that are assumed to be relevantly random. The macroscopic properties of a gas are therefore essentially determined by its environment. In solids, by contrast, autonomy is substantially lost; the intermolecular forces are so strong as to bind individual molecules to fixed group roles. Solids may themselves form "super individuals" (rigid bodies) to repeat the range of possibilities at an aggregated level. Note though that solidity may also create freedom for submolecules, for example, for electrons moving into the Fermi conduction band of a metal crystal. Liquids occupy an intermediate position; liquids molecules spend sufficient time in "small group interaction" that their behavior is much less individually autonomous, though partial individual "looseness" is restored because the small groups constantly dissolve and re- form and there is partial small group looseness (depending on the lumpiness of the liquid).

Agency terms have deliberately been chosen to describe these different physical circumstances to emphasize parallels with the range of possible conditions of biological populations. Populations of lone migratory creatures, such as ocean turtles, probably lie close to the gaseous condition; those with strong group structure, such as gorillas, behave like lumpy liquids and populations of cells now forming an organ are quasisolids. Some populations, notably the slime molds, have the capacity to switch between modes. Thus even if we think of individuals as classical molecules—impenetrable, immutable, local units—we can still open up the question of the many-one relationship. Suppose instead that we adopt the metaphysics of regulatory systems and think of individuals as complex, highly interactive, regulatory systems containing many partially autonomous subsystems (organs, cells) and entering into larger regulatory systems at several levels (groups, societies, ecologies) in various roles. Then it becomes still clearer that the systemic character of biological populations is an empirical matter.

While each individual's life-history contributes to the history of its population, any specific individual's life may range from nearly irrelevant to absolutely crucial for the history and fate of its population or species. It is surely statistically unimportant for the population of robins whether or not some particular robin is devoured by a cat ere nightfall. On the other hand, it may be crucial to the future of a species that the dominant male of its one surviving colony is extremely wary, having once nearly been taken by a predator. In general, the details of specific individuals are not statistically important to the history of the population, which typically far transcends any individual's history. Under such conditions, evolutionary dynamics can be accurately specified in terms of population characteristics alone, say via fitness coefficients. The classic Hardy-Weinberg law, for example, presumes such conditions. This representation is, however, an idealization producing an attractive simplification of treatment when applicable but only ever approximately true and often enough clearly inapplicable. When individuals become important for understanding specific ecological dynamics, the statistical assumptions of an autonomous population dynamics break down. Population history without individual explanation becomes dynamically blind. (Conversely, for complex individuals exhibiting social community and learning, individual explanation without social and historical context is blind.)

There is nothing distinctively biological in this, it is a common feature of all statistical theories. The randomness and homogeneity assumptions of statistical mechanics break down at phase transitions where particular molecules may be crucial in initiating crystalline or gaseous formation, panmixing assumptions break down for suitable chemical reactions, forming spatial segregations that may be initiated by small molecular fluctuations (e.g., in the Zhabotinski-Belusov reactions), and so on. Correspondingly, biological individuals are recognized to often be important in radiation of a population into new territory (founder effects) and in adaptive radiation into new niches, or where populations are small, where socially transmitted learning plays an important role and so on. (These are cases where fluctuations are important or where stochastic models cannot be applied because there are not well-defined long-run frequencies.)

The previous point is only a special case of a more general one now emphasized by many biologists: The phenotype is essential to biological explanation. Phenotypic structure and functioning is essential to understanding the dynamics of selection because it is essential to understanding ecology. Environmental selection forces act on the individuals of a population, on its phenotypes, but the living part of the environment itself is made up of other populations. Changes in each of these can likewise only be understood, or even discovered, by reference to these same factors. As Levins puts it:

> Organisms (a) select their environment actively, (b) modify their environment by their own activity, (c) define their environment in terms of relevant variables, (d) create new environments for other organisms, (e) transform the physical nature of an environment input as the effects of their activity percolate through the developmental network, (f) determine by their movements and physiological activity the effective statistical pattern of environment, and (g) adapt to the environmental pattern that is partly of their own creation. Further each part of the organism is "environment" to the other parts. (Levins 1979, 766)

Furthermore, individuals are complex regulatory systems whose internal regulatory viability is a primary constraint on which genetic variants are able to be presented for environmental selection.[7] So even the possible populational determinable properties, as well as their determinate distributions, depend on phenotypic features.[8]

Nature is a nested set of regulatory systems. These stretch down to at least sub- cellular structures, which are already extremely complex biochemical systems. And they stretch upward to encompass successively cells, organs, organisms, groups, populations, ecologies, and the biosphere. Each systemic level has some degree of control over its functions and states, which can typically be expressed in terms of homeostatic/rhetic conditions, and each is constrained by both the character of the subsystems it contains and the systems for which it is in turn a subsystem. Over time, what was topmost biochemical regulatory structure becomes embedded down in the regulatory structure, placing its constraints on the operation of the new upper layers, and constrained in turn by their viability (i.e., by their homeostatic demands). In particular, what was a topmost regulatory population of loosely constrained individuals thereby becomes a more tightly integrated component in a larger system, thus bacteria to cells, cells to organs and organ-isms, and organisms to structured groups, and these to local and regional ecologies.[9]

In sum, both individual organisms and the groups and ecologies they together constitute are complex, highly interactive regulatory systems (see also Pattee 1973, Salthe 1985, 1993, Ulanowicz 1986, and the asterisked bibliography generally). To understand the dynamics of this complex total system, it is essential to represent the properties and interactions of each of the subsystems. One can locally factor out momentary gene frequency time dependencies, but except in special, simplifying circumstances, by themselves these have little dynamical significance or explanatory value.

The relationship between developmental processes in individuals and selection processes in populations is of central importance to biological dynamics (note 7). Gray 1992 argues that developmental systems, rather than DNA, organisms, or whatever, are the basic evolutionary units. There is something clearly right about this argument; for example, DNA strips without supporting organismic and ecological resources are useless, and just as organisms are reproduced through the biochemical activity of DNA, the organisms collectively reproduce the DNA and help to reproduce the ecological resources (e.g., humus) the next generation must inherit. But I do not wish to hold, as Gray seems to, that there is no causally distinguished role for DNA strips (no functionally distinguished role for genes); for example, they form a functional memory in a way that other resources do not, and they play a causally and functionally distinctive role in reproduction. Indeed, I wish to prescind from the dispute about the so-called units of selection here,[10] except to reiterate its empirical character, and simply emphasize the importance of developmental systems to understanding evolutionary dynamics.

From a dynamical systems perspective then, the distinction between populations and individuals is better viewed as one of degree rather than kind. Other considerations could be listed in support of this approach. Hull 1984 has argued that species (i.e., interbreeding populations), should themselves be seen as individuals. Also, understanding ontogenesis, especially cellular differentiation, may require partial representation of individuals as Darwinian populations of cells, or even of subcellular parts. Conversely, there may be well- defined, if limited, roles for group selection thus endowing sub- (and perhaps supra-) populational groups with a distinctive unity. Once cells, individuals, biological communities, and so on are all understood as complex, highly interactive dynamic systems there is no longer an obvious causal (i.e., dynamical) basis for drawing any sharp difference in kind between them; differences become a matter of empirical detail.

There is then the issue of how to model the individual\population distinction. The choice is essentially between that of modeling it on the distinction between an aggregate or collection and its members and modeling it on the relation of subsystem to system. Despite our natural tendency to take our individual selves as primary, the latter seems the better model. The former can always be recaptured, if applicable, as the limit of zero interaction among subsystems.[11]

An immediate methodological result is the elimination of an a priori or unquestionable status to the evolution/development distinction and the opening up of the issue to empirical investigation. And the interaction structure sketched above does imply a certain displacement of simple Darwinian variation-selection-retention process as the *self-sufficient* basis for evolutionary theory. However, that interaction structure is in itself certainly compatible with orthodox neo-Darwinism. If the account is to be taken beyond neo-Darwinism through the introduction of non-Darwinian self-organizing systems conceptions (à la Brooks/Wiley 1986, Gray 1992, Kauffman 1993, or Wesson 1993), then the relation between development and evolution will become still more important. The whole issue is at the cutting edge of research and will be left unresolved here. All that is required of the distinction here can be accomplished without having to presume a resolution. So for expository convenience, a naive, minimally committed population/ individual distinction will be used and where pertinent (e.g., in the examination of

Piaget's biology in chapter 5), the position taken will conservatively side with neo-Darwinism.

The population/individual structure can now be applied to the description of the scientific community. Science is conducted by a population of scientists, or more properly by a collection of such populations: nuclear physicists, evolutionary biologists, etc. bound together in institutionalized groups and communities in laboratories, universities, learned societies, government agencies, and so on.[12] The scientific behavior of individual scientists includes out-communication (lecturing, publishing, letter writing, etc.), practical activities (designing or building equipment, conducting experiments, etc.), and in-communication (reading, listening, etc.). Each scientist is partially characterized by a set of cognitive commitments by virtue of which they display their variety of cognitive behaviors. For example, a scientist may fully believe a theory, or accept part of it and doubt other parts to varying degrees, or accept it for trial while remaining agnostic, and so on. The same range of variation applies to all the categories of commitment involved: commitments to concepts, principles, data, theories, methods (theoretical and practical), norms, possibilities, metaphysics, valid and plausible reasonings, scientific values, responsibilities and ethics, and so on. A population of scientists exhibits a characteristic distribution of each of these cognitive commitments.

These characteristic populational distributions may, and typically do, change over the course of time. Because of their cognitive ability, scientists can change their institutionalized cognitive environment by introducing new empirical results, new instrumental or theoretical methods, and so on and in so doing create new demands or problems for themselves and the other scientists in their environment. In the case of animals, the response to these problems may be directly behavioral, that is, within their current behavioral repertoires (feedback loop #1). Alternatively, it may require more profound phenotypic change, but species with sufficient phenotypic adaptability will survive, though changed (feedback loop #2). If conditions are appropriate, these changed phenotypic requirements will come to be reflected genetically because of the adaptive advantage of ontogeneses that lead, or lead more efficiently, to the new phenotypic requirements (feedback loop #3). In all cases, a new ecological structure emerges with redistributed populations and behaviors, though this change is likely to be successively more prominent across these feedback loops. For scientists, a satisfactory response to the problems posed to them may also be within their current explanatory and laboratory repertoires (feedback loop #1). Alternatively, it may require more profound neural (cognitive) and behavioral change, but the research activities of scientists whose current methods confer sufficient adaptability will survive, though in changed form (feedback loop #2). If conditions are appropriate, these changes will come to be reflected in changed neurally instantiated cognitive commitments (e.g., new theoretical principles) that explain, or have greater power to explain, the new successful laboratory methods, empirical findings, and so on (feedback loop #3). In all cases, a new institutionalized cognitive environment or ecology emerges with redistributed populations and behaviors, though this change is likely to be successively more prominent across these feedback loops.

Rephrased more conventionally, methods, theories, and data are all subject to elaboration or rejection as well as entrenchment by the scientific community, though the two processes are usually spread unevenly across it. Rejection emerges from observational mis-match with theory, internal criticism of argument, the introduction of competing theories and/or methods, the arguments of colleagues who have shifted to alternative values and practices, and so on. Reactions to criticism, overt or implied, vary widely. Some scientists readily accept a critique, while others resist it; some, accepting it, transfer commitment conservatively to a nearby alternative, while others seek a radical critique, and so on. Elaboration and entrenchment derives from the same kinds of sources when consensus rather than dissensus is operative. The net result, in some approximation, as Campbell 1974 and Toulmin 1972 have described so vividly, is a generic process of variation, selection, and retention that retains the general cybernetic features of the biological process. But we should note that, though charming, this is clearly a simplified "first-order" picture. There is, for example, no sharp distinction between the three feedback processes of the previous paragraph; each may interact with the others. The full picture involves a highly interactive nonlinear dynamic system whose behavior is correspondingly complex.

The specific characteristics and life experience of individual scientists are essential to the explanation of their individual scientific behavior and fate, including the reproductive success of the research programs each supports. But while each individual scientist's life-history contributes to the history of some science or sciences, any specific individual's life may range from nearly irrelevant to absolutely crucial for the history and fate of any particular theory or other commitment within science. Under conditions where the statistical irrelevance of individuals applies, epistemic dynamics can be accurately specified in terms of population characteristics alone; it suffices to represent the time dependence of the populational distributions of the determinable characteristics. Standard analytic philosophies of science, for example, presume such conditions, as do standard evolutionary epistemologies. Indeed, the former assume that acceptance and rejection of theories is determined solely by formal logical rules, that is, by context-free universal rules. In this case, there can be no variation among scientists in matters of rational judgment, a kind of cognitive panmixis and panselectionism with all commitment frequencies 1 or 0. The cognitive states of individual scientists are then essentially irrelevant to an account of the history of science. Though not couched in this terminology, Feyerabend has been among the most forceful at arguing essentially the inadequacy of this simplification.[13] From this point of view, the standard evolutionary epistemologies represent some improvement in representation of the scientific process, since there is room for cognitive variation; even so, their representation is still an idealization.[14] These theories produce attractive simplifications of treatment when applicable, but they are only ever approximately true and often enough clearly inapplicable. Individual scientists may vary from negligible to crucial importance for scientific history, and as they become important for understanding specific local ("ecological") cognitive dynamics, so the suppressed statistical assumptions of an autonomous population dynamics break down. Scientific history without individual explanation becomes dynamically blind. (Equally, given the complex interactiveness of human individuals, individual explanation without reference to social and historical context is blind.)

Just as there was nothing distinctively biological in this point as applied to biological populations, so there is nothing distinctively cognitive about the same point here; it is simply the application of a general feature of statistical theories of interacting populations. Indeed, the previous point is again only a special case of the more general one, that the scientist is essential to explanation of the history of scientific commitments, because scientists' commitments in their local environment are essential to understanding the resulting local ecology of commitment and activity changes (cognitive dynamics of selection). Reproductive success is typically a function of each ecological circumstance. Among these circumstances is swiftness and reliability of a local laboratory procedure, and the degree to which the theory extensions attempted by a local group match the currently deliverable local data. The time-development of global scientific commitments cannot be understood, or even discovered, without reference to these local factors and in particular without taking the cognitive (and practical) capacities of scientists into account.

These points receive a subtle but important reinforcement in the cognitive case. There is the tension between the rational requirement for inter-subjective scientific agreement, expressed through the search for cross-situational invariances of various orders, and the equally rational requirement for criticism based on disagreement at all orders and across all subjects (see below and Hooker 1987, chapter 8 on tensions within rationality). This is the functional correlate to the twin biological requirements on any population of maximal adaptive fitness and genetic diversity. Neither requirement can be allowed to eliminate the other if finite, imperfect creatures commencing from evolutionary ignorance and acting under (genetic or) informational resource constraints are to improve (adaptively or) cognitively. So diversity and complex group interrelations are going to be central to good scientific dynamics. Moreover, Cherniak 1986 draws deep and widespread conclusions from the cognitive resource constraints under which we operate and not for philosophy only, but in a wide range of sciences as well. His arguments may be generalized and deepened with a theory of idealization (Hooker 1994c), the important point here being that finite agents have no rational alternative but to employ context-dependent, risk-taking, often heuristic, reasoning to determine inquiry selection, the standards of care to be applied, and the methods for obtaining cognitive closure. In these circumstances, it is inevitable that agents differ in their information, cognitive resources, reasonable cognitive commitments, and subsequent cognitive behaviors. Any attempt to force formal uniformity on them is literally requiring the impossible. An adequate theory of science needs to respect these constraints. So despite the acknowledged pressure to treat all individuals equally under an idealized rationality, which finds its expression in the standard formal philosophies of science, it is important that this not happen. Moreover, this conception of finite rational agents is appropriate for embedding in a regulatory model and forms the analytic basis of the conception of reason explored in subsequent chapters.

To paraphrase the quote from Levins above: Scientists (a) select their cognitive/operational environment actively, (b) modify their environment by their own activity, (c) define their environment in terms of cognitively relevant variables, (d) create new environments for other scientists, (e) transform the interpretation of an environment input as the effects of their activity percolate through their cognitive developmental network, (f) determine by their activity the effective statistical pattern of cognitive/operational environments, and (g) adapt to the environmental pattern that is partly of their own creation. Further, each part of a scientist's cognitive commitments is "environment" to the other parts. In other words, individuals are complex regulatory systems whose internal cognitive regulatory viability is a primary constraint on which cognitive variants are able to be presented for scientific testing and selection.

Though the description of science deliberately parallels the biological discussion quite closely, there are certainly differences to be acknowledged. As noted in the discussion of evolutionary epistemologies in chapter 1, in a unified naturalistic approach, there is only one proper genotype: the biological one. Nevertheless, because of the ubiquitous applicability of the complex regulatory system model, various functional and process parallels can still be (cautiously) drawn. There is, for example, certainly variation in cognitive commitment across the population of scientists, and the impact of recalcitrant observational data has a selective interaction with cognitive commitments that tends to "tune" them to the data environment. This does suggest a general cognitive variation, selection, and retention process to match the biological one, and this is the commonest view. But at best, the parallel has to be developed very carefully. Scientists certainly retain their altered commitments and teach them to students, for example, but there is no clear-cut reproduction step, and typically no clear-cut lines of cognitive inheritance, to match that in biology. The patterns and intensities of communication in science are different from and greater than those of genetic communication among biological populations, even in sexually reproducing populations. And so on. (See here the generic model carefully developed by Hull 1988a, b, vis-à-vis Gray 1992.) On the other hand, the differences should not be superficially exaggerated. Popper supposed that evolution produced diversity but science produced unity (Popper 1979, 262), but this contrast is not at all obvious.[15] As concluded in chapter 1, it is better to understand the cognitive processes as one facet of the biological process rather than a mysteriously exact but independent copy of it.

At this point those imbued with the assumption that rationality is capturable by conformity to a logical formalism may move to declare this whole discussion irrelevant to the rational assessment of science on the grounds that only the conformity or not of the end-products to the formalism counts. Recall, however, that a premise of this book is the abandonment of that failed anti-naturalist conception of reason (chapter 1). According to the alternative naturalist account of norms (chapters 1, 6), they should equally help constitute an explanatory theory of good scientific practice. So the dynamics of the complex adaptive science system are relevant. To anticipate chapter 6, science will be rational when its institutional processes are so designed that science satisfies the ideal of reason to maximum feasible extent, that is to have efficiently gone as far as is feasible toward having canvassed all relevant reasons for each action or belief, to have assessed them using all relevant assessment arguments or procedures and to have subjected these reasons, their assessment and the grounds for these in turn to all possible criticisms. The hypothetical imperative is "If you wish to proceed rationally, then do A (design institutional roles according to design D, etc.)." Pursuing rationality is obligatory, so doing A is the relevant norm. To anticipate Part II below, a central part of the normative design specification for science is that it supports the construction of cross-situationally invariant representations, since this is central to pursuit of complete explanatory power, a central proxy for truth. The result of this and like considerations will be a distinctive design to the sciencing system, yielding a distinctive cognitive dynamics for it.

We can certainly ask, as Solomon 1992 does, whether science exhibits a "hidden hand," akin to the hidden hand of the economic market, which ensures that local decisions aggregate to global acceptance in conformity with the logical formalism expressing rationality. The answer is in general not, as she discovers; the dynamics can be expected to be, and are, less perfect—so it is important to understand the dynamics in order to understand how to improve the process. More fundamentally, the dynamics of science can typically be expected to be more complex than a hidden hand mechanism permits because that mechanism requires that local differences be washed out to form a single global consensus (cf. averaging exchanges to construct global market prices). This kind of global effect is quite uncharacteristic of non-linear, self-organizing systems, especially of systems, like science, where the dynamics is sensitive to the activities of many subsystems—cf. here the roles played by leading scientists of many different theoretical and experimental traditions, powerful funding bureaucrats and influential publishing editors in any scientific field. A similar result applies in economics. Recall that my concern in this Part is not with building a normative account of science—that is multiply premature—but with developing a general theoretical framework within which to properly formulate that project, and in particular to understand science as a specialized form of biological complex non-linear self-organizing systems. The construction of invariants, for example, is intimately related to the normative goals of science, not by some logical formula, but through its dynamical relation to achieving higher order anticipative regulation and hence higher order internal representation, a general feature of complex regulatory systems (see below).

We can perhaps gain some little idea of the complexity to be expected in science by adapting Kauffman's schematic cellular automata models of genotype dynamics (Kauffman 1993), taking each cell to represent the cognitive state of some individual scientist, the iterated reproduction relations to schematically represent mutual information and influence between scientists and the fitness coefficients to represent the scientific values of the cell states. Kauffman shows that under quite general conditions one obtains a very complex distribution of subgroups of cell states and, importantly, that one can obtain a highly complex critical automata state which is highly sensitive to small local changes which propagate to switch the automation among differing global dynamics. Good scientific practice requires, as just noted, a similar sensitivity, here gained through the creation of dissensus as well as consensus, to enable its members to reorganize its practices and acceptances in the light of new information. As the discussion in chapter 1 of the disanalogy as well as analogy between biology and science in the case of evolutionary epistemology will have reminded, this kind of comparison has to be developed carefully, understanding the cognitive process as a facet of the biological one; nonetheless, in Kauffman's models, and in cognate work (see asterisked bibliography), we plausibly see just the beginnings of a new theory of complex systems dynamics that can begin to relevantly characterize the design and dynamics of scientific practice. Meanwhile, the conceptual framework developed in this and subsequent chapters is designed to prepare the ground for the emergence of the corresponding normative design theory.

Returning to the discussion proper, both individual scientists and the populations and ecologies they together constitute are complex, highly interactive regulatory systems. (See earlier references plus Dyke 1988.) To understand the dynamics of this complex total system, it is essential to represent the properties and interactions of each of the subsystems. One can factor out momentary formal argument sequences, but except in special, simplifying circumstances, by themselves these have little dynamical significance or explanatory value. The complex dynamic system represented in these interactive processes ensures that the dynamical relationship between cognitive developmental processes occurring in individuals and selection processes occurring in scientific populations is of central importance.

From a dynamical systems perspective then, the distinction between populations and individuals in science as well as biology is better viewed as one of degree rather than kind. Further, the arguments of Hull and others that populations should themselves be seen as individuals can be extended to scientific populations. Also, there may be well defined roles for group selection thus endowing various populational groups with a distinctive unity (see note 10). Moreover, the recent emergence of dynamic learning models may lead to a blurring of a sharp cognitive individual/population distinction through the entwining of Darwinian learning systems, typified by genetic algorithms, and traditional inductive systems. This intersection is just beginning to be developed. (See Holland et al. 1986 and the discussion of it in chapter 1, also Edelman's neural Darwinism vis-à-vis growing and/or tuning neural net models, Edelman 1989, Hooker 1994e.) However, it opens up the possibility of replacing the distinction between evolutionary adaptation and individual learning processes with a richer theory of dynamic learning processes in which differences again appear as matters of empirical detail instead of a priori principle.

Once scientists and the various scientific communities are all understood as complex, highly interactive dynamic systems, there is no longer an obvious causal or dynamical basis for drawing any sharp difference in kind between them; differences become a matter of empirical detail. There is then the issue of how to model the individual/population distinction for science. The choice is essentially between that of modeling it on the distinction between an aggregate or collection and its members and modeling it on the relation of subsystem to system. Despite our natural tendency to take ourselves as primary, the latter seems the better model. The former can always be recaptured, if applicable, as the limit of zero interaction among subsystems.

An immediate methodological result is the elimination of an apriori or unquestionable status to any evolution/development distinction for science and the opening up of the issue to empirical investigation. Note that while the interaction structure sketched above does imply a certain displacement of simple Darwinian variation-selection-retention process as the *self-sufficient* basis for evolutionary epistemology, interaction between the historical development of public or populational cognitive commitment and individual commitments is in itself compatible with respecting the constraints of an orthodox neo-Darwinian process model should that prove warranted. If the account is to be taken beyond neo-Darwinism through the introduction of non-Darwinian self-organizing systems conceptions (cf. Dyke 1988, Jantsch 1981, Kauffman 1993, Trappl 1986, and other asterisked references), then the relation between individual and populational commitment dynamics will become still more important. The whole issue is at the cutting edge of research and will be left as unresolved here as it was in biology. All that is required of the distinction here can be accomplished without having to presume a resolution.

2.I.3. A Framework of Two Dichotomies.

Sections 2.I.1 and 2.I.2 provide two distinctions, yielding four descriptive fields. In diagram 2.1 on the following page, their biological and scientific instantiations are given.

A complete description of biology or science (or any similar complex systems field) requires simultaneous, mutually cohering recourse to all four representations. Science, for example, can be characterized generically as an information accumulating, self-organizing, environment-transforming social process functionally based on the information processing capacities and social role capacities of its participating individuals and causally instantiated in their neur-

COMPLEX SYSTEMS SPECIFICATIONS: BIOLOGICAL

Systems Level		Perspective: INFORMATION	Perspective: CASUAL
	POPULATION	Population Genetics	Phylogenesis
	INDIVIDUAL	Functional Ontogenesis	Casual Ontogenesis

COMPLEX SYSTEMS SPECIFICATIONS: EPISTEMIC

Systems Level		Perspective: INFORMATION	Perspective: CASUAL
	POPULATION	Analytic philosophy of science	Cognogenesis
	INDIVIDUAL	Functional Psychogenesis	Casual Psychogenesis

Diagram 2.1 Complex Systems Specifications
There are four distinct kinds of descriptions which may be given of any complex system; each corresponds to a choice of system level of description combined with a choice of description perspective.

al states and causal inter-relations and in the relevant causal states of surrounding equipment, books, etc. Its normative specification then involves further specifying the generation and transformation of information, etc. so as to construct invariants and so forth, in the general manner discussed above (and further below). By contrast, traditional philosophy of science deals only with the semantic information contents of presumed globally accepted scientific statements, and it forms a (degenerately idealized) special case of the information/populational representation. In what follows, each of these four sub-fields will be briefly commented upon just sufficiently to indicate how science can be represented as one aspect, roughly the information concentrating aspect, of biology. These subfields do, however, form the proper general framework from within which to develop full theories of the subject matter of biological science and of theories of science itself.

Causal/populational representation. Describes the relatively macroscopic dynamics of collective properties of a population of individuals interacting within themselves and within their environment. A biological population is a collection of interbreeding material individuals who are (relatively) reproductively isolated. The environment itself is thought of as a collection of populations (all in principle interacting), a collection of inanimate objects together with their relations to all the living populations, and a set of environmental constraints.[16] As noted in section 2.I.2, a population and its environment jointly causally determine one another. The result is a coevolution of population and environment generating a potentially indefinite sequence of dynamic states showing from time to time stable equilibria of more or less duration.[17] In principle, one would like to treat the collection of all populations in the inanimate environment as a single dynamic system whose time scales for change may range from the very short to the very long term. It is only the practical difficulties imposed by theoretical ignorance, finite resources and constraints on the management of complexity that force us to resort to the conventional treatment by parts (populations), levels, and statistical averages across ecological fluctuations.

Turning to the discussion of the equivalent processes operating in scientific evolution, a population at a given time comprises a collection of individual scientists who share a sufficiently large set of cognitive commitments and who are relatively "reproductively" isolated from their wider intellectual environment. Even more clearly than in biology these are matters of degree.[18] As in biology generally, a population of scientists exhibits three significant classes of causal relations: relations internal to individuals, relations among scientists within the population, and relations between individuals of that population and their social and physical environment. The internal causal relations pertinent here are those that causally instantiate the cognitive processes; presumably they hold primarily among neural states. The external causal relations include causal instantiations of in-/out-communications and all relevant practical activities (e.g., experimenting). The environment itself is thought of as a collection of other human subpopulations (all in principle interacting) and a collection of inanimate structures together with their relations to all the living populations, and a set of environmental constraints (nomic or structural). By exerting causal selection forces on a population, the environment brings about specific cognitive adaptations. Conversely, the population will be a source of selection pressures for other populations in its environment, causing cognitive (and other) change in them. It will, in general, also bring about change in its inanimate environment, for example, in the collection of laboratory apparatus. (Recall here the Levins characterization of section 2.I.2 above.) The upshot of these interactions is a co-evolutionary dynamics of populations of scientists and environment, proceeding through a sequence of dynamic cognitive states from time to time and showing metastable equilibria of more or less duration (e.g., causal instantiations of currently held orthodoxies in any scientific field).[19]

The whole complex of processes is a societal dynamics describable with the tools of the social sciences. There is a continuum of processes ranging from the clearly cognitive to the clearly noncognitive.[20] What counts is the broadly representational capacities of the process, not the conventional philosophical or sociological description of it (much less an opposition between them). The aim of constructing an epistemological theory of science is to approximately extract the representationally structured, information-accumulating processes from the larger societal dynamics and focus on their design and dynamics.[21] The simplest way to do this is to introduce functional, specifically information-processing, criteria.

Information/population representation. Describes the relatively macroscopic functional, and specifically information-processing, structure at the population level. As already noted, this field includes as the commonest special case the standard representation of population genetics, in which all that is recorded is the information concerning gene allele expression frequencies as a function of time. So long as this representation is seen for the approximate, statistical abstraction it is, it is a convenient representation for some explanatory purposes. Even when this convenience is unavailable, though, there will still be some (more complex) functional genetic description of the population in principle available. The same applies cognitively. This field includes as the commonest special case the standard representation of science as a formal logical system; in effect it assumes uniformity of rational cognitive commitment (an extreme case of statistical uniformity). So long as this representation is seen for the approximate abstraction from context-dependent variety that it is, it is a convenient representation for some explanatory purposes. Even when this convenience is unavailable, though, there will still be some (more complex) information-processing description of scientific populations in principle available.

Certain strains of neo-Darwinian and philosophical orthodoxy, however, tend to regard these representations as self-sufficient. This view is rejected on grounds already stated. This kind of representation is specifically not sufficient for explanation in all those cases where statistical averaging cannot be assumed. More generally, the specific forms of dynamical laws appearing in the information representation always depend on the causal details of the inter- and intrapopulation interactions, phenotypic capacities, and the like. One can construct mathematical "laws" of genetics or logical "laws" of methodology within the information representation, but one cannot assert that they apply, or understand the limits of their application, or understand the alternative possibilities where their application breaks down, without recourse to the causal details. This is the basic insight pressed home by many historians of science against the orthodoxy of formal universal method and belief, and it is such truths that tend to convince the sociologists, anthropologists, and others that satisfactory explanation can only be given in causal terms. Yet both sides, while having something valuable to offer, by themselves succeed only in presenting incomplete and often biased accounts. Again, what is needed is the regulatory systems framework to unify their contributions and extract a principled cognitive dynamics from them.

Causal/individual representation. Describes the relatively microscopic dynamics of individuals. Each phenotype is a causally complex system of regulatory subsystems, from subcellular to whole-individual, interacting in various ways. Individual cells themselves exhibit a complex internal regulatory structure governing their 3,000+ biochemical reactions; and cells belong to specialized communities governed by their own regulatory responses (e.g., the kidneys or the endocrine system), while the body made up of these cellular communities imposes still higher-order regulatory structures (e.g., eating/digestion patterns). Moreover, while the 'lower' structures obviously place some constraints on the 'higher' structures of this structure, there is also evidence that the reverse processes can also occur, in particular that the mind can exert remarkably detailed and far-reaching effects on physiological processes at all levels and regulatory orders (biofeedback, hypnotic control, self-healing/self-harming, etc.). Finally, through perception and behavior the individual also enters into causal interaction with those larger structures that make up its ecological environment. Here both the environment and the individual will be altered in consequence; in particular, individuals will develop a range of features in response to the evolving interactions, from physiological (e.g., muscular hypertrophy—corresponds to structural memory in the information/functional representation) through neuromuscular (e.g., swimming and group hunting patterns— corresponds to behavior in the information/functional representation), to neurophysiological (e.g., neural modifications subsuming animal behavior patterns—corresponds to cognitive learning in the information/ functional representation).

A causal specification of biological individuals refers to molecules that participate in causal processes in the cell. Genes are not among the referents, being functionally specified; in what follows, *molecular genes* will be used to refer to those molecules that dynamically control the ontogenetic regulatory processes. An individual's molecular genes form an organized regulatory structure that, in the presence of specified environmental signals, causally directs the manufacture and maintenance of phenotypic structure. A program is specified, in a manner more or less sequentially sensitive to environmental conditions, for the sequenced synthesis of organic compounds that are coregulated in their rates of reproduction, the net result being ontogenesis and self-replication.

In psychogenesis too, there is greater or lesser sensitivity to environmental influences. Molecular genes control the development of the neuroanatomy and neurophysiology of basic optical processing; their cognitive function is relatively fixed. (That is, the generalized information about the optical world they incorporate in the arrays of metastable states realized in their neural structures is relatively fixed.) Yet to be raised in an environment deficient in the appropriate range of visual (or visual-tactile) stimuli is to fail to fully develop these structures, and environments placing differing functional demands on vision result in differing processing structures in their organisms. The actual directions in which we point our eyes to gain information is highly plastic or cognitively penetrable; the general form of the resulting saccades is less so (but not zero; artists and military commanders scan scenes differently). This plasticity of visual search patterning increases our ability to learn. We not only learn where to look for items of salience, but we also learn what to look for and how to look for it, and this process occurs throughout life, as witness the continuing development of many adult artists. Like physiological processes more generally, visual processes aim at the development of an inner homeostasis, in this case, by searching until an invariant optical reference structure has been extracted from the stimuli.

The processes of cognition are therefore to be understood as highly complex causal regulatory processes, 'top layers' of the complete set of regulatory structures that form the whole organism and whose formation is its ontogenesis proper. Following Piaget (see chapter 5 below), individual cognitive development, or psychogenesis, is understood as an extension of embryogenesis, that is, as a continuation of ontogenesis. There is no difference in kind, only in degree, namely, in the degree of volatility of structure vis-à-vis that of more basic functional structures. The eye and its retinal ganglia are less volatile than some of the deep visual processing structures, perhaps only marginally for three-dimensionality and more noticeably for object constancy. These in turn are less volatile than the molecular structures that subsume episodic visual beliefs. They all exemplify processes by which environmental information incorporated structurally as metastable states both affects and effects function. Piaget stresses the feedback relationship between cognitive states of the individual and the environment within which it dwells. The capacity to entertain theoretical hypotheses, for example, gradually develops from the capacities to "fill in" hidden objects, notice patterns, and make conditional judgements, and these in turn derive from still more elementary sensory-motor regulatory structures. It is in this way that we understand Piaget's insistence that psychogenesis represents an integral part of embryogenesis. There is, indeed, no reason to believe that ontogenesis proper ever concludes. The causal incorporation of functionally relevant environmental information into the organism is a lifelong process, from initial embryogenesis to adult learning.

Information/individual representation. Describes the relatively microscopic information-processing structure of individuals. Psychogenesis, the informationally specified aspect of ontogenesis, is to be understood as the joint-structured process of expressing genetic information while incorporating environmental information. Exactly how the incorporating proceeds determines the accessible psychogeneses available to a species.[22] Note that, as in biology generally, so also in cognition there is no sharp boundary to be drawn between stored information and regulatory structure; each item of information is incorporated into the system as a component of its regulatory structure (by being incorporated causally as alteration of physical structure). There are only differences in the entrenchments of those items (which can be specified both causally and functionally). As in the case of any other human's, a scientist's psychogenesis continues throughout life.

A scientist's (subsequent) cognitive commitments and behavior are a complex function of cognitive commitments held and impinging stimuli, in particular information received. Because of their training and institutionalized roles, the cognitive states of scientists are typically highly sensitive to interactions with their natural (especially laboratory) and social (especially scientific) environments, although certain cognitive commitments may nonetheless be highly stable, for example, those specifying high-order norms for the research program being pursued and those expressing low-order practical beliefs (note 19). The cognitively interpreted environmental feedback received in consequence of cognitive behavior (e.g., instrument readings, e-mail critiques) results in a specific time sequence of cognitive states, that is, in (scientific) psychogenesis. In short, the complex of cognitive commitments forms a regulatory structure whose exercise, in a specific environment, results in a specific psychogenesis.

Conclusion. A thermostated hot water urn can be given primitive information (functional) and causal descriptions, and an interacting collection of them can be described at both population and individual levels. Biological systems are (much) more complex adaptable systems that exhibit correspondingly more complex versions of the same four interlocked descriptions. The point of this discussion has been both to clarify this descriptive framework and to begin to see science in its terms, as a special case of biological systems.

Both scientists and scientific institutions are dynamic complex regulatory structures whose states and structures are constantly in flux. Switching to the more conventional analytic information representation, data and accepted theories all may change, and changes in any of these in turn lead to changes in method at various orders, and so on. Shifts of this kind may extend further across regulatory orders to encompass abstract analysis, general methodology, and even metaphysics, philosophy, and metaphilosophy of science.

Changes in the regulatory structure of the science accepted by a scientist in turn means that the information sought next, the degree of saliency that it will receive when generated, theoretical alternatives contemplated, and methodological designs developed to search out salient information will all change. These changes tend to be coordinated by communication through groups and across institutions in complex patterns. Thus the new information available to the system and the manner in which it is incorporated is a function of the whole existing scientific system design. This is the characteristic historicity of many nonlinear dynamic systems. Empirical studies of these interactions typically fall within the sociology and/or history of science and are now proliferating. Unfortunately, the debilitating legacy of analytic philosophy has meant both ignoring this dynamics and an absence of corresponding normative institutional design studies. Within the dynamic systems perspective, the key issues remain those of extracting the cognitive dynamics, that is, of distinguishing those processes that serve to systematically generate, organize, store, and transmit information causally related to the environment—and ourselves within it. Only then can a naturalist normative theory be provided which explains their design.

2.II. Cognitive Systems Dynamics for Science

My aim is to portray science as a certain kind of dynamic self-organizing regulatory process. This part develops two ideas about that process. The first idea (section 2.II.1) is concerned with the construction of objectivity, the distinguishing characteristic of the science process. Here construction of objectivity is directly related to the construction of invariants, and this process in turn is first related to the psychodynamics of development and learning and then to system regulation more generally. The whole interactive process, individual + public, is conceptualized as a normatively relevant specialized case of the dynamics of biological regulatory system self-organization.

The second idea (section 2.II.2) develops a distinction between regulatory *refinement* and *ascent*, that is, between refining regulation at a given regulatory order and enlarging regulatory flexibility by developing higher orders of regulation. This latter may be achieved either by refining higher regulatory orders already present and/or introducing new ones, the net effect in either case being to increase the number and variety of conditionalizations of response to stimuli. In science, the refinement and ascent processes interact in complex ways; for example, failure to refine regulation at one order of theoretical regulation may stimulate the search for ascent to achieve a suitable generalization with which to regulate cognitive activity. The ensuing dynamics stands at the very heart of the science process. It provides a unified framework for both method and theory dynamics, so breaking the static method/changing theory dichotomy of the logic-based accounts (chapter 1).

The process of invariance construction is only described in general or schematic terms; as it stands, it is missing a specific dynamics. Trying to provide a completely specific dynamics is impossible; it would be like an engineer trying to write down the Schrodinger equation for a bridge in order to understand its structural dynamics. At best, we need a judicious selection of features that provide insight into the dynamics. Even this is difficult at this time because the reconceptualization of science as a dynamic process is in its infancy. What is offered here is intended as a beginning on the task.

First, the connection to psycho-dynamics of development and learning needs to be elaborated. Here a principled regulatory systems reading of Piaget provides a promising beginning framework (chapter 5), including even the construction of our cognitive ideal of truth as an (operational) invariant. As for learning, the literature on dynamic learning processes is vast yet seems largely unhelpful; it is either concerned with specific behavioral effects, opaque to cognitive mechanisms, or with mechanisms that don't yet connect to more general systems dynamics, in particular to invariance construction. Holland et al. 1986, one of the most promising attempts to bridge the gap from psychology to science, is a case in point; it does not explicitly discuss invariance construction at all (though the construct evidently appears implicitly and partially in assumptions about conceptual frames and perhaps in the reinforcement rules for classifiers). On the other hand, while there are now new models for instantiating learning processes not captured by Holland et al. that show the capacity to develop discriminations that can act as the basis of invariant construction (specifically, neural nets and LV controllers, see respectively Churchland 1989 and Hooker et al. 1992a), these are only just now being embedded in larger architectures where these linkages can be developed. These challenges must be left for another time.

Second, at the public process level the joint refinement-ascent process provides the missing dynamics for invariance construction. However, while it too is only outlined in general terms, it is important to understand that it cannot be filled in a priori; indeed, it cannot even be considered as a contingent but universal set of rules for which any period in the history of science may provide equivalent data points. The dynamics of science, like those of most irreversible nonlinear self-organizing systems, is itself constantly changing as the system increases its complexity, not in a random way, certainly, but in a manner that reflects regulatory refinement and ascent. Unlike its formal logic-based predecessors, the process is highly context dependent, and the contexts are historically dependent, particularly upon past learning. Some day we may have a detailed generalized dynamics of such systems; at the present time, the only way to fill in the dynamical details, as with geology, evolutionary biology, economic development, and other dynamically similar processes, is to provide detailed historical studies and to snatch limited generalizations wherever they present themselves (e.g., the development of representation spaces for invariants in physics; see below). Pursuing these studies must be left to others (cf. Franklin 1986; Galison 1987; Latour 1987), while I press on here with establishing a framework within which they will make sense.

2.II.1. Regulation, Invariance, and Objectivity (Idea 3).

The primary problem for all systems is stability, and this problem mounts in importance and complexity as system complexity increases. A stability is, quite generally, a capacity to restore and maintain invariance, within limits. And for those regulatory systems whose states are far-from-equilibrium fluctuations stabilized by an energy-negentropy flow through them, as all living systems are, the stability problem is especially important, since every system process must be constantly supported by the acquisition of fresh ordered energy. (We call our supplies food.) Simple systems may achieve stability by sinking into low energy states inside potential energy "wells" and remain static; for complex self-organizing, self-reproducing regulatory systems, however, it is the regulatory processes themselves that must be stabilized.[23]

A general way to achieve stability is through negative feedback, where departures from some reference condition are compensated in sufficient quantity to produce a return to it. Typically, positive feedback, which amplifies processes, is utilized to stimulate basic system processes and negative feedback utilized to stabilize the resulting dynamic system. A natural population, typically has positive reproductive feedback (more offspring generate still more) with population size controlled through various negative feedbacks: predation (the more prey the more predators and so the less prey, etc.), starvation, crowding-induced disease and aggression, and so on. Similar remarks apply to feedforward or anticipative systems regulation, which is essential to increasing response effectiveness. Feedforward regulation ranges from chemical production by plants anticipatory of seasonal change to anticipatory prey avoidance behavior by rodents and our procreation of large families to ensure care in old age and our own forward planning of moon flights based on the construction of complex internal cognitive models (see Rosen 1985, 1991). Complex regulatory systems will consist of complex interweavings of positive and negative feedback and feedforward processes (in designs for which there are no good general theories at present).

Feedback specifies a reference level, deviations from which form the signals that are fed back. Stability consists in maintaining the system state at the reference level by utilizing the various input stimuli and internal processes available to the system. Mammals achieve homeostasis (i.e., body temperature invariance), which is a necessary condition for the development and function of a complex nervous system (although many mammals also hibernate). Deviation below reference blood sugar levels initiates the search for food whose consumption restores that parameter to its reference value. In that search, anticipatively driven perceptual systems play a crucial role; these too work by referring incoming stimuli to established, typically learned reference levels. The structure of, and construction of, reference levels and the maintenance of invariance conditions defined by them is functionally central to the operation of complex regulatory systems.[24] Feedforward works in analogous fashion and plays an equally important role (cf. Rosen 1985). In complex creatures, feedforward involves the construction of a reference internal model or representation of the environment in order to provide the basis for anticipatory action. Constructed from constructed invariants, reference models can be modified through higher-order regulatory processes when they lead to inadequate actions. In us, this generalizes; the central role of invariance in cognitive system functioning is manifest in our construction of an objective world and in the conception of scientific objectivity to which it gives rise.

With these few crude remarks, I am going to presume an entire theory of viable systems in terms of internal process stability and invariance construction.[25] But to underline the relevance of this conception to my ultimately epistemic concerns, I shall indicate how the cognitive use of an invariant spatiotemporal framework is fundamental to the process of transcending our own egocentric and, ultimately, homocentric point of view. Notions of object constancy and dimensionality do not emerge in the young child's conceptual operations immediately at birth but are acquired over the first few years of postnatal experience. Significantly, they emerge in tandem with the developing notion of a separate self, initially centered in an awareness of an individual body. We may understand this cognitive achievement as the representation in the brain of an invariant three-dimensional spatial reference framework that transcends any momentary egocentric point of view, and in which all of those points of view are represented as so many varying two-dimensional perspective projections. In this way, the young child is able to organize and unify experiences. Since it too is situated in the three-dimensional space, this reference framework provides the basic means for distinguishing one's own body from other things and hence for representing the fact that one has an egocentric point of view. Prior to that time, such a point of view was not explicitly representable, but it can now be represented as one point of view among others. (This was beautifully expressed many years ago by Bohm 1965, appendix.) The appropriate criterion for what is fundamentally real will then be what is invariant across all points of view. Thus we regard three-dimensional physical objects as real.

This is replicated in classical physics for the species as a whole. In Newton's mechanics, all objects fall under the category of motion, in contrast to Aristotelian theory, where motion was referred to the earth but the earth itself lay outside that category. In Newton's system, we can recover our homocentric point of view by taking any true description and transforming it into a frame moving with the earth. In this way, we may both learn to recognize explicitly that we have a point of view and learn to transcend it so as to achieve a more objective assessment. In the larger framework, our homocentric point of view becomes but one perspective among many. It is only what is invariant across reference frames that has objective reality; in Newton's case, the mass and three-dimensional shapes of objects and their absolute motions. Thus within both physics and cognition more generally, we may think of our fundamental process for escaping unconscious viewpoints as the construction of a higher dimensional space of possibilities within which each particular egocentric/homocentric viewpoint is embedded as a possible lower dimensional projection, a projection from that point of view.

We move to cases of contemporary relevance when we note that the transition from classical to relativistic mechanics provides a parallel construction. We may understand the principle of relativity as beginning from the demonstration that our classical notion of simultaneity is in fact an unconscious projection of our own local space-time coordination, a coordination that has itself no explicit representation within the theory. Relativity theory provides the means to construct the description of the universe from within any one of infinitely many inertial frames of reference (all frames in the case of the general theory), given a description within any one specified frame. We are then able to represent the sum total of all such relationships among frame-dependent descriptions by embedding the representative of each frame of reference in a single invariant geometry: the Minkowski geometry (or more generally a Riemannian geometry). Quantum mechanics can also be thought of as offering a specification of physical systems from any 'measurement' frame. We are able to represent the set of all such possible relationships among 'measurement' points of view abstractly by embedding each of them in a single invariant geometry: Hilbert space.[26]

The development of physics represents the slow achievement of a particular kind of objectivity, namely, the representation of a world that successively eliminates homocentric projections through the construction of more adequate invariance spaces within which to locate events and specify descriptions from each individual point of view. This achievement is relevant to the normative characterization of science since the pursuit of objectivity is intimately connected to the pursuit of explanatory power. Explanatory power is measured by ontological depth and coherent simplicity, together with scope and precision.[27] To have an invariant representation is to permit unified explanation of all the various specific phenomena which fall under it; that is, it achieves a coherent simplification of the domain by reducing the number of independent laws and/or parameters involved in explanation. Further, transitions from one invariance space to the next characterize the greatest historical achievements of deeper ontologies and so increasing explanatory power (thus Aristotelian to Newtonian/classical to relativistic and quantum mechanics). Deeper ontologies may also be achieved within a given invariance representation scheme, as, for example, the shift from phenomenological to cellular disease theory shows. And to increase the scope and precision of an invariant representation is to increase the scope and precision of explanation. Explanatory power comprises a central complex of epistemic values characterizing science (chapter 6); its pursuit is expressed in the drive for objectivity.

These analyses offer only a fragment of a theory of reason in science. The open issues within objective physics itself remain to be explored, in particular the fundamental dispute still going on about the basic nature of rational representation of invariants. This dispute affects our highest regulatory orders and is a good illustration of the subtle capacity of science to regulate such explorations. (This is essentially the dispute between Bohr and Einstein, see Hooker

1994a for a summary.) More widely, the shape of objective theory in other sciences besides physics remains to be completed. One challenge derives from complex, self-organizing regulatory systems themselves, which are of especial importance to the biological and psychosocial sciences, for they are highly context dependent in structure and dynamics, and so their variation is as important to their understanding as are their invariants. At present, we try to construct invariant theories of their variations, with parameters measured suitably invariantly, but perhaps there are fundamental difficulties with this. Tart 1972, trying to deal with altered human states of consciousness, introduces what he calls state-specific sciences. Furthermore, there are other epistemic values besides objectivity that it is also reasonable to pursue (e.g., empirical adequacy, scope and precision, technological applicability, and mathematical completeness), and their structure needs equal exploration. Since any pair of these values may, in appropriate circumstances, compete, there is the further exploration of their reasonable joint pursuit. This latter requires a social allocation of resources and thus forces the introduction of another facet of objectivity: the institutionalized processes through which objectivity is pursued. These are processes designed to spread research effort and to eliminate accidental errors, bias, projection, etc. They are concerned with both promoting critical appraisal and achieving invariance of acceptance across methods and across persons. These invariances express themselves in the invariant representations of ourselves in our world discussed earlier. They apply to both individuals and epistemic groups, and both applications are essential.

To understand the significance of institutional structure vis-à-vis invariance construction, consider one relatively low-order piece of method: the use of Langmuir probes to measure the density and temperature of a plasma (i.e., a hot ionized gas). The central idea behind the Langmuir probe is simple: One pokes into the plasma a small piece of wire connected to some electrical circuitry; by measuring voltage and current in the wire one hopes to deduce the state of the plasma surrounding the wire. Even at this stage, one requires the use of electromagnetic theory to understand plasma/wire interactions, the use of solid-state theory to understand the transmission characteristics of the wire, the use of electromagnetic circuit theory to understand the accurate measurement of the wire responses, and so on. Moreover, the presence of the probe in the plasma systematically distorts the value of the probe responses as a measure of intrinsic plasma characteristics. The really hard constructive work that has gone into developing the Langmuir probe as a useful instrumental methodology in plasma physics research has focused around the development of theory and experimental practice to understand and correct for these two distorting effects. The resulting practices include the use of multiple probes of various sizes and with various electrical biases on them, the extraction of information from transient as well as average probe responses, and so on (see Glastone/Lovberg 1960). These practical and theoretical methods for obtaining useful plasma information have in turn required the extensive further application of the theories noted above (cf. note 43).

It took many years to develop the theory and practice of Langmuir probes to the point where a consistent and useful experimental methodology was available. During this time, theory, theoretical methods, and instrumental practices developed in delicate interaction with one another. New specific applications of electromagnetic theory to the probe sheath were developed. A set of theoretical methods concerned with the formation of systematic usable approximations for the consistent joint application of these theories and for the instrumental construction of data was developed. New statistical methods were developed for the extraction of plasma data from probe responses. And of course, new engineering procedures and experimental probing procedures were developed to meet the conditions laid down by the foregoing pairs of specialized theories. If fundamental electrodynamics had been seriously questioned—for example, had it predicted seriously wrong collision cross-sections for plasma processes—then a fourth element would have been opened up for interactive adjustment as plasma information evolved.

It is only when feedback stability has been achieved among these components (i.e., only when an interlocking web of measurement invariances, preparation methods, and so on has been established that supports a single coherent theoretical representation) that scientists will say that they really understand the laboratory plasma-probe situation. There is no neat methodological priority between method, theory, data, and technology of the kind found in conventional methodological philosophy; rather (in their corporeal instantiations), these components form a mutually interacting evolving system. The achievement of explanatory power to which this methodology leads in turn reinforces the method by providing many more ways in which a given theoretical quantity can be measured and shown invariant; plasma properties, for example, may be measured by spectroscopes, laser scattering, radiation counts, etc. The development of each of these latter methods in turn requires a similar complex, interlocking set of applied theories, theoretical data processing methods, and a supporting set of engineering and instrumental practices. It is only when feedback stability of this kind has been achieved across the full range of reasonably accessible instrumental methodologies that scientists accept that an objective understanding of plasma physics has been achieved.[28] Achieving these stabilities, local and extended, requires many different laboratories exploring various aspects of plasma-probe behavior, comparing different techniques, cross-checking assumptions, etc. In pursuing these activities, scientists and groups of scientists seek to opportunistically increase accessible epistemic values (e.g., kinds, quantity, and precision of plasma information available at reasonable cost), choosing theoretical models and experimental arrangements to suit. So these epistemic values should also be included in the dynamic system characterizing the development of plasma physics.

The overall goal is to arrive at a set of theories, theoretical methods, and instrumental practices that satisfy certain constraints, centrally that there be a collection of theoretical parameters for characterizing systems that are experimentally accessible and yield values that are cross-situationally invariant under multiple measurement procedures, and that specific instrumental practices are cross-situationally invariant across scientists and laboratories and can be similarly cross-correlated to a diversity of related instrumental practices in a cross-situationally invariant manner, of a kind that promotes accessible epistemic values. Science then presents an objective world in terms of the cross-situationally invariant across perspectives, objectively supported through the satisfaction of the institutional and methodological cross-situational invariance requirements just noted. These requirements represent whole-system properties of each science (or complex of sciences), they cannot be attached simply to this theory or that instrument. Rather the whole complex is a mutually regulating system tuning itself to its environment in a complex way so as to achieve local cognitive improvements.

2.II.2. Adaptation/Adaptability, Refinement/Ascent and Progress (Idea 4).

There are two impressive features of science that speak for its being an important path to knowledge: It works, and in this capacity, it gets better and better. It works because objective knowledge is possible, science achieves it, and applied objective knowledge works. The bones of an explanation of these propositions are already developed in the theory of objectivity in the preceding section.[29] And it is easy say what progress consists in; it is progress in objective knowledge, both in its scope and its objectivity. But this is to label the process, a naturalist wants to understand how this process might be seen as a natural effect of organized biological systems. The features focused on here as the basis of a naturalist story are tentative and controversial, and they are not sufficient, but nonetheless they move in the right direction and are about as much of the story as our (my) ignorance will safely permit telling at this time.

Adaptation/adaptability. Pursuing the subtheme of science as a specialized case of more general biological processes, the discussion commences with more general biological considerations. Regulatory complexity has to do with the height of orders of processes that are regulated and with the degree of regulatory refinement available. If we think of fixed features as first-order properties of organism states, then processes are second order properties and processes for regulating processes third-order properties, and so on. The development of organs is a common second-order developmental process, while the hypertrophy of one organ of a pair after the other is damaged or removed instances a third- order developmental process. Animal behavior often displays processes of high order; a dog's searching is a third order process (a modification of running patterns that are themselves second-order processes of muscle tension and relaxation), and the modification of search pattern with search environment and object sought (cf. search at leisure versus with hurrying owner) is a fourth-order process.

The capacity for higher-order modification offers clear advantages, ceteris paribus, whether this be during development or in the adult phenotype. Organisms with this capacity can exploit the freeing up of lower-order properties to adapt to local environmental inputs. The capacity of an organism to alter its first-order physiological properties (e.g., developing calluses or muscle bulk as a result of behavior), is often referred to as phenotypic plasticity, but the more general concept is adaptability, the capacity to adapt adaptations.[30] There are orders of adaptability paralleling the orders of process in a complex regulatory system: capacity to adapt the capacity to adapt adaptations (second-order adaptability, = third-order adaptation), and so on. In the discussion to follow, *adaptability* will be used to refer in a context-dependent way to whatever next-higher-order capacity is relevant. A process that increases adaptability will be one in which regulatory height and complexity of the adaptable system will have increased.[31] So also will the range of environmental inputs have increased over which viability can be maintained (i.e., relevant parameters held invariant).

Phenotypic adaptability has a special value for a species living in a heterogeneous environment, whether it be spatially or temporally variable, for as an organism is subjected to differing environmental conditions, first one adaptation and then another becomes advantageous. Selection for these capacities may produce genetic and phenotypic complexity of a higher order. In homogeneous environments, by contrast, there is no selective premium on adaptability; rather, there will have been selective tuning to near optimality and competition for the selective advantages that marginal fitness gains bring. The result is entrenchment of that status quo displaying most refinement in adaptation.[32]

Behavior is the phenotypic character that most vividly and extensively displays adaptability. In order to adapt to a changing environment, or to their own changing needs, organisms will often have to change their behavior. If the behavior is unconditional, for example, directly genetically regulated (first order), then this will require genetic and developmental changes, a process that is slow and may be expensive in genetic resources (cf. notes 30, 32, 42, and text). So it is that genetically regulated development typically leads to phenotypes whose behavior is conditionalizable to many orders and so highly learnable (cf. note 33). The complex relationship between a population and its environment that Levins describes (section 2.I.2) has its basis in behavior. It is through behavior that animals modify their environment in their own interests; it is with their behavior, especially its adaptability, that they create problems for their predators and competitors, thereby provoking further reactions that create new problems for themselves. And so on. In this way, it is behavior that determines much about the kind and severity of the selection pressures a population will face. Conversely, selection for behavioral capacities is central to the evolutionary process.

For adaptability to be of use, the adaptations have to be systematic as well as sufficiently wide ranging. Increased muscle bulk, for example, requires an increased oxygen delivery system to be usable and so a modified blood circulatory system, while these coordinated changes may in turn have many other repercussions. The systematicity of behavior expresses itself in purposive or goal-directed anticipatory behavior. Animals endowed with highly developed sense organs coordinated by a complex nervous system will be able to purposively adjust their activities according to the prevailing environmental conditions. Such capacities can prove highly advantageous. In a relevantly heterogeneous environment, improvements in this capacity also confer an important selective advantage. Complex central nervous systems are what support complex behavior. Thus we expect genotypes to accumulate that provide ontogeneses for increasingly complex central nervous systems, which in turn support increasingly complex behavioral adaptability. This is evidently so for the mammalian line. As far as we know, the capacity to learn from experience is most highly developed in mammals. Finally, the possessors of behavioral adaptability may advantage themselves by the ways in which they are able to disturb their environment so as to learn its responses and thereby increase its heterogeneity. This adds a reinforcing positive feedback loop to the development of adaptability, since we now have to do with both the generation of new conditions (problems and responses) per se and with the rates of generation of these.[33]

This latter dynamic is one of particular importance for human science, since science disturbs, in fact transforms, its environment(s), generating an accelerating "change machine" (chapter 1). Intelligence has to do with the height of conditionalization of regulated behavioral responses. If we think of a fixed behavior B as a first-order functional property, then a conditional response ("If M then B," where M is some internal condition) is a second-order property, a doubly nested conditional response ("If M, then if M then B") a third-order property, and so on. Generally, M = M(S, m), where S is some stimulus (input) bundle and m some complex mental state. As noted, animal behavior often displays behavioral processes of high order that are, ceteris paribus (note 30), advantageous, but then so too does the behavior of groups of animals (e.g., timber wolves or killer whales hunting), so these characterizations may be extended to any relevant functional subsystem of a complex system. It is then natural to introduce a distinction between cognitive adaptability and cognitive adaptation for the behavior-regulating states m. Cognitive adaptability is the capacity to cognitively adapt (i.e., to modify cognitive states), the regulatory control of cognitive adaptation so as to render it conditional on environmental information and internal state (both cognitive and noncognitive). Cognitive adaptability has an order structure of the same kind as does behavioral adaptability. A process that increases cognitive adaptability will be one in which the regulatory height and complexity of the supporting system will have increased (cf. note 31). So also will the range of environmental inputs have increased over which viability can be maintained (i.e., relevant parameters held invariant).

This is evidently the kind of process that Popper is after when he speaks of the importance of the development of "plastic controls" (see chapter 3). Unfortunately, Popper placed himself in the straitjacket of a logicist theory of reason that did not permit the development of his regulatory ideas. This too is the link to Piaget's emphasis on the autonomy of the active organism brought about by the completion of its endogenous operations at ascending orders of generality (chapter 5). But Piaget's work also shows the same regulatory-formalist tension. Brown/Hooker 1994 presents a nonformalist regulatory theory of reason based on these ideas.

In homogeneous environments, there is no selective premium on cognitive adaptability; rather, there will have been selective refining of behavior to near optimality and competition for the selective advantages that marginal gains in refinement bring. (In more primitive creatures and for many inherited components of our own cognitive system, the result may be genetically entrenched.) Cognitive adaptability has a special value for a species living in a heterogeneous environment that is amenable to behavioral adaptation for that species, for as an organism is subjected to differing environmental conditions, first one behavioral adaptation and then another becomes advantageous.[34] Selection for these capacities may produce system complexity of a high order. Within its workable domain, science is our most successful and powerful means for increasing cognitive and behavioral adaptability. Each successive regulatory order of science added represents an increase in these abilities.

Refinement/ascent. The conventional information representation of science as abstract propositional contents can also be considered as specifying a regulatory system. Theories regulate the development of practices (technologies) and data structures (facts), and methods regulate the development of theories. The case of mechanics provides an illustration; see diagram 2.2.

Here, complemented by method, each order provides a recipe for constructing the orders below it. Given the logic/algebra to be employed in dynamics, for example, together with a characterization of time (say, modeled by the real numbers), then the general form of the dynamics is fixed (e.g., as between classical and quantum dynamics). Generalizing across science, theories are used to provide both the foundation for methods of data generation in their domain and a normative critique of the nature and status of that data. So we may think of theories themselves, at their more modest order, as providing a set of instructions in a program for the construction of lower order methods, technologies, and data. Even data have a residual regulatory character in virtue of their theoretical embeddedness.

The regulatory power of science derives from the height of the regulatory orders it provides. As the theory and method orders are ascended, the range of data structures that can be incorporated

Theory	Method
Order Structures	Pure Mathematics, e.g., Analysis
Logic/Algebra	Geometric Quantization
Geometry: Topology	Topological Dynamics
Projective	
Conformal	Global Stability
Affine	Analysis, Chaos
Metric	Methods
General Dynamics	Deterministic and Statistical Methods
State Specific Dynamics	Instrumental Design and Error Analysis

Diagram 2.2 A Regulatory Structure for Physics
Each level on the left-hand side regulates the constructions at levels below it using the methods at or below its level on the right-hand side.

increases. At first, only the specific empirical structures corresponding to specific empirical generalizations can be encompassed (explained), for example, how this ball falls or this pendulum swings. Then classes of generalizations can be derived through alteration of parameter values in a theory, and thereby their classes of empirical generalizations encompassed (e.g., the law of free fall or the pendulum law). Next, classes of structurally different theories can be derived at successively higher orders (e.g., from Newton's laws), and so on, for example, to Lagrangian and Hamiltonian theory and thence to symplectic structure on differentiable manifolds and so on up. Each ascent allows regulated response to be conditioned one nesting higher: "If the pattern is of kind K then if the conditions are of kind C then . . ." The widening range of data structures encompassed means a widening range of situations to which regulated adaptation becomes possible. This gives science an increasing adaptability.

But no simple linear hierarchy is involved. All of the orders of science potentially interact, and modifications in any one of them may induce modifications in any (or even all) others. This is quite clear going down the structure. On the other hand, new data can lead to the modification or even abandonment of theory, even fundamental theory, and/or of the methods based on a given theory. Theory and method are also intimately interlocked; methods are used to confirm and disconfirm components of theory, while each method is built on and requires a set of theoretical presuppositions. One does not use, for example, a statistical analysis appropriate for independent particles if theory specifies interacting fields. Furthermore, theories interact across domains with other theories and methods; for example, gas and fluid mechanics interact with thermodynamics in heat transfer theory and even in Einstein's quantum models of radiation. (See also Hooker 1987, chapter 4.) Thus we may very appropriately think of all the content components of science as together forming a highly interactive complex regulatory system.[35]

Methods, theories, and technologies may all be refined and extended to new cases. Many refinements to lens making and optical theory and practice over the previous two centuries, for example, gradually led to the improvement in performance of microscopes and telescopes thereby providing important information (e.g., for a germ theory of disease and for cellular biology generally) and engendering further extensions of optical experimental methods (cf. Brown 1987, Hacking 1983). This is the 'normal' situation. Even so, the consequences of refinements can reverberate throughout science, as those of optical instrumentation did. Or consider that our refinement of methods for measuring solar system constants, combined with our refinement of computer modeling techniques, now lead us to believe that several planetary orbits can exhibit chaos.

But science may also change in more radical or revolutionary ways. Consider first Einstein's generalization of the Newtonian mechanical concept of reference frame invariance so as to include electromagnetic as well as mechanical phenomena. This led to a profound transformation of the structure of physics that illustrates a fundamental regulatory process, the forcing of development up regulatory orders by forcing a retreat to less committed (cf. more neotenous) assumptions. Einstein was not the first to explore the idea of modifying classical mechanics in the light of electromagnetic theory, nor even the first to discuss the idea of some kind of relativity principle. But he was the one who clearly grasped that the modifications required reached as high in the regulatory structure of physics as kinematics and were not to be confined to dynamics alone. This proved to be the crucial regulatory insight. The Einsteinian shift stimulated the development of generalized theories of space- time structures and the abstract representations of dynamics on them as configuration spaces (Lagrangians) and on their Hamiltonian co-spaces, which ultimately led to still further generalizations of dynamics (e.g., to symplectic flows on abstract differentiable manifolds), an ascent of the regulatory structure through adding new top layers.

Now consider the more dramatic case of quantum theory. The new principles of quantum mechanics led to quite profound revisions in the theoretical propositions that were previously accepted, and these extended from limitations on the concept of causality down to generalizations about atomic emission spectra, that is, from metaphysics all the way down to simple empirical generalizations. And they led to an equally profound revision of methods, from new abstract mathematical methods for the synthesis of polyatomic systems to the development of wholly new technologies (lasers, superconducting fluids, etc.). Despite the magnitude of these changes, there was also a lot of continuity. The Correspondence Principle ensured that established macroscopic results, along with the same general dynamical framework of trajectories in state space, were retained. Here quantum theory also stimulated the development of more abstract dynamics generalized from classical mechanics in the search for a deeper characterization of quantization, generating geometric quantization theory, topological dynamics, and the like. The introduction of the quantization of action and related principles of quantum mechanics evidently functioned as new components in the regulatory context of classical physics.

So for both the relativistic and quantum revolutions, we find partial shifts at several different regulatory orders plus the stimulation of higher-order theory and method development. This complexity to revolution is not confined to modern science; a similar treatment can be given of the Copernican-Galilean revolution and is summarized in diagram 2.3. The same kind of regulatory representation can be given even of logic and mathematics and general theory of scientific method, where there has been historical change as knowledge improves.[36] But the power of revolutionary change lies in the degree to which it disturbs the existing regulatory order and stimulates development of a richer regulatory structure to replace it. In Galileo's and Newton's hands, the Copernican revolution stimulated the development of empirical science, especially method and high order structures (Harper 1989, 1993; Hooker 1994b, c, and references). As just noted, both the relativistic and quantum revolutions stimulated higher order regulatory structure. And from these, jointly flowed eventually the various quantum cosmologies and cosmological dynamics (black holes, big bang, baby universes, etc.) that currently captivate imagination. However, these revolutions would not have

Regulatory Level	Agreements	Disputes
Philosophy	Conformity to scripture Realism	Reason/revelation relation
Metaphysics	Natural kinds (includes earth, air, water)	Perfection/corruption dichotomy
Order structures	Natural motion (circular) Natural place	Kinematical form (motion reference, decomposition)
Method	Logic (includes use in general scientific method)	Roles of: causality experimentation instrumentation mathematics (models) Quantification
Theory		Cosmological structure Fall: 1 motion or 2? Telescope: applicability, accuracy.
Experiment/Observation	Relevant observations (falling objects, east-west winds, etc.)	Projectile trajectories, dynamics Stellar parallex Winds, tides

Diagram 2.3 Aristotelian and Galilean Principles: A Brief Comparison
Points of agreement and difference between Aristotelean and Galilean science allocated roughly to their regulatory order.

been possible without the plethora of small regulatory changes produced by the 'normal' science that preceded them, for example, Tycho Brahe's refined stellar observation instruments or the measurement of atomic spectra with refined spectroscopic methods.

The key then to understanding scientific development, whether of the normal or revolutionary kind, is understanding the dynamics of this entire highly interactive, context- dependent process. There is no sharp or even unidirectional boundary separating normal and revolutionary science; the distinction is just a first, simplest approximation to the non-linear dynamics of cognitive "phase transitions." These may be taken in analogy to physical phase transitions exhibited by most nonlinear complex dynamic systems (e.g., the liquid→solid transition). The breakdown of simple logical rules of method at revolution is the cognitive equivalent of the breakdown of linear stability analysis near physical phase transitions. But this should be taken to indicate not the irrationality of the transition, but the representational poverty of logic-based method in relation to cognitive dynamics (chapter 1). Feyerabend is right; no specific transition in the history of science will be quite like any other, any more than any specific change in the evolutionary history of genetic regulatory systems is quite like any other. But he is also wrong; this fact does not prevent our understanding them through theorizing them—to the contrary (cf. Hooker 1991a). What is truly revolutionary about science is surely not the sheer magnitude of its capacity for collecting information, nor even its capacity for generating valuable theories, important though these are, but its enshrinement of a process that has steadily pushed us to explore ever higher orders of regulatory structure.

Regulatory systems, we have seen, may show two importantly different dynamical processes, which I have called refinement ('horizontal' extension of regulation) and ascent ('vertical' extension of regulation). Regulatory refinement is the process of increasing regulation within the existing regulatory architecture available to a system. Its correlates in science include increasing precision, increasing scope of generalizations about stably characterized classes of entities, increasing technological applicability and diversity of laboratory procedure, and so on. In short, it corresponds to increasing adaptation and is roughly what Kuhn refers to as normal science. Regulatory ascent is the process of adding new regulatory layers to the system that override and conditionalize those already in place. This results in a more explanatorily powerful and adaptable regulatory system. Einstein's distinctive contribution to the development of mechanics was offered earlier as a simple example. In short, it corresponds to increasing adaptability and is roughly what Kuhn refers to as revolutionary science. (But revolution as merely large-scale change misses the centrality of ascent to scientific revolutions.)

These two processes interact in complex ways, especially for creatures and institutions operating under resource constraints. I call the combined process "superfoliation" (chapter 1). The situation is summarized in diagram 2.4, which provides an information characterization of both processes. Positive and negative feedback relations between the two processes, now to be discussed, are represented schematically on diagram 2.4.

The conditions for regulatory ascent are not well understood in any domain of complex regulatory systems. One process, already noted, is that of the neotenous response to regulatory failure. As the attempt to achieve regulation fails, commitment to specific regulatory processes may be suspended in an effort to free up resources to develop alternative, more adequate regulatory structures. Biologically, the species may arrest the rate of sexual maturation (increased neoteny), thereby freeing up genetic resources for the exploration of alternative phenotypic structures. The development of our own brains seems to be a case in point. The example of relativity theory

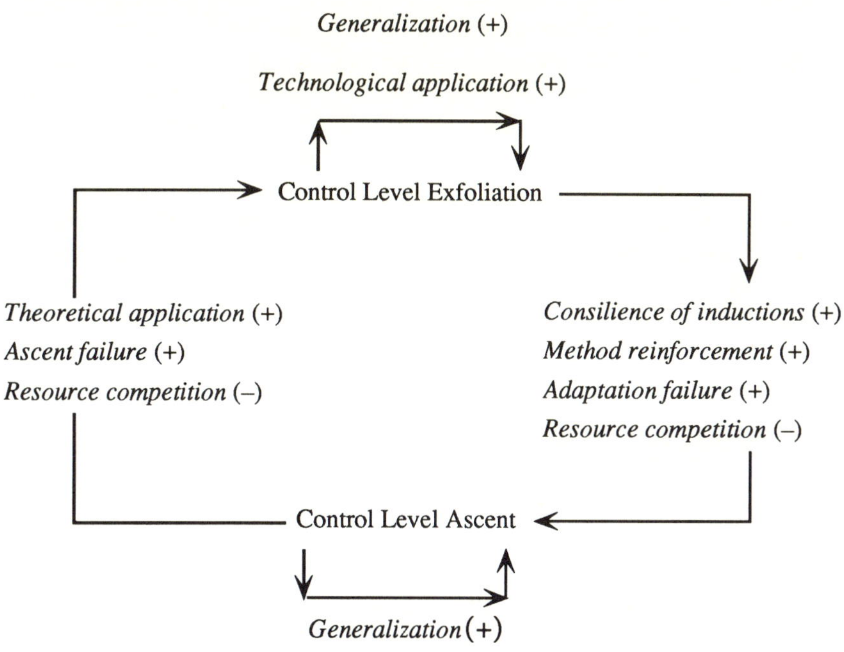

Diagram 2.4 Regulatory System Development for Science
The arrows indicate positive feedback (+) and negative (–) feedback relations between the processes of exfoliation of cognitive commitments (i.e., refinement and extension of existing commitments) and regulatory ascent in cognitive commitments (i.e., shifting categories of cognitive commitment so as to achieve increasingly higher order regulatory control).

just discussed provides a parallel 'backing up the regulatory ordering' in the cognitive domain. These are cases of positive feedback between *failure* of refinement and exploration of ascent. A related process is evidently Whewellian consilience of inductions; several instances of a higher order regulatory structure (e.g., of generalized Hamiltonian dynamics) that have emerged during the course of horizontal extension (i.e., from specific Hamiltonians) may then be compared in order to extract explicitly the higher-order structure instantiated in them. Here is a case where success in refinement leads to success in ascent.

There is also negative feedback between the two processes. Ascent may lead to theoretical representations that make the development of laboratory practices and social technologies extremely laborious and expensive (e.g., high-energy accelerators), or dangerous (e.g., genetic engineering), etc., which may detract from further refinement (or even from both processes). Again, refinement may so alter the larger circumstances in which the system of science is embedded that further ascent becomes impossible. (This will be sadly the case, for example, if we succeed in destroying our socioeconomic systems through ecological disruption, nuclear warfare, medical pandemic, or any other of the many destabilizations of the planetary regulatory system made possible by our technologies.)

Finally, successful refinement will also tend to attract resources to it, for both cognitive and socioeconomic reasons, thereby tending to distract from focusing on ascent. This corresponds biologically to creatures evolving in homogeneous environments; genetic resources are devoted to refinement of, and so entrenchment of, the status quo. Then when an environmental change does come, the result for a highly adapted species may be extinction because of its rigidity. The same is (tragically) true of science, where success means entrenched cognitive commitments and tends to breed dogmatism, the cognitive equivalent of adaptive rigidity. Then when recalcitrant experimental results or alternative theoretical ideas emerge, our dogmatic scientists (or philosophers) are unable to integrate them in any meaningful way. These groups may disintegrate, or the ideas may be suppressed and recalcitrant results explained away by ad hoc means. When the design of the sciencing system is so defective that it cannot marshal the resources to change, it must quite literally await the death of the group in order for the institution to be freed of its lack of adaptability. Famously, and notoriously, Newtonian mechanics operated (unwittingly) for over two centuries in a relatively homogeneous environment essentially characterized by human-sized energy levels and space-time scales. In that environment, it achieved a very high degree of adaptation. The population of Newtonian physicists was able to successfully predict and explain the outcomes of essentially all of the perceived relevant range of experimental situations. Newtonian-derived technologies became increasingly successful and widespread and so on. These are also the conditions for undermining motivation for adaptability, and that is why when the recalcitrant experimental results and concepts associated with relativity theory, quantum mechanics, and irreversible thermodynamics came along they proved such a profound shock for that scientific community.

Thus the interrelationships between cognitive regulatory refinement and regulatory ascent are a complex specialization of those in biology generally. No formal logic-based relationship between them will capture the dynamics of superfoliation. Rather, the whole process must be theorized within a dynamic regulatory systems context, which provides the right framework for understanding the role and dynamics of methodology within rational science; see, for example, the discussion of Kuhn's distinction between normal and revolutionary science above.

Progress. The result of this epistemic dynamics has been the (increasingly rapid) development of the science-technology system over the past three centuries. This has spurred superfoliation; it has produced greatly increased regulatory refinement, but more importantly, it has also stimulated very significant regulatory ascent, the elaboration of increasingly sophisticated theories of many kinds, laboratory, theoretical, and mathematical methods, metaphysics, and so on as already discussed. It is the increasing scale, complexity, applied empirical competence, and adaptability of the science-technology system that we associate with the progress of objective knowledge.

Talk of progress here has to be careful and cautious. Notoriously, Darwinism undercuts any but a strictly local notion of progress in adaptation. A species may make progress in the sense of refining the precision of its adaptation to a particular spatially local ecological niche over a temporally local period when the environment is stable—but not more. Short-term success in adaptation is no measure of long-term success in adaptation. (Nor, for that matter, is short-term failure of adaptation any guide to long-term failure of adaptation, so long as some survivors remain; a favorable environment may again re-appear.) Darwinian selection simply forces adaptation to track the local environment, whatever that may be; there is no direction built into this tracking process as such. For that reason, adaptation is noncumulative. Anticipating, let us call this fragile, directionless/noncumulative, strictly local progress in adaptation refinement, local horizontal progress.

A similar lesson applies, it seems, to cognitive adaptation. As the history of major scientific changes or revolutions sharply reminds us, cognitive adaptation is also not permanent or cumulative. The new principles of Newtonian mechanics (inherited in significant part, but not wholly, from Galileo) contradicted Aristotelian principles at many orders. Yet Aristotelian principles had served humans successfully for two millennia and, as Feyerabend likes to emphasize, precisely because they captured the patterns of the commonsense environment of daily action.[37] The splendid two-hundred-year history of (largely) successes for Newtonian mechanics, producing a wide array of wonderful refinements, did not prevent the demise of the theory in the changed evidential and theoretical environment that began to support relativity theory instead. Short-term explanatory/predictive success is evidently no measure of long-term explanatory/predictive success. (Nor, so long as some scientists survive who have access to a theory, is short-term failure any guide to long- term failure—a favorable environment for renewed application of the ideas may again re- appear; see, for example, the fluctuating fortunes of the wave theory of light.) Horizontal cognitive progress is evidently as local, fragile, and directionless/noncumulative as is horizontal progress generally.

This evaluation is certainly compelling. But it doesn't seem quite right. To see the primary aim of science being that of empirical adequacy to the current environment, as important as that is, is to nearly completely miss the functional significance of modern scientific knowledge, namely the transition from actuality to possibility. The age-old method of learning has been to observe passively, so as not to disturb nature, generalize from what was observed, and test by prediction. It is essentially the only survival strategy available to early human societies, which had access to relatively little energy and information, and we still find such activities today in areas where we have little information (e.g., in some biological classification) and/or little energy (e.g., astronomy). Its only goal could be refinement, specifically generality and precision, whenever the environmental pattern was stable and simple enough.

But alongside passive observation, there has been the equally ancient method of probing to disturb nature and learning from her reaction. Generalized, it is the experimental method. But to properly apply this method really requires information (to know where to direct disturbance and what kind of probe to choose) and energy (to provide the disturbance and to hold other conditions constant); it was only given prominence and systematic application with the Galilean revolution in science at the Renaissance. The secret of its increased epistemological power over passive observation is the ability to create not only regulated conditions, but conditions that would not normally appear (like near-frictionless motion). In this way, one passes from knowledge of what is merely actual, is going to happen given present conditions, to knowledge of what is *possible*, even should it never happen in normal conditions. It is this transition that made possible the overthrow of Aristotelian common sense. Newton's laws of motion tell us not only why as a matter of fact some particular projectile fell where it did, but they also tell us all the possible trajectories that projectiles can have, including projectiles that didn't exist at Newton's time, such as terrestrial satellites.

But what is possible transcends the current empirical situation, indeed transcends the current empirical environment; it speaks about all possible environments, which are natural variants of the actual one. Talk of possibilities in a natural environment provides the clue to a further, more recent elaboration of the epistemic role of possibility. Regulated disturbance (i.e., experimentation) proved adequate to also develop the technological knowledge required for the industrial revolution. But a consequence of this development of the science-technology system has been to develop a third, still more powerful, method, which I shall call the method of designed possibility. While experimental method in its classic phase revealed what was naturally possible (instead of just actual), the essence of designed possibility is in turn to obtain possibilities that are not natural but that become possibilities within a deliberately designed system. Study natural chemicals how you will, only under designed conditions can one make possible special purpose plastics with the tensile strength of steel; study purification and molding how you will, only in near-zero gravity conditions can you find the possibility of very high purity and sphericity. The scientific relevance of this method is made dramatically obvious by human space travel: There were no facts of humans in space to observe objectively in advance, and there were no tentative probings of ways of traveling to the moon to learn from disturbing them; rather, we needed to understand in advance all that was possible, though not naturally possible, for us so that we could do it right the first time. The technological relevance is made equally dramatically obvious by the current efforts being put anticipatively into developing fifth generation computers, new designer materials, etc. Any country not able to plan anticipatively on a 10–30 year future time horizon can no longer compete internationally in new technologies but is restricted to producing local variations on existing products (i.e., to horizontal technological progress).

The shift to increasingly wide ranges of possibilities corresponds biologically to conditionalizing responses in ever higher nested layers, that is, to regulatory ascent or increasing adaptability. An individual or species with sufficient adaptability will not be eliminated by an environmental shift but will be able to adapt to it. The degree of adaptability of an individual or species is measured by the number of environments to which it can adapt, what possibilities it can encompass. (Successful adaptation in an environment also requires sufficient adaptive refinement; to represent this, we could introduce a notion of adaptive power as, crudely, the product of degree of adaptability and adaptive refinement.) While successful adaptation is fragile, local, and noncumulative, adaptability is, by its functional definition, less local and therefore less fragile. (Note that this is a matter of degree, there is neither guarantee nor expectation of either universality or complete resilience.) Moreover, under appropriate conditions (e.g., reinforcement from a heterogeneous environment) and within certain limits, it may accumulate and hence define a less local direction of progress. Let us call this "vertical progress." Difficult though it is at present to characterize in a satisfying way, vertical progress qualitatively describes overall evolutionary development as complexity increases, and describes the mammalian line in particular.[38]

Humans represent environmental possibilities cognitively. The shift to an increasing range of possibilities corresponds quite literally to ascending cognitive regulatory orders. An individual scientist or group of scientists with sufficient cognitive adaptability will not have their cognitive commitments eliminated by introduction of new elements into their scientific environment but will be able to adapt. The degree of adaptability is measured by the number of environments that can be adapted to, that is, by what possibilities can be encompassed. These can be increased by regulatory ascent (i.e., by vertical progress), which at the same time often also makes possible increasing refinement (i.e., horizontal progress). (The cognitive regulatory framework cuts down possibilities even as it opens them up.) In physics, for example, very general representation theorems have been proved this century that effectively show that, given only very general assumptions about time and possible physical states, Newton's and Schrodinger's equations are the only possible dynamics for their kinematical structures and that certain symmetries must hold. And Whewellian vertical conciliational ascent has uncovered a number of fundamental constants with tight mathematical interrelations and consequences for dynamical possibility.[39] Successful adaptability has led us to successful adaptation (though not always, and often not in the short run), and of course it also requires sufficient adaptive refinement to promote basic functioning. Cognitive adaptability is, by its functional definition, less local and therefore less fragile than local empirical adequacy. Each successively higher regulatory order yields a correspondingly wider notion of (relative) non-local vertical progress. Moreover, under appropriate conditions and within certain limits, it may accumulate and hence define a direction of vertical scientific progress.[40]

Horizontal progress specifies no general or global direction to change because the kinds of changes that count as horizontal progress are specific to each environment. Vertical progress specifies a relatively more global direction to change because it covers a range of environments rather than just one. But the character of the changes constituting vertical progress will also change as one passes outside the adaptability set of environments. And should environmental shift bring this about, adaptability will fail, just as horizontal progress fails when its particular stable environment changes. And yet there is something more to vertical progress than just a less local form of horizontal progress. To see this, consider again that each successively higher regulatory order yields a correspondingly wider notion of (relative) vertical progress. Now extrapolate this to the possibility of increasing adaptability until the range of environments encompassed includes all of the possible environments in this cosmos. A creature with this capacity would be the ultimate universal creature, able to maintain its life processes in any possible circumstance. And it is a unique capacity for a given species; though there may be many routes to this condition there is a single final outcome. For cognizers, this capacity would involve having a representation of all possibilities. Indeed, just this latter condition suffices to satisfy the goal of science; it is not necessary to actually be able to survive in every possible environment. This notion is a complex one and, when thought through in detail, may prove difficult to even formulate coherently for various fundamental physics. Nonetheless, it does serve to point out a universal goal for adaptability. It is evidently the idea Piaget had when he claimed that developing a mature cognitive structure was to develop a universal set of operations, ones that could be guaranteed to apply to all possible situations, and hence to truly grasp necessity (see chapter 5).

The ascent from actuality to wider and wider sets of possibilities connects vertical progress to the achievement of objectivity. Recall from section 2.II.1 that the form objective knowledge takes is the representation of each "viewpoint" as one projection among many in a higher-order geometry. It is this geometry that specifies the possibilities, including both the dynamical possibilities (manifold of possible solution trajectories) and the various descriptions of these across possible dynamical observers. What is real is what is invariant across these, and objective knowledge is roughly description satisfying these invariances. In a heterogeneous world, the search for invariance is driven to a higher order than in a homogeneous world. Sufficiently general patterns must be found that remain invariant across heterogeneous situations (environments); these will only be found at a higher regulatory order than the data. Achieving objectivity will require increasing cognitive adaptability and so vertical progress. As heterogeneity increases, so will the requirement for adaptability to deliver objective knowledge and (so) successful adaptation or horizontal progress. The recent understanding of chaos makes a striking illustration of this (see Bak/Chen 1991, Berge et al. 1984, Cvitanovic 1984, Glass/Mackey 1988, and other asterisked references). Thus there are complex relationships between the achievement of objectivity and the achievement of progress determined not by logical relations, but through the cognitive dynamics.

2.III. Science as an Intrinsically Social Regulatory System (Idea 5)

A system of the sciencing kind has to have a complex internal organization to support its regulatory processes. For science, this is given in its institutional structure. The design of science's institutions is crucial to its cognitive capacities and dynamics. The dynamics of those designs is itself an important part of science's cognitive dynamics. The purpose of Part 2.III is to show this by exploring the intricate ways in which the social structure of science is intimately involved in its cognitive characterization and so in its normative epistemic design and behavior. This position opposes the standard view on which cognitive and social aspects stand independently, perhaps even opposed (chapter 1). The unification achieved here is made possible, and necessary, by the shift to science as a dynamic regulatory process.

2.III.1. Regulation, Information, and Institutional Design

One evolutionary strategy for realizing adaptability is (roughly) to produce simpler, more functionally inflexible organisms but to show more phenotypic variability, whether this be synchronically across an existing population or diachronically through genetic mutability.

This strategy in effect tracks environmental change through changing phenotypic statistical distributions but with individually fixed embryogeneses (e.g., many bacteria). Humans by contrast belong to that evolutionary adaptability strategy where the phenotypes are (relatively) complex, with longer lifespans and few progeny, and (relatively) highly adaptable, the population relying for its survival on their socially learned and coordinated abilities. This strategy involves locking in large genetic resources relatively inflexibly to produce a complex nervous system and its somatic coordinations. Environmental change is tracked through adaptable individual and group behavioral development. This strategy places particular demands on social organization.[41]

Humans are finite, fallible general problem solvers, they begin life with a highly adaptable general learning capacity rather than genetically entrenched specific response patterns, but must learn most of the specific skills and information they require. Given that individuals have decidedly finite capacities, the more general problem-solving capacity each individual has, the more immediate is the need for cognitively oriented social coordination, that is, for an "external nervous system." This is so because the learning capacity of each individual is a tiny fraction of that of the collective learning capacity of the social group, and individuals can develop only a tiny fraction of the specific skills that can be possessed by the social group. There is very considerable cognitive reward available to us humans through exercising our collective socialization and general problem-solving capacities—but only if we can coordinate our individual learning, spread the collective result selectively but effectively across the group, focus on it explicitly so as to improve it and transmit it from generation to generation. The social processes by which these four crucial cognitive processes occur are structured through social institutions, that is, sets of coordinated social agreements setting up systematic expectations of self and others (cf. Vickers 1968, 1983). Fortunately, the greater cognitive capacities of general problem solvers also allows them to grasp and develop complex social institutions for these purposes, at least to some extent. Scientific institutions are parallel distributed processing systems which serve to distribute limited skills, information and resources and, by doing so judiciously, also distribute risk and responsibility in manageable ways, the joint net effect of these several distributions being the prosecution of the collective cognitive enterprise in a sufficiently effective manner.

Equally, humans are fallible, prone to illusion, bias, ego-/homocentrism, and so on. In these circumstances, it is essential that human judgments be tested for error and, where necessary, corrected. But in the absence of other epistemic sources than ourselves, our only recourse is to construct systematic processes for requiring and checking that we have inter-subjective agreement. This is, we have seen (section 2.II.1), at the heart of constructing objective knowledge. But there we also saw that these processes can be realized only through institutional coordination. Once again institutional structure is essential to science.

Finally, institutions provide the framework for the scientific education unavoidable for general problem solvers with finite resources. (If we were born with all past learning already accumulated in a cognitive Lamarckism, it would both impose a heavy memory burden, uncritically reinforce accepted but erroneous commitments, and lead to a huge loss of flexibility as options left open when learning from ignorance were closed down.) Our educational processes grow increasingly necessary to support our extending cognitive neoteny (the continuing expression of our biological neoteny), that is, to our increasingly delayed cognitive maturing as research horizons recede before accumulating specialized knowledge and skills.

In sum, institutions permit cognitive division of labor compatibly with coherence of cognitive strategy. Indeed, institutions regulate every important area of scientific activity. They are in these respects typical of the structure of biological communities more generally, except for their degree of institutionalized cognitive specialization.

The cognitive superfoliation process, which is the heart of science is, we have seen, complex. Its subprocesses require coordinated increases in both inter-order and parallel distributed processing structures and increased individual and social complexity. The framework for this is expressed in an institutional design. To support this dynamic process, scientists are required to carry more complex information, even as they are also forced to specialize more and more narrowly within their major discipline. A contemporary molecular biochemist might know less than 0.1 percent of chemistry as a whole and be an expert on only a few gene structures, yet be familiar with the relevant parts of the whole structure of chemical theory and be sensitive to techniques and results in a dozen different areas in physics, mathematics, biology, and neurophysiology. Conversely, chemistry plays intimate roles in biology and geology, while borrowing from physics, engineering, and biology, in turn promoting increasing horizontal refinement or adaptation. So as well as refinement and ascent of regulatory orders, the development of science also leads to increasing cross-theoretical interaction. The result is an intricate set of interrelationships across science, patterned by the context-dependent requirements of local problems and methods, a dynamically restructured version of Campbell's fish-scale metaphor (Campbell 1969, cf. Hull 1988b, 493).

Effective support and regulation of this intricate and dynamic process requires the development of an increasingly complex institutionalized social structure for the conduct of science. When thus normatively designed, call them *epistemic institutions*. A theory of epistemic institutions is a fallible normative theory of how best to design the sciencing process so as to feasibly pursue the ideals of rational knowledge (chapters 1,6). The finitude, flexibility, and fallibilities of individuals lead, we have seen, to science as a collective enterprise in which the division of labor extends cognitive capacity. Conversely, the socially mediated correlation of labor extends objectivity through systematic intersubjective comparisons. Epistemic institutions in effect form the external nervous sub-system organizing science.[42] Out of the activities of this larger nervous system, grows a larger objective knowledge. Thus institutions are both our necessity and our strength. The institutional complexity of science is only compounded by the rapid increase in scale of scientific operation this century and the increasingly intimate involvement of science in the advance planning of technological development and environmental transformation.

Science is an intrinsically social, institutionalized activity. This isn't a surprising discovery sociologists made, nor is it so because a political philosophy or ethnomethodology insisted on including science among cultural traditions; it is so for deeply embedded regulatory, cognitive reasons. Moreover, the sciencing process will not generally be neatly demarcated by conventional institutional boundaries. Science is a dynamic process in a complex regulatory system which selectively incorporates information from the environment as regulatory structure, accumulating it systematically so as to produce objective knowledge. This system extends to wherever the sciencing process occurs (cf. Hull 1988b). It includes subsystem components of individual humans (e.g., visual system, perhaps the immune system), and individuals and institutionalized groups of scientists, but also many other actors across the full societal environment (see above and note 12). Within this regulatory complex there is no principled subsystem boundary which can be drawn marking out the scientific parts and based solely on intrinsic subsystem properties. Rather, the scientific processes need to be demarcated by their functional consequences, specifically by their contributions to transcending cognitive limitations.

Let us pursue the general biological embedding of science through its systems dynamics by considering the regulatory structure of selective forces driving change in the system. Recall the complex relations that exist between a population and its environment as briefly reviewed in section 2.I.2. A nested set of selecting environments can be usefully distinguished in order to understand biological dynamics. Individual phenotypes, complex regulatory systems that they are, form the first selecting environment for genotypes: environment (1). For in their embryonic environmental circumstances, genotypes have to lead to viable ontogeneses; otherwise no reproductively able phenotype is even presented. This requirement is far from trivial and becomes tighter as the complexity of the phenotype increases (cf. Wimsatt/Schank 1987, 1988). The result is selective elimination or reinforcement of genotype changes according to whether or not they cause embryogenesis to depart from the currently entrenched sequence. The actions of viable phenotypes then take place in the three successively more encompassing environments; (2) the rearing environment formed by the operative social group of phenotypes and their technologies, which is the site of initial learning and which for mammals typically includes the extended family and its site, tools, etc., (3) the wider but social community with which the rearing unit interacts regularly, typically comprised of other members of the species in a local colony or village together with its tools and adapted sites (burrows, houses, etc.); and (4) the larger ecological environment formed by the collection of other species in their physical setting. These environments are not sharply separated from one another in practice; rather, they represent convenient abstractions from the complex welter of interactions that is the ecological reality.

Environments (2) and (3) generally increase in importance as nervous system complexity increases, because of the importance of learning during rearing. Environment (2) may also be important in other ways (e.g., because young that are sufficiently abnormal or numerous might be killed), while the determination of colony role for larvae by local colony members in social insects and the selective killing of chieftains' sons in some human cultures illustrates the corresponding importance of environment (3); but I shall not pursue these considerations. In the human case, these environments are crucial for the formation of the basic personality matrix out of which arises those specific activities that we describe as learning, but which is the seat of the prior grounding attitudes of openness—curiosity, honesty, sharing, etc.—without which science would scarcely be possible. Only after surviving these two environments, in whatever condition it does so, is the organism/phenotype presented to the larger environment (4) to survive and pursue a reproductive strategy. Finally, let us recall that other species play important roles in historically shaping each of these environments as well as in their current dynamics, especially for environment (4); we are really dealing with two-way interaction here. It is this structured, dynamic process of multiple selections that drives biological change.

Diagram 2.5 indicates the presence of a corresponding nested set of selecting environments that can be usefully distinguished in order to understand scientific dynamics, comprising (1) the individual scientist, (2) the local experimental research/teaching group (includes theoretical research) that forms the locus of daily activity, (3) the larger institution of science, and finally (4) the wider natural and social environment. Because of the intricate structure of scientific interactions, these environments are not sharply separated from one another in practice; rather, they represent convenient abstractions from the complex welter of interactions that is the societal reality. Though the institutional environment of science is internally very complex, it is still useful to distinguish within it the group sub-environment that is the immediate locus of testing activities (experiments), while the larger scientific environment is that from which the scientist draws and to which results are submitted. Similarly, though there is tight socioeconomic integration in our complex societies, because of the transformative impact of science-based technology on the wider natural and social environment and the feedback to science from technological performance, it is useful to distinguish a specifically technological subenvironment within it.

Individual scientists themselves, complex regulatory systems that they are, form the first selection environment for (causally instantiated) cognitive commitments. Certain proposals, for example, those that conflict with more deeply entrenched beliefs, are eliminated. (Though inconsistencies may be tolerated and 'quarantined', as in quantum field theory, if there are sufficient benefits to doing so.) Conversely, proposals that reinforce currently entrenched beliefs are the more forcefully and persistently developed. Commitments that cannot be sufficiently coherently developed for individual scientists to participate in generating significant development and testing of them cannot even be candidates for scientific acceptance. This requirement is far from trivial and becomes tighter as the complexity of commitments increases. The result is selective elimination or reinforcement of commitments according to whether or not they relate productively to other commitments held. But in the scientific case, this simple picture is complicated by the wide range of commitments across science that may be relevant to any given scientist's considerations and by the deliberate promotion of

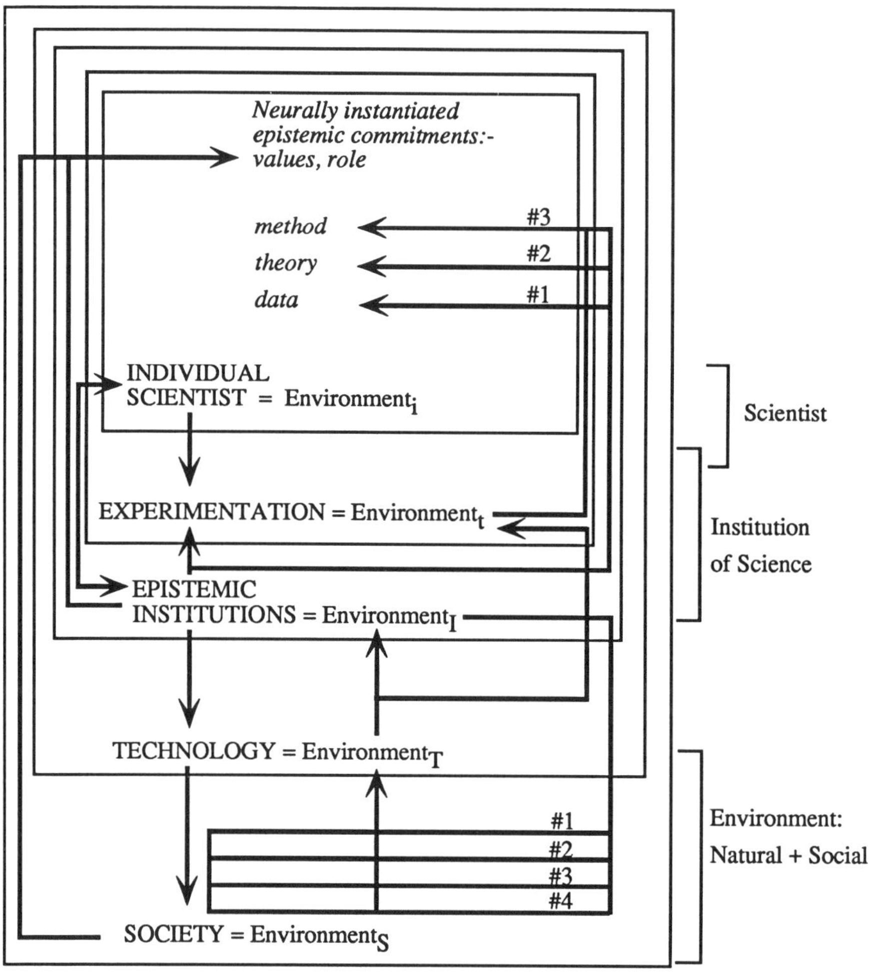

Diagram 2.5 Science as a Causal Feedback Structure
Arrows indicate generic kinds of scientist-altering feedback resulting from changes brought about by scientific behavior.

KEY:

Science-scientist feedback
#1 = data shift
#2 = theory shift
#3 = method shift, etc.
#4 = values, role shift

Society-science feedback
#1 = parameter value change
#2 = category or value shift
#3 = institutional design shift
#4 = science-society relation shift

⟶ : problem-solving/problem-creating process

criticism of extant commitments and the selective support of at least some scientists who try to develop unconventional ideas.

The other scientific environments are equally complex. Something of the complexity of experimental situations, for example, is realized once one understands that not only must the theory nominally under test be applied in the detailed test conditions, which are often complex and require several other theories to characterize, but there must equally be available theories of the testing procedure and instrument performance, consequent data processing methods, and the like. Nonetheless, when these are all coherently conjoined, they form a highly regulated environment designed to focus epistemic attention on a narrow and distinctive class of propositions.[43] Hence they form an important locus of scientific change (including deeper entrenchment).

Similarly, cognitive commitments which should, but don't, lead to successful technological development typically do not survive, and feedback from technological application typically acts as a powerful selective force. Technological development originating from a group of scientists may bring about both direct instrumental change to the working environment of other scientists and also bring about a host of other indirect changes (e.g., economic or communicational ones) to other scientific working environments. Science-technology is itself an internally organized process that reinforces its own dynamic development through positive feedback; see diagram 1.3 of chapter 1 and text. The science- technology system is then connected into the societal system through each of its major institutions, but especially through the generation of advantageous economic change (positive feedback) and sociomorally repugnant practices or proposals (negative feedback).

Technologies amplify processes to produce artifacts. Transport technology, for example, amplifies capacity to travel and/or shift loads, the result being the artifactual network of roads, railways, etc. carrying cars, trains, and the like. Our capacity for technology is itself a cognitively amplified biological capacity to alter the environment so as to provide circumstances to which the organism is preadapted. Nests, burrows, hives, and the like are exosomatic architectures that constitute an effective preselection of the environment for selective advantage to the species concerned. The resulting inanimate environment (cities, transportation and communications technologies, agricultural landscapes, etc.) has been pre-selected for human capacities—at least up to something like the same ignorance limits with which the bird or mouse creates its artifacts.[44] Our educational processes also represent a massive environmental reorganization this time of an institutional character, which is essential to the maintenance of our artifactual environment, which in turn aids science, completing the cycle of positive feedback.

The process of preselection provides a crucial role for phenotypic capacities within the dynamics of evolutionary development. As noted earlier, there is a complex regulatory structure of feedback interactions between phenotypic behavior, environmental structure and dynamics, and selection forces. The massive enhancement of behavioral capacities that science brings to the human species has only served to intensify the action of this regulatory structure. This leads to a number of outstanding and crucially important problems for our species. *First*, it is an open question how pliable the universe is. It is as yet unclear how much of its structure can be transformed into a human artifact at all and, more stringently, in a way supporting human flourishing. *Second*, this very feature poses an important problem: Is it possible for the human species to lock its development into what will ultimately prove a dead-end by choosing a combination of environmental transformation and cognitive-cultural development that ultimately becomes a self-reinforcing but stultifying process? Alternatively, what kinds of environmental preselection will promote continued cognitive regulatory expansion? Nor is this an idle question, for surely it is already possible to identify cultures that have locked themselves into a self-reinforcing staticness in a way that has in the long run been to their detriment.

Third, there is a problem arising from the rate of environmental change, especially simplification, induced by our preselection activities fed back into cognitive development. Cognitive activity disturbs the environment (1) by probing it in inquiry, (2) by preselecting it, and (3) by the unintended consequences of our activity. During this process, however, we also acquire more information about our environment, a more adequate scientific regulatory structure, and more resources with which to pursue science. The net result is an increased rate at which the environment is being transformed into a human artifact. To what rate can this environmental preselection rise before the consequent pace of change outstrips the human capacity to create understanding of it and regulatory systems to govern it? Dimensions to this issue in more popular form are: Has our technology outstripped our ethics? Has our knowledge created a dangerous level of ignorance? Will we survive our own inventions and environmental degradation? But we begin to see these issues more clearly once we place them in the context of regulatory systems dynamics.

Returning to the discussion of environments for science, beyond practicable technologies lies the larger society with at least four distinct ways in which it may impact upon the narrower subinstitution of science (see diagram 2.5). The most superficial kind of change is one in which the quantity of funds, numbers employed, laboratory space, etc. alters (shift 1). These are characterized as a shift in (social) parameter values. Next, shift 2, there may be shifts in the priority society attaches to various research fields, for example, as between nuclear science, space science, and environmental science. Although these are ultimately expressed as parameter value shifts, they go deeper than those, systematically reorienting research efforts and training. Deeper still lie shifts (#3) in institutional design, for example, the recent societal support for the computer science and artificial intelligence field, the growing importance of international institutions to science, or the increasing insistence on institutionalizing ethical review and limits on research. These shifts ultimately have a major impact on the entire character of science-technology. Roughly, each successively deeper shift occurs on a longer time scale with, on the longest time scale, shifts in the general relation between science and society (shift 4). An important shift of this latter kind characterized post-Renaissance Europe and took 500 years to develop, but the result was a fundamental transformation of the scale and institutional character of science (see below).

There are diverse sources of pressure for cognitive change that originate more or less clearly in the noncognitive environment. Perhaps a government directive declares a particular area of research to be a national priority and funds are supplied accordingly. Perhaps the society at large judges particular methods to be unethical, or alternatively, particular theoretical perspectives to be culturally important or controversial. All of these eventualities will create institutionalized sources of pressure for changes in scientific activity. But like the stabilization or buffering of embryological developmental processes (creating a homeorhesis, Waddington 1957), epistemic institutions buffer their members to some extent from distortions to the pursuit of cognitive goals, whether these distortions come by way of the intrusion of prejudice/dogmatism or popular enthusiasms, of emotional, hasty, or otherwise defective methods, or via the sheer deprivation of information and resources. Conversely, though, if those institutions show cognitive pathologies, then their negative impact on scientists is correspondingly intense. At present, for example, they less often ensure that developing scientists are exposed to a wide range of cognitively relevant stimuli and behavioral models so as to ensure that cognitive development is vigorous and balanced. Thus the examination of the institutionalization of science should form a central component of any adequate philosophy of science, but it is excluded by traditional analyses.

There is here then a complex and delicate regulatory relationship between society at large and its science-technology subsystem, each changing the other while each pursues its own interests. However the situation is still more complex. The response of individual scientists to any and all of these pressures for cognitive change may be to change their commitments, conceptions of their institutional roles and/or decision processes, and so on to remove the pressure. But it may equally be to resist that pressure. One way to do this is to increasingly isolate oneself from such interactions, but scientific institutions quite properly resist this strategy in turn by insisting that scientists interact or face exclusion. Another strategy is to exploit the ineliminable ambiguity and insufficiency of evidence to argue the reasonableness of continuing present commitments. This strategy raises delicate questions about the actual and rational designs of the unavoidably imperfect collective judgment processes that can be institutionalized, questions of the kind raised (and largely dodged) by Lakatos 1970 over the choice of scientific research programs (but increasingly studied; see Solomon 1992). Finally, as emphasized by Vickers 1968, individuals may instead seek to change the roles or decision processes of the relevant scientific institutions so that the resultant scientific role expectations (role norms) for them conform more closely to their own present judgments. This is an important strategy, essential to intelligent institutional designs. Its presence in science is manifest, for example, in the creation of new research groups, journals and learned societies.

So the science-technology system finds its institutional design changed both from within and without. At the same time, it acts with self-organizing capacity to buffer the epistemic process from external distorting change signals (e.g., political pressure), and this capacity is an expression of the support for its institutionalized processes, which it normally receives from its individual members and its society. Like any other ecology, individuals and the institutions to which they belong undergo a delicate coevolution. The most convenient picture is probably one in which the science-technology system is thought of as a subecology of a larger ecology of institutions, with individuals participating in several institutions simultaneously. Understanding this complex and delicate regulatory relationship is the more urgent today as scientific activity becomes more closely entwined with society at large and governments become more active in the transformation of scientific institutions.[45]

It will often be rational for individuals and the various institutional groups of which they are members to use differing methods and even pursue differing values. An individual scientist S may rationally accept theory T and accept experiment A as the best next performed while S's laboratory may rationally require S to carry out experiment B instead and S's discipline may also rationally support development of theory T', incompatible with T, and a corresponding critique of both experiments A and B. These differences are real and important to the dynamics of science. We cannot understand them within the formal conception of rationality as logical rules but the decision-theoretic model of rational action is well able to model the variety of insitutional situations involved, providing a natural setting in which to theorize rational institutionalized action because of its specific recognition of context-dependent beliefs and values entering rational action (chapter 1). At all levels of institutionalize decision making aimed at increasing epistemic utility, methods can be modeled as conjectural and risky resource distributing strategies for theory development and testing.[46]

In sharp contrast to the formalist logic-based conception of rationality, the interleaved patterns of consensus and dissensus that emerge naturally within a decision-theoretic representation of epistemic institutions can be, and typically are, complex. The achievement of collective epistemic gain does require the formation of a consensus, but the critical assessment necessary to objectivity also requires promotion of dissensus. Consensus is of a very specific kind and is compatible with a very large amount of epistemic disagreement. For consensus on what to do, given a background of accepted cognitive commitments C, the participating scientists need to be agreed: (1) that it is worthwhile to investigate alternatives to some given theory T, (2) that among all of the logically possible alternatives to T a finite set of investigative paths are those that are to be followed first, (3) that certain general methodologies (or certain general constraints on specific methodologies) are to apply to formal and experimental exploration and the mutual communication of results, and (4) that certain general procedures are to constrain the ways in which each of the members of the group will change their own research strategies in the light of the results communicated to them. Notice, however, that the scientists need not agree on any particular alternative to T as the one most worthy of investigation. Indeed, they need not agree on any ranking of the alternatives in terms of either likelihood of truth or preferability of pursuit. Even more fundamentally, they need not agree on what precisely is at fault with T or even that it definitely has some major flaw. Finally, they need not agree exactly on particular experimental methodologies, on methods of statistical inference, or on the heuristics of theoretical research programs. Indeed, they need not even agree on what the relevant experimental data are at any given point. Despite all of these possible sources of disagreement, it is still reasonable that the scientists conclude, both individually and collectively, that there is a net epistemic gain, even a large net epistemic gain, to be had from a cooperative investigation of alternatives to T. It is precisely the purpose of an intelligent epistemic institutional design that it encourages cooperation to proceed, even while also encouraging appropriate critical dissent.[47]

Individual scientists demonstrate their rationality by arranging things so that all individuals follow their own commitments, within institutional constraints, while collectively they exhaust the plausible research strategies. Collectively, the remarkable interdependence of individual epistemic preferences that such arrangements demand requires an institutional explanation focusing on the way in which scientists search for, and adapt to, the institutional roles they perceive themselves to play and the way in which these roles are designed in relation to the overall epistemic enterprise. To the extent that the designs of those roles systematically supports the epistemic enterprise, we may speak of epistemic institutional rationality.[48]

Epistemic institutional rationality has to do with the generation and reinforcement of patterns of consensus and dissensus of the sort illustrated above. Too much emphasis on consensus formation leads to overhasty and uncritical, therefore dogmatic, acceptance; too much emphasis on dissensus leads to paralysis through idiosyncratic speculation and methodological fragmentation unconstrained by critical assessment and synthesis of others' work. In the former case, the epistemic pressures emanating from the three enveloping environments in which scientists work (see above) are felt too strongly in one respect (the pressure to agree) and not sufficiently strongly in another (recognition of genuine difference needing investigation and explanation). In the latter case, the reverse obtains (or individuals remain indifferent to one another; that is, they feel no pressures). Felt pressure emanating from the social environment to agree with some ideological line is only the worst case of a continuum of epistemic institutional irrationalities here. The institutional framework itself must both give scientists cognitively as well as socially appropriate roles to play and reinforce the values driving those roles through suitable rewards and punishments (e.g., rejection for fraud, promotion or prestige for successful new theories *and* for successful criticism of extant theories, data, or methods). Studies by Galison 1987, Hull 1988b, Latour/Woolgar 1979, Latour 1987, and others are beginning to elaborate the extremely complex dynamics of these institutional processes, though we are not yet in a position to subject them to cognitive theorizing.[49]

Their support of either consensus or dissensus formation hardly makes epistemic institutions unique among human institutions. Tribal and political rivalries, for example, generate and reinforce dissensus; various forms of group allegiance, suppression and coercion generate consensus. Often the two processes will occur together in the same cultural group; but whether separately or together, they are often not systematically related to epistemic development. (Consider policy-generation processes in democratic political parties.) On occasion they can be more closely related with learning (e.g., when a tribal council seeks consensus on seasonal migration by pooling accumulated experiences and current observations), but not essentially so, and in many institutions, especially where power or profit dominate, truth may be deemed irrelevant or deliberately disavowed.

What is relatively unique about science are the patterns of consensus and dissensus formation supported: patterns that are systematically relevant to epistemic development. The institutional procedures for submitting a new proposal to public critical assessment through journal publication, conference presentations, and the like are designed to maximally widen possible dissensus formation and so regulate that formation (the loci of dissensus), and the resultant interactions among dissenters, as to focus selectively on the epistemically relevant features. By comparison with most social dissensus processes in our culture more and relevant (informed, skilled) actors are drawn into the dispute, yet their interactions are also far more complexly and relevantly structured. The feedback structure of this institutional design underpins the construction of the epistemic invariance that grounds the achievement of objectivity, the product of institutional rationality (section 2.II.1 above). This is possible, however, only because, and insofar as, rational epistemic institutions also so arrange their epistemic activities, especially experimentation and its extension in technological development, that the entities in some domain must, if knowable, play an appropriate causal role in the generation of our representations of them (Brown 1987, 1988; Campbell 1986).

Rational scientific acceptance requires at least the following: (1) The process of acceptance is one for which the relevant features of the world played a relevant causal role in actively bringing about the local consensus on acceptance, which is the institutional basis for its wider scientific use as acceptable. This is a necessary condition for being genuinely open to learning the truth (in the surrogate form of our best conjectures thereon). The reason why we object to the intrusion of politics into science is because that intrusion cuts scientists off from causally relevant sensitivity to reality. (2) The process of acceptance is one where public, shared epistemic values emerge as a clear counterweight to both sectional and wider social interests and in which what is accepted is critically but fairly assessed. This applies to both local acceptance and wider institutional use; it ensures, for example, that the evidential bases for local acceptances are always open to retesting. (3) The relevant community that is to choose is clearly specified and its role in the choice process is clearly specified, and in both cases not constrained beyond what our currently best science specifies as epistemically relevant (e.g., exclusion of the blind as visual observers). (4) The acceptance process itself is continually (in practice, regularly) open to correction. This process should be capable of fundamental reconceptualization as well as fine tuning. So much is merely prudent in view of the subtleties already raised. But each review has a cost, so that there needs to be an institutional design also for the process of initiating reviews, and this too should satisfy the foregoing clauses. (5) If acceptance is to be a rational process, then a minimum requirement is that the institutional design solutions intended to satisfy the first four of these clauses should themselves be argued critically against competing alternatives. That is, these clauses should be metaconsistently applied to their own process of institutional design, which they define.

In this context, note that the decision-theoretic model of rational agents provides the opportunity to unify the treatment of science methodology and public policies, including science policy, as various cases of institutionalized strategic action. This is not only cognitively valuable but of practical importance in contemporary circumstances. For the practical consequences of science have led to a growing debate about its social control, which has struggled with the incoherence generated by the dichotomy between theoretical and practical reason emerging from the formalist conception of reason (Hooker 1987, section 7.9; Brown/Hooker 1994). The five conditions for methodology sketched above, for example, have counterparts in a theory of objective public policy contents (see Hooker 1989c).

Of course there is operative in all the foregoing design characterizations a large ceteris paribus clause to cover the imperfections of human individuals and institutions, the vicissitudes of history, etc. But these are complications; the distinctive design principle is that characteristic of a social learning system, that is, of a rational epistemic institution. However, the complications run deeper. In science, we exclude mental incompetents and the relevantly immoral from the epistemic community, and this already requires careful theoretical design reflection; beyond that, the case becomes even more complex. The community for science is increasingly "soft"; very few people can really dispute nuclear physics reasonably, for example, so science is not in this sense a public enterprise, and this despite the impact of nuclear science on the wider community. Moreover, the competent scientific community is shrinking steadily as scientific complexity increases (cf. the contemporary chemist above who can have read less than 1 percent of even the important papers in the discipline). Only epistemic institutional design holds the scientific process together; otherwise, it would fragment into a thousand specialties, each too small to sustain an objective process of scientific acceptance. One has the sense of some difficult epistemic regulatory problems emerging here for the human species. But at least here they can be given a principled formulation in this regulatory systems context as a first step to addressing them.

What is, in the last analysis, progressive over the longer term is the superfoliation of the regulatory structure itself. In the prosecution of science, we are engaged in a magnificent and potentially unlimited superfoliatory process, which takes us "beyond ourselves" as we increasingly reflect the cosmos in the regulatory order of planetary epistemic organization. Magnificent as this sciencing process seems to us, at least from our feeble parochial standpoint, and as much as we seem here to have caught on to a cosmic process that genuinely transcends us, I close this exposition with a cautionary historical note. All individual humans, and the species as a whole, are in general ignorant (in advance) of the consequences of their own disturbances. The risk carried by inducing change is that it may rebound and eliminate the initiators. Application to the current human situation is all too obvious, with the massive threats to our planetary environment induced by the advent of nuclear warfare, industrial environmental pollutants, and resource consumption and population explosion. Everywhere across the globe, ecologies are being simplified and destabilized, their resiliency decreased. The real issue for us is whether our scientific understanding, and our larger institutionalised control of our own processes, can develop at a pace fast enough to control our disturbing impact. Recall the Conant/Ashby theorem that a controller must have access to at least as many states as does the system to be controlled (Conant/Ashby 1970; cf. Ashby 1970), and the nonlinear increase with complexity of information required for system coherence, and you have the race against ignorance we are in. The secret of sustainable advantage through adaptability is to be able to relate the changes induced to the adaptabilities possessed. It is precisely this relationship that is in doubt for us humans and which at prior times, thanks to our ignorance, we have not even been in a position to sustain. The third- and-fourth order institutional feedback control loops required here are an urgent developmental priority.

Chapter 3

Reason and the Regulation of Decisions: Popper's Evolutionary Epistemology

with Barry Hodges[1]

Introduction and Overview

Popper's philosophy of science is widely and properly regarded as the logical successor to empiricism. The basic idea motivating empiricism is that science can be objective only if it is dictated by observation and logic alone, because only then can it be independent of all human decisions and so independent of all human errors and normative or evaluative judgments. Empiricism essentially eliminated decisions by either logically forcing them (they followed deductively or inductively from the evidence) or excluding them from the realm of the cognitive as purely pragmatic and conventional.

In this light, and in preparation for the discussion to follow, let us call "rigid decision processes" those context-insensitive procedures whose outcomes are uniquely determined purely algorithmically and call their outcomes "rigid decisions." Here context will include the historical state of the background knowledge assumed, the institutional roles and responsibilities of those involved, their skills, the technological, monetary, and other resources available, and like factors. By contrast stand those decision procedures that are context-sensitive, heuristic, and risky (may not lead to a uniquely best, or even a satisfactory, outcome) and based on nonformal judgments. Let us call these "flexible decision procedures" and their decisions "flexible decisions." (Hooker 1994c argues, expanding on the lead of Cherniak 1986, that finitude alone requires that flexible and not rigid decision processes are central to our being rational.) Now we can turn to Popper's position.

We believe that the problem of the rational control or regulation of decisions lies at the heart of all Popper's work. We understand his intellectual career as exhibiting a developing strategy for addressing this problem, which passed through three broad stages:

Stage 1. The Conjectures and Refutations Account. In his earliest period, Popper took over the general empiricist approach to decision as part of his empiricist legacy; the elimination of decisions in the growth of knowledge was assumed essential to guaranteeing objectivity and achieved, once again, by either logically forcing them (solely deductively) or excluding them as conventional. However, he showed an explicit awareness of the necessity of dealing with human decisions.[2] Unfortunately for Popper, as Part 3.I will show, this move fails; it turns out that many of the forced decisions are in fact not capturable within the confines of deductive logic and that many excluded decisions are cognitively substantive. In effect, all these decisions are rationally uncontrolled decisions lying at the heart of his method.

Stage 2. The Evolutionary Epistemology Emendation. Popper's response to this failure of control was to add an evolutionary epistemology to his stage 1 account and attempt to show that all decisions could now be forced through the "decision processes" of natural selection. This moved Popper away from the mixed logically-force/externalize strategy and toward more reliance on control or regulation, but it remains a primitive conception of rigid control. Part 3.II examines Popper's evolutionary epistemology and concludes that this attempt also fails, and for exactly the same reasons as the first: The requisite decisions still escape the net of the enlarged rigid decision procedure.

Stage 3. Regulation by Flexible Decision Processes. The third stage is concerned with the introduction of a notion of *plastic controls* (Popper's term) and is examined in Part 3.III. Though this stage is only embryonic in Popper, it is (in our judgment) the crucial one, because it moves simultaneously away from rigid decisions and toward primary reliance on the human regulation of decisions using flexible decision processes. Further, it points toward a more productive conception of objectivity in terms of the institutional regulation of decision. (Popper unfortunately still externalizes this conception, in that he never applies it to the empirical sciences but only to the logic of social situations.)

Our treatment will follow Popper in this overall development but will concentrate on the evolutionary epistemology, where Popper sows the seeds for an evolution of theory of science beyond his empiricist limitations.

3.I. Logical Empiricism and Popperian Method: Formalism and the Control of Decisions

3.I.1. Logical Empiricism

We shall begin by briefly reviewing empiricism; it will set the problem frame for our examination. One of us (Hooker) has argued that empiricism generally, but logical empiricism especially, provides an inadequate theory of science and that a, or perhaps the, principal reason for this is its attempt to confine scientific rationality to logical formalism. And further that Popper's theory of science suffered from similar defects, indeed that metaphilosophically he was essentially an empiricist.[3]

Empiricists distinguish sharply between reason and experience and regard the latter as the primary (indeed, usually the only) source of knowledge. Fundamentally, the empiricist program is to employ reason as formal logic to develop a general knowledge of the world out of the collection of particular experiences that humans have of it. Though logical empiricism is primarily concerned with the *epistemological* problem of the basis and scope of scientific knowledge, much of its discussion concerns philosophical translation into a formal language of science L from natural or colloquial scientific languages and with syntactic and semantic analyses within L. Logical empiricists essentially denied that there were any substantive underlying presuppositions to the translation requirement since only a collection of necessary truths concerning logical structure was held to be involved and these were held to be devoid of empirical content because logically true. The language of science thus became a necessary framework for the expression of any intelligible content, but was itself held to be contentless. So we have a powerful two-component system: insistence on first-order translation combined with second order denial of any significant presuppositions to translation.[4] But this allegedly innocent beginning in fact establishes the whole character of logical empiricism. For this essay, the key feature is that translation was designed to ensure the elimination of human decisions from science.

The ideal for logical empiricism was that its epistemology be directly "read off" from formal logical metatheorems. One metatheorem for the simplest version of L, for example, says that every sentence of L is logically equivalent to a finite truth function of elementary observation sentences, thereby ensuring that all admissible knowledge claims can be reduced to logical derivation from an observational base. In this way, the characterization of empiricism itself would become an exercise in pure reason. Though one cannot attain this formal ideal for more than the simplest version of L,[5] it nonetheless continued to guide empiricist construction. For logical empiricism formal logic both captures the structure of reason within science (as the structure of valid argument within the object language) and within philosophy (as the set of formal metatheorems about the object language). This makes logical empiricism a powerful and elegant position. (The powerful role for a formal conception of reason exhibited here deserves attention for itself, since it lies behind much of western philosophy; cf. Hooker 1991a; Brown/Hooker 1994.)

Nonetheless, this elegant attempt did not succeed. Its simplest version, for all its power and elegance, is not a persuasive theory of science. A large critical literature can be summarized as follows: (1) Theories are not definitionally reducible to finitely, observationally verifiable assertions; (2) scientific method is not rationally confinable to entailment from the facts; (3) observation is not a fundamental, transparent category, but a complex, anthropomorphic process, itself investigated by science; (4) the history of science is not just the accumulation of observed facts and their rational organization, and historical inter-theory relations do not fit the accumulative model; (5) accepted observationally-based facts do not belong to an eternal, theory-free category but are theory-laden and subject to theoretical criticism; (6, 7) science is not isolated from the human individual and from society in the manner presupposed here; (8) method cannot be reduced to logical rules alone, and it is quite reasonable that it not be universal either across scientists at a time or across history; (9) logic does not have the privileged status given it here but is itself open to broadly empirical investigation; and (10) there is not the gulf between the normative and the descriptive that is built into this position. The introduction of richer logical structures to L, especially the predicate calculus and inductive logic, did something to ease objections 1 and 4, but it never really blunted them, and it left objections 3 and 5 untouched. And while it removed objection 2, replacing it with a confirmationist/inductivist program, this promptly added two more traditional objections to replace 2: (2′) Inductive logic is impossible (inadequate, incomplete), and (2″) it is impossible to inductively distinguish true laws of nature from accidentally true generalizations.[6]

Rather than rehearse all these objections, consider just the problem for empiricism of establishing its observational or empirical base. Empiricism requires that this be guaranteed reliable, providing empirical truths untainted by human error and judgment; otherwise the entire logical edifice built on it will be equally contaminated. But this account of observation is in deep difficulty. One obvious problem is that perception is not in fact always reliable; carelessness and distractions, illusions and hallucinations all reduce accuracy. Nor is there any way to read off from perceptual experiences themselves which observational reports are reliable and which ought to be discounted; it requires good theoretical guidance to do that. This difficulty (for empiricism) is a model for the next problem: All the scientific evidence we have points to the view that perception is itself an activity essentially cognitively similar to theory construction; the mind forms the "best" model it can of the scene before it on the basis of memory, stored information processing methods, and current information input. In both of these cases, observations, which are the end products of this process, cannot have any privileged cognitive status and so cannot provide foundations for knowledge.

These features of perception constitute problems, because they require the introduction of decisions and thereby rob science of its autonomy from evaluative judgements. If perception is not always reliable, then it must be decided whether this particular observation, taken in these circumstances, is reliable, and in what respects and to what degree. And it must be decided what tests to apply to check both the judgments made in this particular case and judgment and decision policies about such matters generally. And it must be decided of those tests how reliable they are; they will certainly involve perception at some point and may have other fallible components. And it must be decided of the humans involved in all this testing whether they are reliable in carrying out their tasks. (Or in each of these cases, it may be decided to take another's judgment as surrogate for direct testing, for example, that of an "expert"; but this decision too derives from a judgment, or rather a bundle of them.) Then again if observation itself results from a theorizing-like process, this already involves judgments and corresponding acceptance decisions. Which are the relevant alternative theory-like perceptual frameworks to consider? This decision must be a matter of judgment. By what method is one to be chosen as that to be used? By what tests is the most plausible observation to be constructed within that frame? These decisions too must be matters of judgment. And so on. Popper himself introduced an even simpler version of this criticism: Scientists, being finite and ignorant, must therefore make decisions about which aspects of reality are worth observing (cf. Popper 1972, 46). Such decisions can be foolish or reasonable, depending upon the circumstances obtaining. Instead of being a clean objective report, an observation suddenly becomes immersed in a sea of judgments and corresponding decisions of many different kinds.

Ultimately, all the other criticisms converge on this same issue. Pursuit of the general problem of the status and roles of judgments and decisions in science constitutes a highly destructive line of criticism of empiricism (and of Popper; see below) which, pursued far enough, would lead us far away from not only an empiricist conception of knowledge (though not of course away from the view that experience must play an important role in knowledge formation) but also away from a formalist conception of reason. Given that human bodies are essentially instruments and given the fundamental role of theories in the design and evaluation of measuring instruments and experimental methods, for example, it is not at all likely that a purely formal account could be given of the range of important decisions involved in accepting an observation.[7] The alternative is to write them off as noncognitive, but this just avoids the issue by fiat. And it is now well understood that a purely formal analysis of inductive reason poses grave problems. The fundamental reason for this is that rational inductive inference is context dependent, the context being primarily fixed by theory.[8] But we shall pursue the generalized critique of reason no further here in order to focus on its application to Popper.

3.I.2. Popperian Method and Logical Empiricism

Despite being trained in logical empiricism, early on Popper became a severe critic of that doctrine, and his position represents an influential step along the developing path of western philosophy of science. In his first book, *Logik der Forschung* (1934—see Popper 1980), he rejected induction outright and argued against logical empiricist semantics. He then presented a conception of scientific method and of the nature of science very different from that of logical empiricism. In subsequent books (see Popper 1972, 1979), he modified his views, but only to move still further away from the terms in which an empiricist conception of science is stated (see Stokes 1989 and Part 3.III below).

In this section, we focus on the relation of early Popperian method to logical empiricism (for references, see notes 6, 13). Despite his criticisms of logical empiricism, it turns out that he essentially remains a logical empiricist metaphilosophically. It will come as no surprise that his position turns out to be even more clearly founded on nonformal decisions than is empiricism. But Popper too is in deep difficulty dealing with those decisions. We shall argue that these difficulties can be traced back to the logical empiricist constraints.

Popper's critique of empiricist semantics (see Popper 1980, 94–5) is in essence that already given of empiricist observation. His way of putting the matter was to argue that observational classification is just as general as any theoretical classification. If I say, "Here is a glass," I commit myself to the existence of an object that will shatter if knocked or dropped, whose composition is of silica not hydrocarbon (i.e., plastic), and so on. These claims in turn are not simple observational claims, and confirming them would require yet other observational and theoretical terms whose content was just as complex. So there turns out to be no semantically privileged observation language. Popper speaks of an endless wave of implications spreading out from the use of every term in science, whether theoretical or observational so-called, with its curtailment at sensible places a matter of judgment (cf. Popper 1980, 47-8).

As for induction, Popper rejects it, arguing that it is irrational to pursue its methodological aim. Empiricists aim for securely or assuredly true informative belief. But, Popper argues, *security* of belief and *informativeness* of belief are two incompatible goals, at least for finite creatures beginning in ignorance (as we do), and since it is rational to aim for informativeness, it is irrational to aim for security. The more informative a conjecture is (i.e., the more possibilities it rules out), the less secure it is (i.e., the less likely it is to be true).[9] So we must choose between aiming to obtain secure but minimally informative scientific claims and aiming to obtain deep scientific understanding with informative but insecure bold conjectures. Aiming at security requires a methodological conservatism, for example, sticking to what is already accepted knowledge and what can be added from the most secure immediate experience. But what is secure here? This strategy tacitly requires uncritical acceptance of what we have inherited, culturally and genetically. Culturally, our historical experience tells us that our inheritance can be highly misleading; Aristotle's dynamics fit much common-sense experience but is wrong. And our biological aetiology not only undermines any grounds for asserting the security of our inheritance, it also supports the alternative aim. When born in evolutionary ignorance, the only rational strategy is to conjecture boldly (i.e., with high information content) concerning our real circumstances, including, for example, the possible defects of our natural bodies as observing instruments, and hope to hit on insight and consequent behavior that will promote survival before we are eliminated. This applies everywhere from the mouse crossing the field to us facing the technology-induced disturbance of planetary ecology. Our only hope is to theorize still more boldly in an effort to understand our complex circumstances before we undermine our own survival or cognitive capacities. (One might add something similar about our moral, social, and political capacities.)

From this perspective then, nothing is certain or guaranteed. There is risk in relying on anything. From an evolutionary perspective what is risky is cognition generally, and we do well to query everything critically. But we must learn to survive, so we need to take the multiple risks of learning about all of it. This line of Popper's invites extension. Agents with finite resources have constantly to make trade-off decisions between what some ideal rational method might enjoin and the value that following it can deliver; the result is constant, rational resort to risky heuristic methods. Thorough perceptual search and complete exploration of deductive consequences might seem attractive methods—cf. the total evidence principle and deductive closure as rational principles—but only to a god with unlimited resources. To a human facing a charging lion or a scientist facing the limitations of laboratory life, they make little sense; in both cases, the law of diminishing returns from pursuing them sets in savagely. What are required instead are risky but intelligent judgments about how far to pursue these ideals, and in which directions, in the practical context to hand, so that the best possible decisions are made (cf. Cherniak 1986, Hooker 1994c).

The appropriate method of science then will run exactly counter to the empiricist security-oriented conservatism. And contrary to empiricism, the theory of science will be critically fallibilist about observation itself, looking to science itself to improve its own observational processes both through critical theory of their imperfections and through substituted instrumental technologies. Some obvious questions now arise. What is this method? How does it lead toward the truth? It was clear how empiricism modeled the progress of science, viz., as an accumulating base of logically independent (hence mutually consistent) observations plus induction on these to increasingly general and accurate theories. What is Popper's model of cognitive progress? These are the standard questions, which Popper himself recognized. But there are others equally natural in the context. Does scientific progress include progress in method itself, in learning about learning? On what kinds of judgments are the decisions in all these processes based? It is here that we shall uncover Popper's hidden empiricism.

Popper proposed a methodology of free creation of highly informative scientific theories, or bold conjectures as he called them, followed by the deduction of testable consequences from them; these latter were then to be the basis for subjecting theories to the most severe testing available. All those theories that failed their tests were deduced false as a result and rejected, while the survivors were tentatively retained for further testing and practical application. This is of course the very barest sketch of Popper's methodology, often worded so as to knowingly avoid the difficulties, and the subtleties introduced to avoid them in turn.[10] But it is accurate within its limits and suffices for our purpose here, which is to bring out both the differences with empiricism—now achieved—and what it shares in common with it, to which we now turn.

Popper's methodology is as formal in its own way as is that of empiricism. Everything turns on the properties of classical deductive logic. Both prediction and falsification are interpreted as species of deduction, taking the general form respectively of *modus ponens* and *modus tollens*. Conversely, it is the fallacy of (i.e., deductive invalidity of) affirming the consequent that is the ground for Popper's methodological insistence that agreement between observation and prediction provides no positive support for a theory. Other methodological concepts, such as informative content and severity of test, are also defined in deductive logical terms.[11] Moreover, Popper places the process of theory creation outside the realm of rational processes precisely because he does not believe that it can be modeled as a valid deductive inference. (This position entails a sharp discovery [theory creation]/justification [theory corroboration] distinction; many empiricists also appeal to such a distinction, and equally to avoid the difficulty of explaining how we come by theories; see note 8.) Deductive logic then is not merely the tool of, but the arbiter and structuring principle for, rational method. Popper's position is in this respect close to empiricism in conception, but more austere because of its confinement to deductive logic.

Nor is this the only parallel with empiricism. For Popper, as just noted, the creation of a bold theoretical conjecture is nonrational because it is nonlogical. That is, the decision that present circumstances suggest exploring a particular explanation of them, perhaps a counterintuitive one, is nonrational because it is not dictated by deductive logical inference. This creates a sharp normative/descriptive dichotomy. Theory creation, he insists, belongs to psychology rather than rationality; to suggest otherwise would be to confuse descriptive and normative. Of course the decision to adopt Popper's methodology itself cannot be one dictated by deductive logic, and Popper also insists that this and like decisions are purely conventional, not rational, decisions (see Popper 1980, section 11). In short, all those decisions that are not dictated by logic, or perhaps logic and observation, are held to be nonrational and pragmatic. This re-creates the empiricist external/internal distinction, and on the same grounds. Indeed, in other works (e.g., Popper 1966) he follows empiricism in identifying the restriction of the objectively rational to internal decisions dictated by logic, or logic and observation, as the ultimate grounds for defending individual liberty against authoritarianism of all kinds. These parallels between the Popperian and empiricist frameworks could be extended.[12] Despite his severe philosophical criticisms of empiricism, Popper largely shared the same metaphilosophical ideals and assumptions about the philosophical enterprise, in particular his stage 1 treatment of decisions. Empiricism, we have seen, did not succeed in realizing its own ideals. It is time to see how Popper fares at this task. And here the criticisms of empiricism are now ready to hand.

3.1.3. Critique of Popperian Method

Popper's conception of science can clearly escape the first five criticisms leveled against empiricism. This is the strength of his position, for these criticisms are the 'internal' ones, the only ones that count for all those who accept the confining of method to formalism and a strict normative/descriptive dichotomy (in short, a generally empiricist metaphilosophy). But for just this reason, his position is much more vulnerable to the remaining five criticisms. Meanwhile, let us remain in Popper's tradition and turn, but only briefly, to the internal adequacy of his account.

The methodology sketched above is quite properly called naive falsificationism in the literature (Lakatos 1970). The idea that a theory makes an observable prediction that is then observed to be false, so falsifying the theory, is simplistic at every turn. In almost all cases, to obtain an unconditional, scientifically useful observation claim, a theory needs to be conjoined both with other observation claims and with other theories. Typically, a great deal of the rest of science is involved in each experimental test, though in practice much of it remains implicit in an accepted 'background'. Other theories, beside that officially under test, are needed to characterize various aspects of the experimental conditions, the operation of observing instruments, and even the processing of raw signals into observable reports (i.e., data). The situation is complicated by the fact that often the theory under test also plays one or more of these latter roles. But the consequence of all this is that a prediction/observation clash tells us only that something is false somewhere throughout this vast array, not where to locate the error or what to sensibly do next. Logic is impotent to specify these further, crucial methodological moves.[13]

There are further methodologically relevant ambiguities that logic will not resolve. Popper's logicist notion of content, for example, is restricted to a one-dimensional ordering in terms of classes of entailed negations of Popper's basic statements. But content has multiple methodologically relevant dimensions (e.g., scope, precision, coherence, and explanatory power), an increase along any of these dimensions reasonably being taken as an increase in content. These dimensions are logically independent of one another; we have theories of broad scope and relatively low precision (e.g., phenomenological thermodynamics), or of high coherence and low scope (e.g., symmetry group theory of fundamental particles), or of broad scope and high precision but low explanatory power (e.g,. biological taxonomy), and so on. It follows that methodological pursuits of increased content along these dimensions may compete with one another, especially in the presence of finite capacities and resources. It is hard to see how logic could dictate decisions among these pursuits. (For discussion of these and the criticisms to follow, see Hooker 1981a, 1993b.) It is equally hard to see how logicist method can recognize that in most circumstances it will be collectively rational to pursue many or all of these improvements simultaneously, though each by different scientific groups. Formalist methodology produces universally binding requirements, when it produces any at all, confining every rational agent to the same action: a methodology of rigid decisions or none at all.

There are a variety of other technical difficulties in specifying Popperian method that also typically issue in ambiguity. In typical contexts where many theories are conjoined to derive testable consequences, including the roles of background knowledge, the Popperian content is a joint property of all relevant theories conjoined; how much belongs to each? Again, a bold conjecture will often conflict with some part of the background into which it is introduced. What has to be revised, and how? And how then is severity of test to be defined (is probableness to be assessed against original or revised background)? And so on. But we do not pursue these here, except to note that deductive formalist method also cannot acknowledge as reasonable the widespread scientific practice of retaining but containing errors (anomalies) and even contradictions until some illuminating resolution of them is found.[14]

Impotencies and ambiguities aside, why does Popper's methodology achieve the truth, or at least head toward the truth? The answer seemed obvious in the case, of empiricism, since induction was to produce most-likely-to-be-true theories on the basis of observational foundations. But in Popper's case the problem is urgent, both because Popper has no foundation for knowledge, unlike the empiricists, so there are no guaranteed truths in Popper's system anywhere, and because of the kind of method that Popper proposes: a negative method of error elimination. Particular cases of error elimination by themselves carry no guarantee that they will be replaced by the truth in any respect. If there were only a finite number of falsehoods on a given subject, then a perfectly working error-elimination method would be guaranteed to eventually reach the truth. But an imperfect, finite error elimination-method working in a field of infinitely many errors has no such guarantee available. Combined with absence of foundations, it demands an explanation of why we can expect truth to emerge. (See also Grünbaum 1976a, b, c.) Furthermore, each bolder conjecture is, according to Popper, a priori less likely to be true. So progress in science must be constituted by a sequence of increasingly initially improbable theories. These features together make it difficult to defend Popperian methodology, within its own internal constraints. Popper tried to offer a formalist notion of progress toward the truth, increasing verisimilitude, but it too fails on its own terms.[15]

Finally, turning to 'external' difficulties, we confine ourselves here to just one problem, that of relating Popperian method to the history of science. Popper offers a universal, eternal method, but the actual history of science shows methods that are context dependent, where context specifiers are discipline, problem, theory, and historical period.[16] Again, Popperian methodology emphasizes the importance of novelty, of new, hitherto unanticipated results to scientific progress. But novelty is a historical matter, not solely a logical matter. What is novel at one time is not so thereafter, yet its explanatory significance, for example, understood purely in terms of its logical place in the structure of science, will not be altered by the passage of time per se. While notions of content, explanatory power, etc., defined in purely logical terms can be deployed to argue for the desirability of risky theories, these characteristics cannot help us to see why, for example, passing severe tests is desirable or failing them undesirable. This is so because logico-structural features are not altered by testing, whereas testing is a temporal, epistemological matter, and the severity of a test (i.e., the improbability of its prediction) is a function of the background knowledge characterizing each historical test context. (The reader will recognize that these problems are not peculiar to Popper but characterize any formalist specification of method in terms of logical rules.)

The attempt by Lakatos to alleviate these difficulties (Lakatos 1970) is instructive. For Popper, theories appear and disappear as units or wholes. Lakatos introduced a two-component structure to theory, a core of principles, and a surrounding belt of auxiliary assumptions. A research program is a sequence of theories with a common core and a changing belt, adjusted to meet developing observational evidence. Research programs are retained as progressive as long as their belts can be adjusted/elaborated in a principled way and they continue to be empirically successful, but they are abandoned as degenerative when sufficient anomalies accumulate that can only be met by ad hoc belt adjustment. By this simple device of adding research programs to the Popperian structure, Lakatos is able to achieve the following: (1) Explain how it is that scientists can utilize the methodology of falsification while yet refusing to abandon the theory as a whole, viz., by exploiting the ambiguity of where to pin the falsity in a falsification so that the core remains untouched; (2) show how science can have an intrinsically historical structure to its methodology; (3) show how Kuhn's notions of normal and revolutionary science may be incorporated within the general Popperian framework.

Unhappily, Lakatos's methodology is equally formalist, though with an extra structure. On the one hand, it suffers from the same rigidity of rules as does any set of formal rules. In contemporary physics, for example, we can specify experiments whose outcomes would deeply penetrate the so-called core (e.g., discovering that PCT symmetry was violated). On the other hand, the extra structure introduces additional decisions uncontrollable by logic. Lakatos's position raises in an acute form the question of why scientists persist with or abandon a research program. Any research program can unpredictably see its fortunes reversed, and there is no decisive criterion for making these most momentous of decisions (a fortiori no formal criterion). Consider the two-hundred-year history of Newtonian mechanics successfully rising to new challenges, to be followed by the relativistic revolution. Lakatos says that it is the community of scientists that make such decisions, but who decides who is to be accounted a scientist for this purpose? Who decides what criteria will dominate in a given decision context? And so on. Thus to all Popper's uncontrolled judgments underpinning rational science, we add still others, and again ones made by the scientific community.

There are various moves Popper has made to try to ameliorate these problems. *Conjectures and Refutations* (1963—see Popper 1972) emphasizes problem solving and criticism as the essential features of rationality. This proves a relevant response to criticisms of traditional empiricism for its simplistic conception of the history of science, criticisms made, for example, by Feyerabend (1978a) and Kuhn (1962). For problems may be dissolved as well as solved, and even abandoned, and certainly reevaluated as to their importance, so there is not the empiricist-style constraint to a strictly accumulative history of science. There are a variety of 'internal' difficulties with reliance on problems, starting with how to identify and weigh them.[17] But the principal difficulty with thus weakening the conception of rationality is that it throws yet more weight onto conventional decisions, made by the community of scientists. What is to count as a problem? Which are worth trying to solve? When can a problem be dissolved rather than solved? And so on. By enlarging the role of social decisions Kuhn raises yet more acutely the question of what distinguishes science from other socio-political activities. Kuhn had already argued, for example, that the problem of changing research programs ultimately had no rational resolution, simply a social one. Although Popper emphasizes criticism as the distinguishing feature of rational science, this is not by itself sufficient, for it must be criticism aimed at the truth, and not made, for example, for bureaucratic or political advantage (cf. Brown 1987, 1988). Even this entry of the social, and with that also individual psychology, raises objections 6 and 7 against empiricism in an acute form for Popper as well.

Still later, in *Objective Knowledge* (1972—see Popper 1979), Popper introduces an abstract World 3, of ideas and logical structures, distinguishing it sharply from Worlds 1 and 2, respectively the realms of nature and mind (psychological states). Objectivity belongs to World 3 and the structure of objective science is found there. But Feyerabend (1974) argues that this addition represents in effect the Lakatosian degenerating phase of the Popperian research program, that World 3 merely labels Popper's desire to provide an objective account of knowledge but does not actually solve any of the outstanding problems (choice among tests, and so on). Saying that an observation report is in World 3, for example, cannot make it objective, it only represents it as if it were so. The shift to World 3 is a shift within the formalist framework, and we are evidently left still in need of a substantive account of rational procedure. We shall shortly explore related charges concerning Popper's evolutionary epistemology.

The common themes connecting all these difficulties, both internal and external, is the methodological poverty of rigid decision processes determined by logical rules and the lack of any alternative theory of rational decisions. Confinement to this methodology derives from the formalist empiricist metaphilosophy Popper tacitly adopts, which suppresses the importance of flexible decisions as the ubiquitous informal components of any methodology.

Consider the matter of perception and observation reports. Popper certainly doesn't face the difficulty of defending observation as autonomous from conjecture or theory and cognitively secure, as empiricists do. As noted, Popper insists that observational reports are themselves fallible theories, not guaranteed truths. However, they are still fundamental to his methodology, for they are the basis on which theories are falsified and corroborated and so must be able to be cognitively defended. And the key consequence of their fallible, conjectural status is that, like all theories, there must be some decision procedure for accepting some and rejecting others. Popper says that one tests observation claims by deducing further observable consequences from them and checking them, but this clearly generates a potentially vicious regress. Ultimately, the only way to halt the regress is for a decision to be made by the scientific community that enough testing has been done and the original claim is accepted or rejected. (Popper offers here the metaphor of science as an edifice built on a swamp whose supporting piles are driven deep enough for current stability but never touch a solid bottom—well, "deep enough" is a matter of decision.) A further community decision is constantly also required, viz., whether to reverse an earlier acceptance/rejection decision.

Indeed, Popperian theory is riddled with yet additional ineliminable decisions: which theories to test, which tests to apply, where to lay the error when theory and experimental result clash, when to stop testing, and so on. The whole of doing science is filled with such decisions. Diagram 1, chapter 1, provides a simplified schematic outline of the decisions surrounding theory testing. (A more complete analysis of the decision structure of formal Popperian methodology is provided in Hodges 1990.) These are the kinds of decisions that must be made by every scientist, though of course their substance will vary from one context to another—which introduces a further class of decisions. Few or none of them can be decided formally.

Popper's method limits us to bold conjecture, but why so simple a specification? We make progress, to be sure, by constructing a theoretically well motivated bold new conjecture, but in the history of science, this has been neither a sufficient nor necessary route to deep innovation. There are a large number of strategies each of which contribute to the likelihood of deep insight, for example, developing a new technology or technique, searching for new results in a hitherto unexplored domain, pursuing a theory known to be inadequate until the nature of its adequacies are more clearly understood, examining the evolution of a concept (say motion) in order to bring to light hitherto unexamined presuppositions (cf. Einstein), exploring generalizations of abstract mathematical frameworks, putting oneself through relevant extended training or acculturation experiences, developing new forms of communication, and so on. To single out just one strategy from among all of these as the only rational one is just to ignore the complexity of the historical dynamics of science. It is to refuse to recognize more than one operative cognitive aim or value, something that Popper cannot do:

> . . . what lends science its special character is not the elimination of extra-scientific interests but rather the differentiation between the interests which do not belong to the search for truth and the purely scientific interest in truth. But although truth is our regulative principle, our decisive scientific value, it is not our only one. Relevance, interest and significance (. . . relative to a purely scientific problem situation) are likewise scientific values of first order; and this is also true of values like those of fruitfulness, explanatory power, simplicity and precision. (Popper 1976, 96–7)

To these we may add security, technological control, predictability (controlled or not), and so on. But distribution among these alternatives is a matter of flexible decision. It is typically the case that not more than one or two of these goals can be simultaneously pursued by an individual; only the institutional structure of science draws the various pursuits together. Articulations of, and trade-offs among, these proxies for truth, equivalently among the research strategies adumbrated earlier, require yet further judicious decisions.

Beyond these decisions, there are all those further decisions forced on us by our finitude, such as choosing among relevant tests just those whose techniques can be feasibly mastered by a given group of scientists, those whose consequences for theory are calculable at the time, those that can be economically afforded, and so on. Decisions of this kind apply at every methodological step. And beyond that again, there is the battery of decisions concerned with the institutional realization of science: Should X or Y be hired as our technologist? What sort of training does Z require before tackling this problem? Should we start a new journal? Aimed at which audiences? And so on. These decisions are just as essential to doing science for, contrary to any impression created by focusing on abstract logical method, it is here that collective cognitive capacity is either realized or aborted—and that includes the capacity for thorough critical appraisal.

None of these groups of decisions is dictated by formal logical considerations. But if science is to be rational, all these must be rational decisions. For Popper, rational decisions are restricted to those dictated by deductive logic, which is impotent to assist here. But this means that at the very basis of Popperian method are substantive decisions that lie outside of rational control. Popper's stage 1 strategy fails on its own terms. When faced with any instance of this difficulty, Popper turns to the empiricist alternative: Exclude the decision as conventional, a matter for psychology and sociology but not normative rationality. But this opens the way to purely sociological or political theories of science and thereby removes its cognitive significance. And that was a result Popper himself fought to avoid.

3.1.4. Control of Decisions: The Popperian Methodological Dilemma

After Popper, through Kuhn, Feyerabend, and all the others, the appeal to decisions by the scientific community widens rapidly—so rapidly, that all of these latter have been accused of abandoning reason. Why? Only because of the tacit assumption that what cannot be reduced to logical method is nonrational. This consequence is instead better taken as a reductio of this conception of rationality.

But would Popper have taken it so? We suggest not. Popper remains wedded to the formalist conception of reason, as his projection of rational processes on an abstract Platonic heaven makes plain. Popper also remains wedded to promoting the life of reason, and not only in science but in sociopolitical life generally. Indeed, it is part of Popper's greatness as a philosopher that he has pursued the consequences of his vision of the rational life across subject matter and history. And we shall later find in Popper more positive suggestions for a theory of reason than his articulation of formal criticism, but these are submerged under his formalist concerns. Popper's first attempt has failed to control decisions, largely because the control provided by logic is too weak. How might the control be strengthened? One strategy would be to develop an account that accepts the necessary and legitimate role of human decisions in science, but that also offers some more adequate form of their rational evaluation and control. We suggest, however, that this is not Popper's solution, that instead his move to an evolutionary epistemology can only be understood in terms of what we have called his stage 2 strategy: Natural selection eliminates what does not deal adequately with the world, and it does it quite independently of the eliminated creature's fantasies, wish projections, prejudices, and the like. Could it be that natural selection is intended to provide a decision procedure that forces all decisions, including selection of theories, in a way that would resolve Popper's problem and restore his empiricist metaphilosophical ideal of eliminating all decisions? Let us turn now to Popper's evolutionary epistemology to see if this move is more successful.

3.II. Popper's Evolutionary Epistemology: Analysis and Critique

3.II.1. The Natural Selection of Theories

> **3.II.1.1. Evolutionary Continuity.** "From the amoeba to Einstein, the growth of knowledge is always the same: we try to solve our problems, and to obtain, by a process of elimination, something approaching adequacy in our tentative solutions." (Popper 1979, 261)

This familiar passage characterizes Popper's evolutionary epistemology. (Hereafter, Popperian evolutionary epistemology will be abbreviated PEE, and italics in the quotes from Popper to follow will be Popper's, unless explicitly stated otherwise.) It emphasizes the continuity of evolutionary processes, in contrast to Bradie's popular division of evolutionary epistemology into two strands: EEM, the evolutionary epistemology of mechanisms (e.g., evolution of the brain) and EET, the evolutionary epistemology of theories (Bradie 1986). EEM is concerned with a causal process involving genetic information, while EET is concerned with a logical process involving semantic information and argues the formal similarity of the process to the way species evolve. There is much in Popper that supports this division, and Bradie claims him as a principal ally of the distinction, but Popper's stress on continuity seems in stark contrast.

From the outset, PEE focuses on the development and defence of generalized evolutionary processes. Popper distinguishes, for example, between three "levels of adaptation": genetic adaptation, adaptive behavioral learning, and scientific discovery. Here each level is claimed to be a special case of the preceding level, the purpose being to assert a "*fundamental similarity of the three levels*" in terms of "the mechanism of adaptation" (Popper 1975, 73). Hereafter, we shall repeat Popper's use of *level* where he would use it; so long as the reader remains aware that this is done without prejudice to the need for a more careful analysis (cf. chapter 1), little harm should result.

At each level, the mechanism works on a basic *inherited structure*. The structures are passed on or *transmitted* to descendants by *instruction*. The bearers of structures are exposed to *problems*, arising from their environment, and emit *tentative trials* as attempts to deal with the problems. In the production of these trial solutions, variations or mutations occur, thus generating new structures, ". . . in response [to problems], variations of the . . . *instructions* are produced, by methods which are at least partly *random*" (Popper 1975, 74). This relative independence of the specific nature of a trial from the environment is the equivalent of Popper's rejection of induction as rational, and the locating of the authority of knowledge in criticism rather than source (see Popper 1972, 24 ff.); it is the logical equivalent of the Darwinian blindness of mutation to environmental success. Variations arise only "from within the structure" rather than from the environment; the environment may elicit or allow a trial to be emitted, but its influence on the *form* of the variations comes into play only in selection.

Not all of these new trials survive to be passed on to later representatives of the structures, for the next stage is a selection process in which the badly adapted are "killed off":

> Those of the new tentative trials which are badly adapted are eliminated. *This is the stage of the elimination of error.* Only the more or less well adapted trial instructions survive and are inherited in their turn. (Popper 1975, 74)

In earlier Popperian language, both biological and scientific variations are tested by their consequences.

Popper calls this overall process, which is held to occur at all three levels, adaptation by "the method of trial and the elimination of error." The elimination of error within this process "is also called '*natural selection*.'" Natural selection thus understood "operates on all three levels." (Popper 1975, 74)

This is a generalized version of the orthodox Darwinian variation and selective retention (VSR) process. Problems, which are "problems in an objective sense," arise out of a mismatch between a structure and its environment. As a response, the structures try out tentative solutions, or trials. The kind of thing that realizes the structure at each level is different, as are the trials. Popper tells us, "All *organisms* are constantly . . . *engaged* in problem-solving; and so are all . . . *evolutionary sequences of organisms*" (Popper 1979, 242).

Trials may be "new reactions, new forms, new organs, new modes of behaviour, new hypotheses" (Popper 1979, 242), depending on the level of analysis involved. At the genetic level, the structure that continues is the "gene structure of the organism"; at the behavioral (second) level, it is "the innate repertoire of the types of behaviour which are available." From the point of view of a phylum, for example, an individual organism is a "tentative solution" to the survival "problems" of the phylum. The individual organism "is thus related to its phylum almost exactly as the actions (behaviour) of the individual organism are related to this organism" (Popper 1979, 243). At the level of scientific knowledge, the "structures" are "*the dominant scientific conjectures or theories*" (Popper 1975, 74). Thus the growth of knowledge, specifically of scientific knowledge, is also included in the VSR process.

This generalized process of adaptation, the method common to Einstein and the amoeba, is subsumed under the "general schema of problem-solving by the method of imaginative conjectures and criticism" (Popper 1979, 164; cf. 243), with P = problems, TT = tentative trials, and EE = error elimination:

(VSRPEE): $P_1 \text{—>} TT \text{—>} EE \text{—>} P_2$

Popper also calls this "learning from our mistakes" (Popper 1979, 266), where the place-holder *our* can be replaced by organisms, animals, human beings, or any structure undergoing such evolutionary change. This VSR problem-solving process is the "fundamental evolutionary sequence of events." The sequence is "not a cycle," though it has a recursive element, because "the second problem is, in general, different from the first: it is the result of the new situation" (Popper 1979, 243). At each level, each application of the trial and error process, especially successful ones, may change the environment within which the structure is being transmitted and thus result in "new pressures, new challenges, new problems" (Popper 1975, 75).

For Popper the structures that remain after selection not only incorporate, but also encode, environmental information; a physical organ in an organism's body, for example, may be seen as a "theory."[18] Encoded information of this kind is accumulated across the relevant evolutionary sequence and forms the basis for the homologous process at the next level "up." At each level, this information then becomes part of the inherited structure, for example, as genetic information, instinctive behavior, or "the dominant . . . theories" (Popper 1975, 74). Thus "no organism is born ignorant." For human beings, the inherited basis from which human knowledge grows includes: various inborn needs and drives (Popper 1979, 23–4); "theory-laden perception" (ibid., 35–7); Kantian-like categorical regularities (a priori for the individual, as inherited species information, but not a priori valid, for example, ibid., 24,[19]); and the capacities underlying rational thought, including consciousness itself and the capacity for symbolic thought (Popper 1979, 137–9). This last aspect is especially significant, because it is here that the "natural selection of theories" takes place. The inherited basis not only makes cognitive evolution possible, it provides the initial selection of what aspects of the environment are problems (cf. Popper 1979, 258–9) and also the beginning theories and expectations against which problems appear as mismatches between the expectation and practice.

Now a crucial question arises. How are we to think about the relations between information-accumulating VSR processes at different levels? According to PEE, they are homologous, certainly, all instances of the schema VSRPEE. But as just indicated, these processes also stand in relations of basis to supported process; how is that relation to be understood? We may see the relations as characterized by *continuity* between levels, or by a *division* of kind between levels.

Popper speaks about the process of error-elimination as natural selection, as if it were a literal extension of the biological process. In this case, the continuity and interactiveness of processes at different levels is accepted. But then there will be no sharply delineated relation of support available; processes at one level may be both cause and consequence of those at others. Hierarchies become approximations, not strict; there is a single dynamical system undergoing complex change. (Compare the Levins quote at chapter 2 and Dyke's rejection of principled hierarchies in favor of LIMAs: level interactive modular arrays; Dyke 1988.) If, on the other hand, a sharp distinction between levels is drawn, then an equally sharp distinction between their evolutionary processes would follow naturally; the kinds of entities involved and the basic terms of the VSR process may change sharply across levels. In particular, for our purposes, the processes of selection might differ essentially.

There is much in Popper to support both readings. As well as affirming continuity, he is quite explicit that such differences exist. Here we find a first tension in PEE between commitment to a continuous, integrationist, inevitably naturalist emphasis on the unity and/or interconnectedness of evolutionary processes and commitment to fundamental difference among processes at different levels within a veneer (a nearly vacuously thin veneer) of VSR unity. It is time to look at these differences.

3.II.1.2. Differences in Error Elimination. The differences between levels that Popper offers us are signaled by an increasing "flexibility of response" available at each level as we go up the hierarchy. They emerge in two closely related ways. First, in relation to trials, there is a lesser degree of rigidity with which the underlying instructions specify the nature of the resulting "organism," resulting in turn in an increasing area of freedom within which the organism can emit trials, as new variations. Second, the nature of error elimination exemplifies an increasing "looseness" or increase in indeterminism. This is because the means of error elimination, especially with the "higher" organisms, is increasingly a matter for the development of "controls" over the trials *within* the organism rather than the elimination of the "emitter" of the trials.

> Error-elimination may proceed either by the complete elimination of unsuccessful forms (the killing-off of unsuccessful forms by natural selection) or by the (tentative) evolution of controls which modify or suppress unsuccessful organs, or forms of behaviour, or hypotheses. (Popper 1979, 242)

This represents a shift in the locus of selection from the exterior of the organism to its interior. The simplest organisms *reactively* produce reflex behaviors that result in their being killed or not. All the action, so to speak, is external to them. But adaptable organisms, at least behaviorally adaptable ones, can increasingly run trials within themselves and there *anticipatively* select the one judged best for the circumstances.[20]

In particular, at the behavioral level, Popper distinguishes between what he calls, following Mayr, "*closed behavioral programs and open behavioural programs*" (Popper 1987, 151). A closed program specifies the behavior of an organism "in great detail" (ibid.), leaving little room for the development of new behavioral possibilities (expansion of the behavioral repertoire) through the emission of new trials at the behavioral level; that is, there is less room for variation to occur at that level. The open program, on the other hand, "does not prescribe all the steps in the behaviour but leaves open certain alternatives, certain choices" (ibid.). (We note for later reference the re-emergence of decisions here for Popper.)

The distinction between populations of phenotypes exhibiting these two kinds of programs may be linked to two evolutionary "strategies" or dynamics, the one characterized by increasingly refined but fixed *adaptation* and the other characterized by an increasing capacity to alter adaptations as appropriate, that is, by increasing *adaptability* (see chapter 2, section 2.II.2). For environments that produce stable, relatively homogeneous selection pressures, there is selective advantage in a closed program that produces a phenotype optimally tuned to survival in those conditions, and conversely, there is some selective advantage to open programs in highly variegated environments (e.g., those that change rapidly), because in such situations there is advantage in the development of flexibility of response.

A major significance of this difference, for Popper, is that it provides a plausible route into the development of consciousness as the capacity par excellence for open program choices. For consciousness provides the capacity to envisage alternative choices and then to evaluate them, a capacity made necessary by program openness: "consciousness originates with the choices that are left open by open behavioral programs" (Popper 1987, 151). This conception of consciousness may be fruitfully compared and contrasted with that of Piaget (see chapter 5).

Adaptable organisms can emit behavioral trials and can learn from those trials which ones work and which don't. This may lead, for example, to the development of exploratory behavior (Popper 1975, 76) through the use of trial and error in behavior—cf. Skinner's operant conditioning (Skinner 1953). Resulting information may then be stored at this level, in further information-storage capacities that have developed via natural selection, and thus be available for future use by the operant organism itself, rather than only to the species as a whole through the reproduction of the characteristics of survivors. At a further stage of development, there enter vicarious trials, in which the open program allows for the organism to play possibilities through "tentatively—on a screen, as it were—in order that a *selection* can be made from among these possibilities" (Popper 1987, 152). At one stage of development, such selection may be made merely through "warnings," for example, a vague memory of pain (or pleasure; ibid., 151). At a further stage, however, such situations of choice may become the beginnings of goal-directed behavior, with the evaluation of the vicarious trials undertaken in connection with "the end state of the imagined behaviour" (ibid., 152), for example, a goal. A crucial part of such development of choice within open programs will be the appearance of "preferences," for example, behavioral dispositions to choose in certain ways that express internal evaluatory procedures (see below).

This internal development of variation, *evaluation*, and selective retention (VESR) processes will, according to Popper, lead to a new kind of externalization of the evolutionary process. For the final stage in this development is the appearance of symbolic operations and language. Popper regards this as "the human step": the "evolution of language and . . . World 3 of the products of the human mind" (Popper 1987, 152). Popper identifies a World 1, "the world of physical objects or of physical states," for example, a book as material object; a World 2, "the world of states of consciousness, or of mental states," such as feelings of liking for a book; and World 3, a shared realm of "abstract meanings and contents" (Popper 1979, 240), "the world of *objective contents of thought*, especially of scientific and poetic thoughts and of works of art" (ibid., 106). Through language and the development of critical evaluation of abstractions, the VSR process is removed from the biological world and transferred to a VESR process in an abstract symbolic world accessible only to consciousness.[21] So "something new has emerged on the human level" (Popper 1979, 262) namely the "emergence of mind" (Popper 1987, 150).[22]

Popper's introduction of his ontology of three Worlds has been a controversial one, but his evolutionary epistemology requires some such notion in order to provide for the operation of natural selection with respect to theories. While Popper asserts a continuity with animal "knowledge" and genetic information in the same breath (Popper 1979, 261 and elsewhere) and that World 3 is an unintended by-product of human activity (ibid., 117), similar to a spider's web (ibid., 112) or a bird's nest (ibid., 117), it is clear that for Popper this development marks a major change to the way in which evolution operates; the emergence of "something new" at the human level signals the "transcending" of the old natural selection (cf. Popper and Eccles 1977, 210), through the creation of a new arena of conscious and symbolic trials, which is open to intersubjective criticism. The stage of *exosomatic evolution* (Popper 1979, 248) has begun.

This specifically human step is characterized by the operation of trials and the elimination of error in a symbolic environment of abstractions, World 3, where we can allow our hypotheses to "die in our stead." We can criticize, and eliminate hypotheses instead of being eliminated ourselves.

> It allows us to *dissociate ourselves* from our own hypotheses, and to look upon them critically. While an uncritical animal may be eliminated together with its dogmatically held hypotheses, we may *formulate* our hypotheses, and criticise them. (Popper 1987, 152)

But this in turn involves a number of changes to the location and the nature of the fundamental evolutionary process. Not only are the trials here vicarious, but they are publicly exposed to criticism in an intersubjective realm; they become "objectified": objects open to investigation by others, as well as ourselves, through their presence in World 3.

> On the level of scientific discovery two new aspects emerge. The most important one is that scientific theories can be formulated linguistically, and that they can even be published. Thus they become objects outside ourselves; objects open to investigation. As a consequence they are now open to *criticism*. Thus we can get rid of a badly fitting theory before the adoption of the theory makes us unfit to survive: by criticising our theories we can let our theories die in our stead. (Popper 1975, 77–8.)

The objectification of the trials in the symbolic World 3 allows for the elimination of the "unfit" by means of "criticism," which occurs on the basis of the descriptive and argumentative functions of language, for without the "development of an exosomatic descriptive language—a language which, like a tool, develops outside the body—there can be *no object* for our critical discussion" (Popper 1979, 120). And it is "only in this third world, that the problems and standards of rational criticism can develop." As he says elsewhere: "Thus in bringing about the emergence of mind, and World 3, natural selection transcends itself and its originally violent character" (Popper and Eccles 1977, 210). Might we also want to say that it transcends itself and its originally *mechanical* character?

This highlights the degree of difference there evidently is for Popper within the common VSR process. Clearly this last step of exosomatic evolution is crucial to PEE, so we examine its account of natural selection.

3.II.1.3. Error Elimination in World 3 and Plastic Controls. How does evolution of theories occur in the human world, where they die in our stead? According to PEE it is still through "a process closely resembling what Darwin called 'natural selection'; that is, *the natural selection of hypotheses* . . . a competitive struggle which eliminates those hypotheses which are unfit" (Popper 1979, 261). And while early on Popper frequently used natural selection as simply an illustrative metaphor for the elimination of unfit theories (1980, 108; 1972, 52) and does so even in places in *Objective Knowledge*, in the latter he also states quite clearly that he intends the description to be taken literally: natural selection "is meant to describe how our knowledge really grows. It is *not meant metaphorically*" (p.261, our italics). Recall that Popper's natural selection "operates on all three levels" (Popper 1975, 74), including within itself the processes of error elimination in World 3. But at the same time, recall Popper's claim that the operation of the trial-and-error process with scientific knowledge is not only the same as, but also different from, its operation at earlier levels.

The major difference lies in the process of error elimination. We noted in section 3.II.1.2 how Popper shifts the locus of selection inward, from an action of the external environment on the organism to a controlled or regulated action within the organism (see note 20 and text). The notion of *control* is the key to Popper's account of error elimination at the higher levels. It signals the appearance of Popper's embryonic stage 3 strategy, and deserves closer scrutiny. According to PEE, control is exercised through increasingly "open" developmental programs, especially those yielding increasingly "loose" behavioral programming and a corresponding increase in consciousness and such paraphernalia of consciousness as vicarious trials, preferences, and evaluation of trials with reference to "ends." This increasingly symbolic process culminates in its transfer to the exosomatic World 3, structured by rational argument (Popper 1979, 239–42; see also section 3.III.3 below). Despite the emphasis Popper gives to the logical character of argument in World 3 (see below), he also holds that the open developmental programs supporting rational agency produce "choice."

> The selection of a kind of behaviour out of a randomly offered repertoire may be an act of choice, even an act of free will . . . . A choice process may be a selection process, and the *selection* may be *from* some repertoire of random events, *without being random in its turn*. (Popper 1987, 147)

The "preferences" that model internal control of behavior will be a complex of in-built directions associated with inherited structure operating at various levels, adaptive behavioral learning and "objective" knowledge, learned through the use of these looser symbolic, rational "higher functions." Preference and choice are the human expression of the increasing looseness that control exhibits as one ascends Popper's evolutionary/cognitive hierarchy.

Within this hierarchical conception:

> Each organism can be regarded as a hierarchical system of *plastic controls* . . . . The controlled subsystems make trial-and-error movements which are partly suppressed and partly restrained by the controlling system. (Popper 1979, 245)

Controls here define the internal part of the error-elimination process. But what are "plastic controls"? Popper introduces the term as a contrast to "cast-iron control" (Popper 1979, 232) by which he means a control that has no freedom in its action: It either applies or it doesn't—a simple on/off switch. A plastic control, on the other hand, is a "selective control" such as "an aim or a standard" (ibid.) that allows for compliance with it in different ways or to different extents, these judgments being made especially according to argument, coming to a decision by professional judgment, for example, within the constraint of a set of guidelines, as opposed to a rigid deductive decision procedure.

> The control of ourselves and our actions by our theories and purposes is a *plastic* control. We are not *forced* to submit ourselves to the control of our theories, for we can discuss them critically . . . . (Popper 1979, 240–1)

The evolution of plastic controls allows for learning; the notion of plastic control combines the restrictiveness of a control (elimination of error) in a "subtle interplay" with the freedom of deliberation, ". . . a kind of *maturing process*" (ibid., 234) which allows for the combination of a determination by the control within the framework of an area of freedom, or indeterminism: "freedom plus control" (ibid., 232). This subtle interplay consists largely of a process of "feedback" (ibid., 239) between controller and controlled, which allows the control to be modified in the light of an encompassing goal. In short, the process of error elimination may itself be learned. This is a significant point, because it signals the failure of the strategy of rigid control. For this reason, Popper's formalist commitments ultimately lead him to ignore it (cf. discussion of Rescher in chapter 4). We return to these issues in Part 3.III.

In World 3, elimination of trials goes by a particular type of plastic control—that of rational criticism: "We expose our World 3 conjectures to selection by *conscious* criticism" (Popper and Eccles 1977, 122 and elsewhere).

> . . . progress in science . . . depends on *instruction* and *selection*: on a conservative or traditional or historical element, and on a revolutionary use of trial and the elimination of error by criticism, which includes severe empirical examinations or tests; that is, attempts to probe into the possible weaknesses of theories, attempts to refute them. (Popper 1975, 78)

The natural selection of theories within World 3 is explicitly identified by Popper with his methodological falsificationism, the method of conjectures and refutations (see Popper 1979, 260–1). It is given here a new gloss—scientific knowledge grows through the introduction of conjecture (new theories = "trials") on the basis of a broad background of knowledge ("instruction": the "conservative" element) and subsequent refutations (the "selection" element) are achieved through the familiar logical structure of critical rationalism. Refutation is explicitly identified with the process of error elimination in the application of the general problem-solving schema to scientific knowledge; it occurs through natural selection in World 3. Within this new realm, the processes of trial and error and natural selection, hence of evolution, now take place. But elimination of errors within World 3 constitute "new standards of selection" (Popper 1979, 240), which form a plastic control, since we are "freely choosing" between theories (ibid., 241).

What provide the control, the elimination of errors, are our critical arguments. What provides the structure of this plastic control is the purpose of the activity, the goal to which the "behaviour" (of science) is directed: the "regulative ideas of truth and of validity" (Popper 1979, 239). It is precisely because arguments are structures that can be directed toward the truth that they loom so large for Popper. However, this may sound a little strange: For Popper, control of innovation (new theories) is by logic, a formally exact and eternally fixed structure. How is a structure as rigid as deductive logic a *plastic control*? How can its eternal necessity square with the evolution of plastic controls? This is another facet of the tension between continuity versus difference across levels within PEE.

3.II.1.4. Unity or Schizophrenia? Popper's Problem of Evolutionary Method. The introduction of control reinforces continuity across evolutionary levels as a central characteristic of PEE. The achievement of control operates at all levels, producing physiological alteration in simple organisms, physical tools, and artifacts in more complex organisms, and subsequently intentional problem solving. With further phenotypic development, these controls are expressed in the appearance of human public language and objective rational criticism. Based on the evolution of adaptability, Popper's open program leads to the development of a hierarchy of plastic controls, which allow for the development of such tools as science itself. The method of science is the upper level of the hierarchy of the processes of problem solving by the general pattern of trial-and-error and the inheritor of accumulated information from lower levels. The trial and error process is identified at the level of the growth of scientific knowledge with the apparatus of Popper's methodological falsificationism. The natural selection of theories is error elimination through refutation in the logic of falsification. This continues the PEE theme that throughout the evolutionary sequence there is continuity in the information gain from level to level and that it is the same VSR process "all the way up"; genetic mutation and theoretical conjecture is essentially the same process, as is natural selection and rational error elimination. All that changes is the steady increase of complexity, expressed first in physiological problem solving terms and then in terms of rationality and freedom.

But here we find yet another facet of the tension between continuity versus difference across levels within PEE. The biological level offers the "mechanical" nature of natural selection without decision, the rational-symbolic level concentrates on free decisions. Biological selection is causal, material, blind (to outcomes), local (ecologically), and non-progressive, while rational selection is (respectively) logical, formal, intentional, global, and progressive. How is this difference to be understood and reconciled within PEE? The re-appearance of this tension suggests that perhaps Popper has not got here the solution that he wanted.

Recall from Part 3.I that the central problem of epistemology for Popper has always been the problem of the growth of knowledge, the problem of accounting for the way in which change of theories actually can be constrained or controlled so as to become progress in the aim of science, that is, to constitute objective knowledge. But for Popper, decisions are taken to unavoidably introduce subjectivity and so must be eliminated. Part 3.I concluded that his first, or stage 1, eliminative strategy failed. At that point, we noted an alternative: Accept decisions and offer an account of their rational evaluation and control. And we can now see that a foundation for this can be found in Popper, for the very plastic controls that are the basis for the conduct of scientific method are also the basis for that looseness that allows for human freedom but then necessarily demands decisions. This makes it even more difficult to see how Popper could avoid an account of decisions. (This is another facet of the tension within Popper, an instance of the age-old tension between freedom and the necessity of rationality.) But Popper, still under the spell of empiricism, chose another route: his stage 2 strategy; PEE is Popper's attempt to resuscitate the eliminativist strategy through using the properties of natural selection to force all decisions. But the account is plagued by tension, by the difference between causal natural selection and logical rational selection; and according to Popper it is precisely at this (dis)location that decisions reappear. This looks ominous for this strategy. We turn now to examine the success of PEE as an account of progress in objective knowledge. We shall confine our attention to the key element, natural selection.

3.II.2. Selection of Theories in a Symbolic Environment

3.II.2.1. Analogies, Disanalogies, and Popper's Problem. Standard attacks on PEE, indeed on evolutionary epistemology in general, are in terms of disanalogies between biological and cognitive processes (Thagard 1988). Such disanalogies are typically offered in all the areas of variation, selection, and retention. But to attack Popper's account on the basis that disanalogies exist seems superfluous, since Popper has already done it himself, for example, by pointing out how the nature of error elimination will differ across levels with cognitive selection rational and intentional while biological selection is neither. What we are concerned with then is not a listing of similarities and differences per se, but whether or not we are being presented with an adequate account of the growth of knowledge. Rescher's substantive emendation of evolutionary epistemology described in chapter 4, for example, is motivated by the disanalogy just noted between undirected biological selection and truth-directed rational selection; see also case 1, section 3.II.2.4 below. Yet Rescher's concerns about teleology, nonblindness of variation, and Lamarckism as further important disanalogies marking out a non-naturalist cause/reason divide are shown to be misplaced (chapter 4, section 4.VIII). Similarly, Popper claims that whereas biological evolution produces increasing diversity, cognitive evolution produces increasing nomic unity (Popper 1979, 262), but this is at least unobvious (see chapter 2, note 15). Each argument from disanalogy must be argued on its specific merits.

This returns us to our Popper problem proper. The interesting question is not that of disanalogy, but to what extent the putative natural selection of theories offers a solution to the problem of decision within objective knowledge. Does the natural selection of theories succeed in eliminating (forcing) human decision? This question cannot be addressed until the potentially obscure notion of natural selection in World 3 is clarified. Here Popper does not help us, so we turn to Ackermann's elaboration of Popper's position.

3.II.2.2. Ackermann on Selection in World 3. The natural selection of theories is, according to PEE, a development of the capacity for vicarious trials and their elimination by the system of plastic controls. It is clear then that in speaking of natural selection, PEE is not talking about the elimination of theories by their encounter with the real, physical world. Theories consist of propositions and can only stand in logical relations to other propositions (Popper 1980, 43). To suppose otherwise would be to contradict Popper's division of things into Worlds 1, 2, and 3. An event such as the deliberate burning of books does not eliminate the theories they contain but is simply a World 1 event, the removal of "paper with black spots on it" (Popper 1979, 115). "Theories are not the sort of things that can be killed or knocked out by a physical environment. Such talk is mere metaphor" (Holland/O'Hear 1984, 209). But neither can Popper offer just a metaphor; he needs a real selection process.

Well then, where would the evolution of theories take place? It must be in a symbolic environment somewhere within World 3. Ackermann 1976, in a chapter called "Popper against Subjectivism," offers the beginning of an account of what this symbolic process might look like (but more will be required later). He characterizes World 3 in these terms:

> World 3 is the world of human knowledge as expressed in public language . . . . Stating one's ideas in language removes them from the realm of the subjective and places them into the objective arena of public discussion. This arena cannot be controlled by individuals [e.g., for their own benefit] . . . . (Ackermann, 1976; 55)

This is a reasonable, if sketchy, characterization of Popper's World 3, and it offers a proposal for understanding its strategic significance: The third world is Popper's answer to the relativist, subjectivist threat to his philosophy of science.

But Ackermann suggests that this is not sufficient to give World 3 any real importance for Popper and that its importance lies in its being the PEE arena where all the symbolic variation and selection activities are going on: ". . . the evolution of scientific knowledge proceeds through the evolution of world 3 structures" (Ackermann, 1976, 56). And more specifically: "a proper scientific history is an internal history of transformations in world 3 structures" (ibid.). This is where our hypotheses can "die in our stead", where Popper's "exosomatic evolution" occurs.

> The significance of world 3 objects is that while they are objective and in the public domain they are also off line in the sense that we can examine them intellectually for viability while temporarily disengaging them as hypotheses for use in action. (ibid.)

This is *where* indeed, but what about *how*? What does Ackermann do with the Popperian ambivalence as to the nature of selection? He *repeats* it: "The selection pressure on world 3 is, in Popperian terms, the pressure supplied by objective arguments and refuting data" (ibid.). But this selection pressure is a matter of human decision. Ackermann contrasts selection in biological evolution with World 3 evolution: "Selection pressure isn't operative in world 3, however, except by human *decision*" (ibid.) This replicates the tension within Popper's own account: selection is by "refuting data" via "objective arguments," but this is still ultimately a matter of "human decisions." Again, we are presented with two accounts of selection in an evolutionary process, a causal account in a biological setting and a conscious, rational account in a scientific setting. Ackermann offers the same metaphor:

> Scientific knowledge progresses by the proposal, criticism, and falsification of falsifiable scientific theories in world 3. . . . The metaphor is evolutionary. New theories are like new species. The unfit are weeded out by natural selection [sic].
>
> By Popperian analogy, a theory is a scientific experiment that either does or does not survive the results [sic] of some given observational datum. Falsifiability is the criterion used to decide if it is viable, that is, could live in the scientific environment. (Ackermann, 1976, 57)

Notwithstanding its evident metaphorical status, he treats this analogy as if it might contain a genuine solution to Popper's problem, taking pains to develop a richer account of it. We shall briefly examine Ackermann's account here, not because we hold that his elaboration of PEE has any special status, but because his account provides a convenient way to clarify the nature of selection in World 3 and leads to a richer analysis of decisions in science.

The two major features that Ackermann introduces concern population and environment. First population. Ackermann criticizes PEE for its oversimplified conception of biological natural selection, and he sets out to rectify this weakness by introducing population notions, particularly the idea of many variants of a theory occurring across the population of scientists, giving the population greater capacity to adapt to anomalies. This makes, from an interesting theoretical perspective, the commonly made point that single-datum falsification is too simple a basis for a plausible account of rational scientific strategy, a point developed in a related context by Lakatos 1970. However, we do not pursue the details further here, because this enrichment does nothing to alleviate the dichotomy within PEE between the causal biological and the rational scientific processes. Indeed, the enrichment introduces further ineliminable human decisions to the scientific process, of the kind already noted when discussing Lakatos in section 3.I.3 (choice of which variants to develop, to subject to which tests, etc.).

So we turn next to Ackermann on the selection environment. Will his emendations there alleviate the causal/rational dichotomy that plagues Popper's version? Ackermann considers what the PEE analogue of a selection environment might be:

> To what does a theory adapt? I regard the ecological niches or living space of a theory to be defined by . . . data domains. . . . A data domain will be defined as a range of data that can be gathered by certain methods. (Ackermann 1976, 60)

The environment in which a theory has to survive and to which it will adapt, if "lucky" is a set of data defined by the method by which it is gathered: use of telescopes and so on. This is scarcely a clear statement (see below), but let us accept it momentarily to make the point that the account is still too simplistic: *methods* covers a far wider range than just chunks of machinery. Scientific method includes procedures of various kinds (e.g., laboratory procedures), techniques of all kinds (e.g., mathematical inference, survey sampling, etc.), modeling of various kinds (physical, computational, mathematical), principles of various kinds (e.g., use of total evidence in statistical inference, treatment of infinite magnitudes, removal of inconsistency, etc.), and so on (cf. Hooker 1989b, 1987 chapters 4, 5). Let us agree to include all these aspects under *method*. Then even with this emendation, and still setting aside the definition of data domain, there remains a major problem.

Note that for Ackermann, environment = a set of observational data, a set of basic statements. Now we know that Popper has a problem with anchoring his basic statements, the empirical basis of science, in experience; ultimately, they are cut off from experience, there is a sharp distinction between "motivated reports" and "logically justified reports" (Popper 1980, 110). For this and other reasons (e.g., ignorance vis-à-vis finite resources), for Popper the empirical basis is ultimately decided upon by conventional decision; it is determined in terms of where we decide to stop the process of "justification" (Popper 1980, 45–8, 93 ff.; cf. Hodges 1990 and Part 3.I above). And in fact it becomes clear that by *methods* Ackermann means not only the technology available and used, but also judgments (presumably by the scientific community) as to the legitimacy of various "means of gathering data" (Ackermann 1976, 60). Such normative judgments are as wide as are methods. We must not only judge where to use a telescope (not from downtown Los Angeles) and how properly to use it (how to correct for parallax, distortion, etc.), but also judge which laboratory and other procedures to use, which observing strategies to use, which principles to apply, and so on. For each of these components there must be a judgment as to the legitimacy of its use. Further, we must judge that in combination they can yield relevant information and that they can all simultaneously operate within their domains of validity while doing so. By focusing on chunks of machinery, Ackermann has missed most of this, but it is a critical part of an account of scientific method.

This discussion shows that introduction of this World 3 analogue for environment leaves the basic causal/rational dichotomy not merely untouched, but reinforced. The symbolic selection environment for theories is anything but natural. It is instead a highly artificial construct out of a myriad human decisions of a thoroughly normative kind. It is this *artificial* environment that provides the selection pressures, the mechanisms of natural selection.

It is also a highly complex environment, as we have seen, being constituted by many interacting components. Indeed, by focusing on the machinery chunks, Ackermann distracts us from the subtler but no less important complexities. First, methods are in intimate interaction with theories. When can one clean apparatus with detergent and when should carbon tetrachloride be used? That depends on applying chemical theory to the contaminated surface and the procedural requirements for adequate purity. What is an adequate statistical sampling methodology? That depends upon statistical theory, and there is still competition and controversy in the foundations of statistics and hence in justified method (see Harper and Hooker 1976). What is a better electron microscope? That depends upon applying theoretical physics to analyze the various processes involved. We find an interactive closure: There is no way to characterize this environment independently of appeal to the very theories that are to be selected in it (see also Hooker 1987, chapters 4, 8).

Second, there is interaction with higher-order components within the cognitive domain. What criteria for judging methodological adequacy are appropriate? What, for example, makes the use of a consistency principle an adequate one? What sorts of criteria are to be used to decide this question? What kind of philosophy of science should we use? Is an evolutionary account of the growth of science an adequate one? What methods should we use to decide this question? Method is itself a question that is part of the domain of knowledge. Science is, in substantial part, the data domain (= environment) for philosophy of science. But philosophy of science provides the framework of judgments that ultimately differentiate to create the symbolic selection environment. Once again, we find a complex interactive closure, but now of another kind, between selector and selected.

Ackermann has usefully enriched Popper's basic metaphor, but he is still oversimplifying. If we are to achieve an adequate characterization of science, we must recognize its immense complexity. It is not our task here to complete this characterization of science, but any account of World 3 as a system in which scientific advance takes place ("an internal history of transformations in world 3 structures" ibid., 56) must not understate the complexity and intimate interrelations of these structures.

Our main concern, however, is to make the point that Ackermann's elaboration has only served to drive the gulf between biological natural selection and scientific natural selection still wider. For Popper, the former is a complex causal process, while the latter is a complex rational (and conventional) process. How can Popper call them both natural selection, sliding across the evident gulf separating them? Our answer: by exploiting ambiguities inherent in his specification of World 3. To prepare for the next step in the analysis of natural selection in PEE, we examine some of these ambiguities.

3.II.2.3. The Contents of World 3. World 3 is "the world of *objective contents of thought*," contrasted with the world of "states of consciousness, or mental states," World 2, and World 1, the world of physical objects (Popper 1979, 106). Popper cast this in terms of "existent realities" and "Platonic Forms"—certainly an unnecessary, and for many an erroneous, move. A chief offense is that it creates a cause/reason dichotomy where before there was only difference: Reason is a particular, functionally specified regulatory process; like all regulatory processes, it is causally realized. But for Popper, reason and cause are banished to separate worlds, and ontologically, the two have nothing to do with one another, while functionally chaos is prevented by the imposition of a rule requiring that World 3 structures are always reflected in Worlds 1/2 causal sequences. This is Popper's "principle of transference." Popper is ambivalent about its status, variously calling it "a heuristic," "a bold conjecture," "a fact" within the same paper (see, respectively, Popper 1979, 24, 6, 68 note; cf. Hooker 1981a). The ambivalence highlights the fact that this is an old and unresolved philosophical problem; e.g., Popper's principle echoes Kant's "double government" theory (cf. Butts 1984). We again anticipate later argument by remarking here that it is not necessary to proceed in this way: To create an adequate theory of objectivity, it suffices to construct not a distinction between worlds, but a principled distinction between cognitive and other social institutions and social processes (see Part 3.III below). But for the moment, let us accept Popper's projection of system design distinctions on to the heavens and inquire as to the contents of World 3.

The initial criterion for entry into World 3 provided by Popper is to be an "objective logical content" (Popper 1979, 157), but this is not at all illuminating. The crucial features that appear to be operating here are the notions of abstractness and public accessibility. But this includes far too much for the selection environment Popper needs. It certainly includes "*theoretical systems*," "*problems and problem situations*," "*critical arguments*," the "state of a discussion," and the (logical) contents of "journals, books and libraries" (Popper 1979, 107), which all sounds fair enough. But the entry criterion for World 3 is the wider one of being an objective logical content, not the narrower one of being part of (current) objective knowledge; so World 3 also contains discarded or hitherto untested theories and data, pairs of incompatible theories, and incompatible data pairs and similarly, myths and stories (Popper 1984, 252) and so on to "all the products of the human mind . . . . [including] the world of human creation in art" (Popper 1984, 252), for example, "poetic thoughts" and "works of art" (Popper 1979, 106), including visual arts (Popper 1984, 252–3) and music (Popper 1979, 254; 1984, 252). This makes World 3 a decidedly peculiar and puzzling device for an attempt to capture objectivity: the scientific is there, but then so is a lot more, including domains generally regarded as epitomizing, even glorifying, subjectivity (poetry, etc.). It is convenient to relabel World 3, World 3–0.

Given Popper's intention to capture the objectivity of scientific knowledge, it would be surprising if he remained content with this bulging World 3. And there is another tougher criterion implicitly present, which acts to narrow the contents of World 3. Significantly, Popper also includes "taste" as the plastic control structures of composers, "their system of musical evaluation," and the same for painters (Popper 1979, 253–4). So a thinner World 3 emerges; here the basic criterion of entry is still to be an objective mental content, but objective now (roughly) means abstract and public, and also capable of being rationally *criticized by argument* (Popper 1979, 136–7). One can see how this moves in the right direction.

But this thinner World 3 is still uncomfortably fat; it still contains, for example, incompatible pairs of theories, or data (all these are *capable* of being criticized) and all art works subject to taste. Indeed, it contains not only individuals of these kinds but whole systems of them. It contains, for example, scientific theoretical systems, including: theories reaching across all the various orders of science; methods for that theoretical domain, including both good and bad methods; theories of method for science, allowing for the evaluation of good and bad methods (and presumably metatheories of theories of method, and so on . . .). And beyond science, there will also be all manner of nonscientific theoretical systems, including aesthetic, sporting, and ethical systems, together with all the relevant paraphernalia of these systems as above (e.g., their aims and systems of evaluation of aims). There is no reason why it should not also include the abstract mental contents of all prescientific belief and current commonsense belief, for we can rationally criticize myths and stories. Let us label this slimmer but still obese ontology World 3–1; it contains, roughly, rationally criticizable mental entities.

We are given little idea of the structuring of this realm, except that obviously it contains some systems (e.g., "theoretical systems"). It cannot be allowed to be an undifferentiated mass. It is not plausible, since logic is the key structuring tool in World 3, and among many other things, it would play havoc with Popper's long-time attempts to achieve a successful criterion of demarcation between science and nonscience. So World 3 must be regarded as being differentiated or systematized into many subsystems—scientific, artistic, social, ethical, religious ("myths"), with each area itself being systematized into theory and data, metatheory and methodology, and so on. Does this slim World 3 down any further or simply organize it? Are some items to be rejected because they resist systematization and are thereby placed beyond rational criticism? Consider here isolated recalcitrant data versus isolated theoretical conjectures or the "state-dependent sciences" of altered states of consciousness (Tart 1972). This is an important issue, but one that seems wholly obscure when posed in Popper's ontologizing terms. To leave open this possibility, let us label the now-systematized ontology World 3-2.

Now, Popper *says* that this is what World 3 contains. But he mostly works with the concept as though it were essentially restricted in membership to the problems and the resources of science. To do so as a matter of a priori principle seems insupportable; what is it about science vis-à-vis other denizens of World 3–2 that would give it this special privilege? Rather, it is this question to which PEE is itself designed to contribute. However, to accommodate Popper's narrower interest here, let us distinguish that part of World 3–2 that comprises just science-related content and label it World 3–3; it contains, roughly, rationally criticizable, systematized, scientific mental contents. (It is far from clear that this distinction can be made in any principled way within Popper's ontology—from an institutional design and process approach, what counts as science depends upon what actually contributes to truth-tracking dynamics context by context, but we let that pass here.) Formally, we now have World 3–0> World 3–1 > World 3–2 > World 3–3.

Theories are to be selected by natural selection: this means that some system will be moving into "fit" with another. We have seen that there are two major possibilities here: (1) a theoretical system moving into fit with the real world (= its environment), or (2) a theoretical system moving into fit with a system of abstractly represented data (= its environment). We have seen that (1) does not represent a real possibility for Popper. Alternative (2), following Ackermann, is a better statement of the situation. The process of natural selection of hypotheses occurs somewhere within World 3 and involves a process of increasing fit between one subsystem of the whole and another, at least. Any system A that is moving into fit with another B is doing it via the selective retention in A of some representation of selected features of the B entities, this is the basis for the view of evolution as accumulation of internally represented environmental information (e.g., Plotkin 1982). But generalities of this kind are too vague to found a theory on; to really understand what natural selection of theories comes down to, we need to look in detail at just how and where within World 3 this process of fit is occurring.

3.II.2.4. Selecting Theories: The Design of Method. For science to be progressing toward the truth, its accepted corpus of theories should be coming to represent the world. What would this mean? We will consider three cases, and see that for each possibility, the natural selection of theories will not yield knowledge without artificial intervention—an artificial selection that is central to the problem of the design of method.

Case 1: Selection by the Physical Environment. Consider first the obvious option for an evolutionary epistemology (although one that Popper cannot adopt) in which the system of theories of science—call it S_T,—moves into fit with the world itself; call this the world system or S_W (i.e., alternative (1) above). This means that survival of theories in S_T would be determined by criteria of fitness with respect to S_W as environment. But there is a problem with this: the problem of progress.

It has always been a major criticism of evolutionary epistemology that there is a significant and fatal disanalogy between science and biological evolution: Whereas scientific knowledge grows in a way that is globally progressive, species change in evolution has no overall direction, and is not globally progressive (cf. Nitecki 1988). The supporting argument restates the causal/rational dichotomy: Scientific change is the result of the intentional, conscious, and rational application of some set of general or global criteria driven by the desire to reach a specific global goal (knowledge). Species change, by contrast, is an unconscious, nonintentional, a-rational increase in fit of a local population to a local environment, blind to any global goal. So while there may be local progress in local fit to a temporarily stable local environment, there is no notion of global evolutionary progress available. Theories "survive" because they satisfy global criteria that apply "over the whole range of science," whereas genes survive as a result of "satisfaction of local criteria" (Thagard 1988, 108).[23] That is, fitness is simply "a function of the extent to which an organism is adapted to a specific environment." But "we have no general standards for progress among environments" (Thagard 1988, 107–8). There is then a fundamental gap to be bridged by any doctrine that, like PEE, claims insightful use of the term *natural selection* for both biological and scientific processes. How could this be overcome? Surely only by *careful design* of some further system of retention of theories over global environmental change, some way of bringing *global* criteria of selection into play. But it is hard to see how this could be anything other than some further process of selection, and *artificial* (i.e., *designed*) selection at that.

This latter we consider the crucial insight. But we set it aside for the moment (it will recur), because case 1 cannot arise for Popper; for him the "environment" for S_T is not the real-world physical environment, but rather a *symbolic* environment, that is, another system within World 3. This system is what Ackermann calls "data domains"—what for Popper is the system of accepted basic statements. Call this system S_D. This formulation will provide another set of problems for Popper, but before we examine these, it is necessary to look at an implicit assumption: that S_T is moving into fit with some *single* subsystem of the complex system that is World 3.

Case 2: Selection by Unrestricted Interaction in World 3. World 3 is a complex system of subsystems. Whatever subsystems of it that S_T is interacting with, it is certainly not only with S_D. Richards, for example, generally sympathetic to the evolutionary epistemology enterprise, states:

> Ideas are selected and retained by men for a variety of explicit and implicit reasons: power, passion, inertia, derangement, stupidity, and reason are all determiners of the ideas men enjoy. . . . when a mechanism modelled on natural selection depicts acquisition [or selection] of ideas primarily for reasons other than reason, then it is not a model of knowledge acquisition. (Richards 1977, 500-501)

Others agree with this judgment (see Rescher 1977, 142 and Ruse 1986, 46). That is, theories have more features in terms of which they may be selected than purely cognitive features, and are interacting with more subsystems of World 3 than just S_D—theories interact with social systems, political systems, and value systems in general (see Feyerabend 1978b, Latour and Woolgar 1979, Galison 1987). And note, these are all legitimate Popperian members of World 3. But in terms of the sort of system that we want S_T to be, and the sort of cognitive progress we wish it to achieve, natural selection as a matter of S_T moving into fit with some or other of these subsystems is not discriminatory enough. It includes too many selection processes to specify the process of science. This theory of scientific natural selection is not selective enough.

We may certainly still speak of selection criteria in all cases, cognitive and otherwise. The noncognitive criteria can also be just as global as the cognitive ones are taken to be. Someone may, for example, believe, and pursue, all and only propositions that stimulate fear in them. But selection of hypotheses by such criteria represent (to the best of our current judgment) bad methods for doing science, because the selection criteria are irrelevant to its cognitive aims. Even so, these *bad methods* must be in Popper's World 3, as we have seen. Indeed, they can belong to World 3–1 and even World 3–2, for there is nothing about "inter-subjective criticizability" and "systematicity" that says anything about cognitive correctness or even relevance per se. (The point applies still more widely, as Boon suggests: Common accounts of rationality in theory choice based on good reasons are "too broad because having good reasons for actions is not restricted to scientific activities" Boon 1987, 161.) Nor as we have just noted, does the fact that the propositions thus selected belong in World 3–3 prevent them from being selected by bad methods. World 3 is not a physical environment, but neither is it exclusively a cognitive environment—it is much more than that, including in fact all life environments.

If this has often been overlooked in discussions of evolutionary epistemology, it is only because an implicit initial selection process, restricting theories to cognitive features or selection to cognitive criteria, tends to occur before such questions are discussed. So we create the fiction that hypotheses are intrinsically and exclusively cognitive objects. But it isn't so, at least not without human intervention within the natural process. Natural processes left to themselves will not result in any overriding principle of selection according to cognitive criteria.

But restriction to cognitive criteria is not sufficient, for this will still permit the use of bad methods, that is, of cognitive methods that do not aim at the truth. Anti-induction is in general one such method (exceptions are cases like that of Christmas feeding; see note 8). So are "wilder" methods, such as believing all and only propositions endorsed by some science guru G. All these methods may be purely cognitive in form, but they are (in general) bad methods for doing science. They also need to be excluded. When this too is done we shall presume here, for the sake of argument, that it will also provide rationally acceptable assurance (*not* guarantee) that science will then progress toward its cognitive goals.

Let us label World 3–3 when further constrained to satisfy these cognitive requirements, World 3–4. This prior process of culling out World 3–4 is a process of selection, but not a natural one. To locate World 3–4 within World 3 (i.e., to specify its entry criteria), PEE needs to specify how all noncognitive and bad-cognitive selection criteria are to be eliminated in such a way that the resulting selection of theories in their data environments ensures our aiming at the goal of scientific knowledge. But this is just the problem of scientific method re-stated and is an instance of a quite general problem fundamental to human societies, which by their very existence attempt to utilize natural processes in ways that will realise *specific values* (cf. Rescher's situation, chapter 4). And note, this is exactly the problem Popper is getting at with his talk of plastic controls and of looseness in the means of error elimination in science; this is exactly where *choice* (and thus *decision*) and the *artificial* come in. Constructing World 3-4 is tantamount to providing a theory of science as a system of cognitive plastic controls.

By definition, the selection of theories within the resulting World 3–4, although governed by reason rather than cause, can be left as natural, that is, survival in World 3–4 may be permitted to be the ultimate measure of cognitive acceptance. This ambiguity in the nature and location of selection as between the physical environment S_W and Worlds 3-x, x = 0, . . .4, has hitherto gone uncommented. It defines the range of selection models available in principle to PEE and locates a basic sleight of hand in Popper's appeal to natural selection as underwriting objectivity.

However, natural selection in an environment that still includes all noncognitive selection pressures i.e., selection in World 3-x, x = 0,. . . 3, offers no principled grounds for believing that the result will be knowledge. The resulting succession of theories will in general exhibit cognitively directionless change. (There may be locally directed adaptations, but these will fluctuate largely among social, political, and economic selection pressures.) Campbell makes the point:

> . . . for us social humans, the belief assertions or public truth claims we make have important utilities for us other than optimally guiding our own . . . behaviour vis-à-vis the objects that are nominally the referents of the truth claim. [We must attempt to] . . . understand how the social system of science inhibits self-serving dishonesty, and as a result, allows "external reality" to have somewhat more influence in the social winnowing of beliefs for consensus formation. (Campbell 1987, 152)

See also chapter 2. So selection in these environments must be artificially structured to engage just cognitive concerns. We are then really faced with only two alternatives: *natural* selection in an *artificially selected environment* (World 3-4) or *artificial selection* in a *natural* environment (World 3-x, x = 0, . . .3). [Note that here, as often elsewhere, we use *natural, select* because they are Popper's terms; their appearance should not be taken as our endorsing this usage or the theoretical views which underwrite it.]

This returns us to the centrality of *the design of the global selection and retention process* (cf. the close of case 1). Our problem is to design the selection process. There has been a standard response to this problem, pursued by empiricists and early Popper alike: Formal logic provides the design. This answer is inadequate and nowhere more so than for Popper's approach. It leads, as we saw in Part 3.I, to the impossible problem of the purely logical control of human decisions so that objective knowledge can still be pursued. Scientific method is just such a control structure, intended to guide or constrain the outcomes of scientific processes so that they give us reason to believe that they achieve, or at least approach, the cognitive outcomes we want.

But this is where we started. It does not matter whether the required cognitive design is represented as the construction of an artificial process in World 3 or the designed construction of World 3–4; the two constructions are equivalent. Though it is posed in a different setting, either construction amounts to the traditional problem of method in science: how do we control theory acceptance and rejection in such a way that subjective, biased, self-interested choices are not allowed and all theory selection is made on the basis of principled cognitive criteria? And so it reraises exactly the original problem of human decision in Popper's method. Popper may have tried to avoid this conclusion by trading on the ambiguity between World 3 and World 3–4, but examination quickly shows that the basic problem remains.

So our examination of both cases 1 and 2 has inevitably led to the problem of the design of method. But even if the correct design could be achieved 'naturally'—and it can't, since it too is part of our developing fallible knowledge—even then the problem of method would *still* arise. To drive home this point we shall next examine the nature of the selecting environment required by PEE.

Case 3: Fitting in with Data Domains. Let us assume that Popper has been able to work the trick of constructing World 3–4. There is still the further problem of isolating within World 3–4 the subsystem of basic statements S_D constituting the data domains that we have agreed with Ackermann must be the environment for S_T. This is the old problem of the separation of theory and data reappearing (as these problems seem to be in the habit of doing for Popper). A further problem lies in the recognition that, if S_D is the environment for S_T, then according to orthodox evolutionary theory, S_T is also the environment for S_D; which will move into fit with which? Does data refute theory or theory criticize data? This is a notoriously unsolvable problem for Popper (cf. Hooker 1978 and Part 3.I above). Here these problems will only be noted, because we wish to focus on another. We will assume that Popper has been successful in constructing some further subsystem of World 3–4—call it World 3–5—which contains only S_T and S_D, with S_T moving into fit with S_D. Yet a major problem remains: the recurring problem of designing a method that will achieve *global* rather than *local* progress in the "evolution" of S_T.

The process of S_T coming into fit with S_D may be locally progressive, but changes take place in data domains, and the processes of natural selection will lead only to the knowledge system locally tracking these changes, just as in the natural ecology, population and environment proceed through a correlated but not intrinsically directed sequence of states. Will the result for science then be globally progressive in terms of our aim, that is, gaining objective knowledge of the real world? It will be, according to the account developed so far, if, but only if, the data domain/environmental changes themselves are progressive. That is, these changes *themselves* should track the way the world is. So how do data domains change? Changes in data domains can arise, on Popper's account, in one of two ways.

First, they may be seen as "motivated reports" (Popper 1980, 110), that is, caused by World 1 factors, producing World 2 mental belief states representative of "the world." In this case, the corresponding propositions are proposals that cannot enter into World 3 as objective data claims (i.e., enter World 3–5 and constitute part of S_D) without a process of testing (see "scientific objectivity," Popper 1980, 44 ff.). Where to stop this process is (from the point of view of "objective" World 3) an *arbitrary* conventional decision, so these propositions are arbitrary with respect to World 3. This is unacceptable; the knowledge system will track these arbitrary changes and will not be progressive.[24] What is required instead is, roughly, the *right design* to the testing process. That is, the ways method, theory, judgment, institutionalized communication etc. are put together need to be so designed as to ensure, or at least increase the likelihood that the propositions that are allowed through this filter into World 3–5 will accurately represent the world (cf. Campbell quoted above). The resulting World 3–5 environment will be an *artificially selected* one, the designed one we earlier saw to be essential to a globally progressive theory of natural cognitive selection. But the design of this environment, we also saw, is just the problem of method repeated in a new guise.

Second, Popper could argue that data domain changes result from changes in method. This is an aspect of the problem of the interaction between organism and environment noted above; as was suggested in the discussion of Ackermann, data changes do regularly occur for this reason, for example, whenever better statistical theories of hypothesis testing emerge. But what of the changes in method themselves? If these are again by conventional decision, as all theory-led methodological changes will ultimately have to be for Popper, then the resulting data changes are again arbitrary with respect to World 3 (World 3–5). The environment lacks the requisite cognitive design. The problem would then be as before: What is a rational design for the cognitive control of these decisions? And this is again the unresolved problem of method.

Alternatively, if Popper wanted to argue some form of rational assessment of such decisions (i.e., as no longer "conventional"; see Part 3.III), then the environment that provides the selection pressures for theories is provided by this assessment method; it is an environment that has been *designed* in this way, an artificial environment. It is designed to produce methods that will in turn select warrantedly acceptable data and theories. The question will then be asked, is it well designed for this purpose? Does it, indeed, succeed in providing appropriate selection pressures? And we are back with the same problem of method.

We note in passing that the shift toward a problems formulation, which has characterized Popper's development, noted in Part 3.I, does not alter this design problem. If anything, it makes it worse. Popper could be interpreted as intending that the environment-analogue within World 3 is a "problem situation": ". . . the fittest hypothesis is the one which best solves the problem it was designed to solve. . . ." (Popper 1979, 264, see also 244, note 53). This is not unreasonable; what constitutes a problem situation is going to be a complex issue, but it obviously is going to delimit a relevant data domain in relation to a particular (selected) group of hypotheses; the process of evaluation will then make methodological decisions about falsification vis-à-vis data revision and so on. But if problems are to be so resolved as to track truth within a problem context and across problem contexts,[25] then these domains equally clearly amount to the construction of the same kind of artificial cognitive environment. To which may be added all the problems of identifying and characterizing problems and their cognitively systematic interrelations (see note 17). Again, an adequate theory of science requires the construction of an artificial environment (a succession of artificial environments); the question of method is *how well* it does this.

So on any account of change in data domains (environment for S_T) available to Popper, we reach the same conclusion: For the increase of fit in S_T to be *globally* progressive, the changes in its selection environment must themselves be progressive, and this can only occur via an artificial process somewhere.

In sum, selection by World 3 environments will not provide us with knowledge via *natural* selection of theories unless those environments are artificial environments systematically designed or selected for that purpose. If they are to be systematically designed, it must be by some method. Whether or not there is progress in knowledge will depend on how well that design method has in turn been designed. Of course, this design method is itself an object (somewhere!) within World 3, and its design will again be a matter of decision from within World 3. If this decision is itself arbitrary in World 3, then the problem is simply repeated. But our point is not this threatened infinite regress, but rather that the division into World 3 and Worlds 1/2 simply perpetuates Popper's decision problem. Rational science ultimately depends on the quality of the decisions involved, and for this we need a critical theory, not an arbitrary ontology.

3.II.2.5. Conclusion. The natural selection *metaphor* has a certain persuasiveness while the idea is that the natural selection taking place is through interaction between the physical environment—the real world—and a system of hypotheses, so that reality is thought to directly dictate the content of science. It is nonetheless a superficial persuasiveness, because in itself it still doesn't offer any notion of progress. In any case, this imagined interaction is neither sustainable nor what Popper wants. But once the account moves into World 3, specifically into a World 3 sufficiently complex to deal with the analogy satisfactorily, it repeats the problems that face method in general, and Popper's methodological falsificationism in particular.[26]

Why then does Popper struggle so to retain natural selection for a variety of selection that his own account would seem to immediately and clearly suggest is very different? Because the metaphor superficially promises to repair the damage caused by the failure of his first strategy to eliminate decisions. Popper's solution attempts to make the most of the continuity between human and animal knowledge that he stresses in *Objective Knowledge*; he attempts to make all knowledge literally part of the same process, thus hoping to transfer the solidity and respectability of the mechanical (automatic, nonhuman decision) selection processes of natural selection to quell the threatened subjectivity, whimsicality, and relativism of human decision processes. The human world of subjectivity is to be connected to the natural world of objects, with a consequent transfer of the beauties of objectivity; we all know about *objects* and natural selection (i.e., natural selection in World 1), and World 3 tries to make theories into objects, connected via the emergent evolutionary human consciousness to the very bedrock of nature itself.

The problems with this fanciful construction are, as we have seen, immense. The persuasion involves a double exploitation of ambiguity. First, there is the ambiguity of symbolic and natural environments. Ironically, when this is reified in the distinction between Worlds 1/2 and World 3, it creates a dichotomy that cuts off the very continuity that is at the heart of the appeal of the natural selection metaphor. Second, there is the ambiguity between cognitively designed and general symbolic environments (between Worlds 3–x, x = 0, . . .3, and 3-4, 5). These concealed ambiguities suppress the issue of rational design for method, the key problem. They are manifested in Popper's ambivalence concerning the nature of the selection process; too much precision on this point would destroy the fragile bridge of metaphor linking the world of natural processes to the human World 3. Objectivity is to be controlled for Popper by various human-designed and operated methods, although he will continue to use the metaphor of natural selection (presumably in the hope that no one will notice the peculiarities of this "natural").

In sum, the natural selection metaphor leaves Popper with exactly the same problems with which he began. It makes no substantive difference to the problem of human decision faced by his earlier stage 1 conjectures and refutations account of error elimination within science. The stage 2 strategy also fails. But it is a fruitful failure. Natural selection creates, as one of its highest achievements, the possibility of looseness, in particular within science. Such looseness or plasticity allows the creative power of science, but it also demands choice: human decision. Talk of natural selection of theories is no help at all; natural selection creates the capacities, but it also creates the need for choice in their use. If we want the decisions concerning the selection of scientific theories to result in something we would recognize as knowledge, then these decisions have to be good ones. In terms of selection, what we must have, at some point in the process, is well-designed *artificial selection*. This means designed intervention in the natural systems, something so characteristic of human activity. This designed intervention is simply the old problem of method. But there has now emerged the interesting idea that we might give an account of method in terms of the design of a system of cognitive plastic controls, that is, the design of a certain kind of dynamic process. We turn now to explore it.

3.III. Toward a Regulatory-Systems-based Reassessment of the Significance of Popper's Philosophy

3.III.1. Problems and Lessons from PEE

Even were there to be no basis for an alternative account of decisions in Popper's philosophy, still Popper's philosophy would have considerable merit. It would have merit first because of the effective critique of empiricism that Popper provides (see Part 3.I above). Correlatively, another merit is Popper's development of the systematic importance of criticism in the rational conduct of science and of his supporting discussion of the notions of severe tests and non-ad hoc adjustments to theory. But we should also credit Popper with doing more than he can properly acknowledge. He made us aware of the conflict between aiming at security (probability) of truth and aiming at high content (relief from ignorance), for example, which set us at liberty to accept that there may be many (proxy) goals or aims characterizing the pursuit of truth and that for finite creatures born in ignorance, these may be partially conflicting as well as mutually reinforcing. We shall add other lessons of this kind in subsequent sections.

Beyond these merits, there is the Popperian emphasis on the adventure of science, on the constitution of an inquiring life and of a nondogmatic society in which inquiry is free to proceed. Indeed, there is for Popper an ethics of rational life, a distinctive set of commitments that are presupposed in the determination to pursue Popperian inquiry. All these are, one might say, the classical merits of Popper's philosophy, and they are large. (And we are indebted to him for them, and we adopt them.) But they are also well known, and here we shall pass over them quickly, pausing only to note that in the last of these, the idea of an ethics of rational life, we shall find the roots of an embryonic third strategy.

Popper had a problem he couldn't solve, namely how to provide a methodology for science that was uninfected by human decisions. But in trying to solve it, he was led to emphasize the continuity of knowledge from its roots in elementary biological organization through to its most abstract and sophisticated products at the hands of humankind. Again, there are important supporting concepts developed, such as the universality of problems, the feedback loop connecting problem solutions to the generation of further problems, and the notion of extrasomatic constructions, which are part of the technology of problem solutions, from burrows and nests through to scientific theories that "die in our stead". These are merits indeed, but they are not distinctive to PEE, and almost every evolutionary epistemology has some version of them.

On the other hand, as we have seen, they do lead to a distinctive problem for Popper. If scientific decisions are to be directed toward the truth, then this directedness must be exhibited in the selection process (theory of method) in some way, but for Popper this proves impossible. Instead of seeing the whole process of science in dynamical terms, Popper's metaphilosophical empiricism forced him to regard it schizophrenically: Scientific theories may evolve but the rigid rational decision structure of science remains eternally fixed. Popper's difficulties with this problem suggest that one must move beyond the focus on theories as the entities that evolve. Rescher points the way here by showing how if one shifts from a focus on theory selection to a focus on method selection, then one can give a much more satisfactory account of the development of science (see chapter 4). But why stop there (as Rescher does, repeating his own version of Popper's problem)? Why not see all the elements of science as dynamic and attempt to formulate the whole cognitive process as a single dynamic regulatory systems process, perhaps an evolving system of cognitive plastic controls? Here again we will find a proposal for resolving this problem with its roots in PEE but with consequences that lie well beyond it. So let us turn then to Popper's embryonic third strategy: the explicit acknowledgment of decisions and their rational social regulation through flexible or plastic decision processes.

3.III.2. The Control of Decisions

Popper is faced squarely with the ineliminability and ubiquity of human decisions in the prosecution of science. He partially recognized this from even his earliest writing (1934, see Popper 1980), namely in all these decisions written off as conventional. Subsequently, faced with an increasing proliferation of decisions that could not be forced (i.e., were not rigid), his treatment of the nature and status of conventional decisions slowly shifted, in a way that suggests a threefold division of "convention" into:

(1) Convention as a *noncognitive decision* fixing a choice for which there is no relevant matter of fact, for example, the decision to adopt a particular formal language for science (Carnap) or to adopt a critical approach to knowledge (Popper).

(2) Convention as solely *social practice*, the way in fact we do things around here, for example, as the manifestation of an existing practice of a scientific community, from laboratory procedure for polishing lenses to the decision to fund much scientific research through military budgets. On this conception, social practices result from unappraised group decisions, whether explicit or implicit, and social conven

tions are the choices that would appear in adequate theoretical models of these group processes as decision processes.

(3) Rconvention as *rationally assessable decisions* concerning the designs of social institutions and their institutionalized processes, especially those that provide control over decisions, for example, the creation of scientific research institutes, the adoption of Policy Planning and Budgeting Systems, or the acceptance of Popperian basic statements.

Note that Rconvention, sense (3), is not in fact properly convention at all; we introduce this alternative only because it figures in Popper (we shall argue). It has a superficial similarity to convention in senses (1) and (2) but is crucially different; we prefix it to emphasize that.

Popper continues to speak throughout his work of these proliferating "conventional decisions" in sense (1), but in practice, the way in which he actually uses the term tends more and more to that of sense (2). Sense (2) appears wherever Popper appeals to the scientific community and its "implicit decisions." It can be seen, for example, even very early on in Popper 1980, in his discussion of the empirical basis. "Basic statements are accepted as the result of a decision or agreement, and to that extent they are conventions" (ibid., 106). Or "from the logical point of view," basic statements are "accepted by an act, by a free decision" (ibid., 109). If there is disagreement, we test them until agreement is reached. This process of testing "has no natural end" and where we stop (i.e., which basic statements are accepted) is decided by consensus; we stop "at statements about whose acceptance or rejection the various investigators are *likely to reach agreement*" (ibid., 104, our italics). Such agreement is clearly not reached on the basis of incorrigible perception (not available to us); rather, it is ultimately reached on the basis of socially negotiated consensus—the implicit "decisions" represented by the actual, brute practice of the group's commonsense perceptual practices. (Note here the continuity Popper asserts between scientific knowledge and commonsense and prescientific knowledge, Popper 1979.) In the final analysis, the acceptance of basic statements goes by social convention (the way we see things around here), conventions whose current design will have emerged from their historical developmental path.[27] It may be initially plausible to see philosophers of science making conventional decisions in sense (1), but far less so when it is practicing scientists actually making substantive decisions about how the world appears to be.

Or consider his remarks concerning the necessity for an accepted "background knowledge" in the evaluation of a theory: "People invclved in a fruitful critical discussion of a problem often rely, *if only unconsciously*, upon . . . a considerable amount of common background knowledge". (Popper 1972, 238, our italics.) Although qualified with *often*, it is clear from his following remarks that he means *always*, that such assumptions are a necessary condition for a "fruitful critical discussion": "While discussing a problem we always accept (if only temporarily) all kinds of things as *unproblematic* . . . ." (ibid., 238). As with the acceptance of basic statements, it is only social naivete that allows Popper to see this unconscious reliance as anything other than a set of social practices of the relevant community determining which components of this background knowledge are in fact accepted, as opposed to those that are not. Clearly, such acceptances do not simply express individual preferences; they are again the result of a socially negotiated consensus, proceeding according to the familiar patterns of social consensus formation (see Etzioni 1961; Galison 1987), that is, conventions (2).

But socially negotiated decision processes can already move us into Rconvention in sense (3), depending on their character. While Popper never explicitly speaks of Rconvention, the sense is often implicit in his treatment of decisions; once we recognize an existing social practice as an implicit decision, we are faced with the (entirely Popperian) necessity of *criticizing* it, of asking whether this particular practice is, indeed, the *best* way to proceed: Are there (might there be) other, better ways in which the aim of the practice could be realized? To this end, the practice itself will be theorized, and improved practices suggested. A convention (2) about accepted background knowledge, for example, cannot rationally be left that way; at some point, the pattern of accepted background knowledge to any problem must itself then become a problem: ". . . any particular part of it [the background knowledge] *may* be challenged at any time, especially if we suspect that its uncritical acceptance may be responsible for some of our difficulties." (Popper 1972, 238)[28] As soon as this occurs, as soon as we begin to look critically at the pattern of acceptance of background knowledge, we are moving into the realm of Rconvention, sense (3): This is the theorising of a social practice (a convention (2)), in order to critically assess and improve it. It becomes the revision of existing practice by theory-guided intervention, a characteristically human (and, in spirit at least, Popperian) path of improvement, both of technology and concepts. This is the rational regulation of decisions, that is, Rconventions in sense (3).

The shift from (1) to (3) represents a radical shift in conception and status for conventional decisions and one with massive ramifications for not only epistemology but the theory of public policy and culture generally. Under the strict empiricist conception, conventions not only were not to be discussed as cognitively significant, they could not be so discussed because their content was very "thin"; the ideal empiricist convention would be that between the choice of two formal languages for science where it was a metatheorem that their expressive powers and logical truths were exactly equivalent. This is notion 1 of convention and, since they have no cognitive substance, they are not open to rational improvement. By contrast, under concept 3 of Rconvention, we can *learn* about the appropriateness or value of decisions and, as Popper would say, thereby improve them. This is the notion of decision we believe is both correct and of crucial importance to a rational account of science.

In this respect, concept (2) is a halfway house, for under it, all such decisions are merely part of the happenstance of our culture. Though they have content, it is not rationally appraised content. Nonetheless, the step from (1) to (2) is an important one, because it recognizes that conventions have a substantive content, that social practices—up to and including the mathematicians' formal practices—are not a contentless framework for something else but themselves represent an important part of the content of life. Indeed, they represent an important expression of knowledge. Though not rationally appraised, they are rationally apprais*able*. Our system of accountancy for publicly listed companies is only one way of doing things financially; even so, it is not only the numbers that constitute the content of public accountancy reports that matters; the processes of accountancy themselves and the form of accountant's reports also represent a substantive commitment to our institutionalized financial self-regulation in ways reflecting our knowledge of the financial system. Dyke 1988 has provided a seminal preliminary theorizing of the nature of substantive social convention from this point of view, but with conventions already seen as entrenched, historically developing components of the design of dynamic social systems. This conception seems to us to be along the right lines but already directs us beyond version (2).

Similar things will have to be said for the social institutionalization of science, the research roles, communication patterns and ultimately research methods themselves; they too reflect substantive (if implicit) decisions to pursue knowledge in certain kinds of ways rather than others, and these decisions in turn reflect our knowledge of knowledge pursuit, our substantive metascientific or philosophical theory of the nature of objective knowledge. Connivance between empiricist-influenced philosophy and sociology of science has resulted in decisions about the institution and conduct of science being construed as merely empirical sociology of knowledge, in contrast to any substantive normative account. The clue to the inadequacy of this view already lies in the way these decisions presuppose or express knowledge about the procedural and institutional possibilities. No doubt such knowledge began historically as tacit and relatively disorganized, but once we understand the nature of social conventions, we are able to bring them increasingly under cognitive evaluation, to deliberately explore their design decisions. This welds sociological analysis to normative analysis exactly in the manner required by the regulatory systems naturalism defended here. Public policy-making and evaluating becomes a rationally assessable cognitive process and knowledge policy-making (i.e., scientific method), is one part of it. Reflexively, policy science is itself part of science and includes the science of science as part.[29]

The shift in the conception of conventional decision from (1) to (3) shifts decisions into the realm of substantive, theorizable practice, which like any other theorizable domain, can be rationally (including critically) assessed. For Popper, scientific knowledge is an extension of commonsense knowledge and, ultimately, of animal knowledge; well, animals not only do particular things on particular occasions, they also have ways or policies of behaving and these are more or less successful. Humans are rational animals, and their ways or policies of behaving are rationally assessable through being theorized, including policies for the pursuit of truth. Thus method dynamics or the evolution of method becomes central to the cognitive dynamics of science, and reason is expanded in scope from deciding on theories to also deciding on methods.

Now let us assess this shift from a theoretical perspective. Social institutions and their policies and practices are the fundamental ways in which we regulate or control our lives in relation to our diverse individual needs and our coordinated public environment. From this perspective, the function of a society's cultural institutions is to regulate decisions by its members. Sociologists have properly made much of this regulatory function of institutions and their accompanying values (cultures), for example, in analyses of the control of deviance (though societies may also selectively produce deviants). When these regulatory activities become publicly conscious and intentionally critically assessed, they become part of rationally assessable social processes. This is the fuller context within which Dyke's analysis of natural convention operates (Dyke 1988), the conception of society as a norm-developing and using dynamic system. In turn, we shall use the expression, *the regulation of decisions* to refer to this conception of organized decision making.

Thus the purpose of scientific institutions and scientific method is to jointly regulate or control the decisions of scientists. (Consider Lakatos's research programs from this perspective.) But now we have made a leap in our theorizing of science. For no longer are logically unforceable decisions written off as external and embarrassingly noncognitive. Rather, decisions are an integral part of scientific roles within the scientific institutions, and it is the designs of those institutions and the methodologies they practice as their policy that regulate scientific decisions and therefore that constitute whatever rationality they possess. This reconceptualizing of science accomplishes what Popper could not do, namely bringing decisions within the scope of a theory of reason, by doing what Popper would not do, namely enlarging our conception of reason so as to include the critical design of scientific institutions and their cognitive policies.

If the nonformal nature of reason is thus reinforced, there is another consequence concerning the scope of reason that we emphasize here. Reason qua regulatory design (control of decisions) now extends from particular scientific decisions to the adoption of scientific method and to the institutional design of scientific roles and procedures generally. Let us call this the "socializing of reason." It is important because it removes the false and stultifying dichotomy between social analysis of science as descriptive and non-normative (or even anti-normative) and the philosophical theory of scientific method as normative but non-descriptive. Rather, the norms of science are realized as institutionalized social procedures and decisions (see chapter 2). Thus just as scientific theories themselves are simultaneously normative and descriptive of their domains, so too are theories of scientific method and scientific procedure generally (see chapter 1). The rational design of sciencing as a cognitive process is crucial to our fallible learning about reason itself.

We suggest that Popper's major contribution to an adequate theorizing of science lies here, in the way in which his flawed efforts to provide a theory of the rational regulation of decisions within the scope of empiricist metaphilosophy instead points the way toward a more adequate naturalized conception of them as nonformal regulated decisions in a dynamic evolving system. In the next section, we explore the hints that Popper himself provides for a positive conception of science of this kind.

3.III.3. Plastic Controls and Social Rationality

In the same work in which Popper introduces his biological extension of his cognitive problem-solving process, he also introduces the conception of plastic controls as a model for human decision, especially for rational deliberation. We have briefly described these at section 3.II.1.3 above. This kind of control is to be neither mechanically deterministic nor statistically random but conforms instead to the "idea of combining freedom and control" (Popper 1979, 232; hereafter, page references in the text will all refer to this book and quoted emphases will continue to be Popper's, unless explicitly stated otherwise.) And since "deliberation always works by *trial and error* or, more precisely, by *the method of trial and error elimination*. . . . This suggests that we might use in our new theory some mechanism of trial and error-elimination" (ibid., 23-4). So the basic mechanism of PEE is to be put to work here also. Though the analysis will lead beyond PEE, the importance of the notion of plastic control is that it provides a much richer model of decisions than that of rigid decisions, and one that supports the sense (3) conception of Rconventional decision and hence opens the way to an adequate theorizing of science in terms of the rational regulation of decisions.

To approach Popper's conception of the nature and place of plastic controls, we begin with language, the basis of World 3 for Popper. In the context of discussing the evolution of language, Popper distinguishes four semiotic functions; two of these are possessed by animals also—the symptomatic/expressive and releasing/signaling functions—and two are unique to humans—the descriptive function, introducing truth values and, most recently evolved, the argumentative function, whose paradigm is critical discussion.

> Like the other functions, the art of critical argument has developed by the method of trial and error-elimination, and it has had the most decisive influence on the human ability to think rationally. (Formal logic itself may be described as an "organon of critical argument.") Like the descriptive use of language, the argumentative use has led to the evolution of ideal standards of control, or of "*regulative ideas*" (using a Kantian term): the main regulative idea of descriptive use of language is *truth* (as distinct from *falsity*); and that of the argumentative use of language, in critical discussion, is *validity* (as distinct from *invalidity*). (237)

Now:

> The higher levels of language have evolved under the pressure of a need for the *better control* of two things: of our lower levels of language, and our adaptation to the environment, by the method of growing not only new tools, but also, for example, new scientific theories, and new standards of selection. (240)
>
> So the evolution of the higher functions of language . . . may be characterised as the evolution of new means for problem-solving, by new kinds of trials, and by new methods of error-elimination; that is to say, new methods for *controlling* the trials. (240)

All this is intended to give PEE process a more generalized and intimate role than merely bringing about the physical organ that supports abstract mental processes.

> My theory . . . consists of a certain *view of evolution* as a growing hierarchical system of plastic controls, and of a certain *view of organisms* as incorporating—or in the case of man, evolving exosomatically—this growing hierarchical system of plastic controls. (242)

Here is a dynamic process in which interaction opens up every component to critical scrutiny, even the aims of science:

> Just as, in a system with plastic controls, the controlling and controlled subsystems interact, so our tentative solutions interact with our *problems* and also with our aims. This means that our aims can change and that *the choice of an aim may become a problem*; different aims may compete, and new aims may be invented and controlled by the method of trial and error-elimination. (253)

Indeed, consciousness itself is seen to be but one more regulatory structure in intimate mutual interaction with all others (251). We support this integrationist, naturalist sentiment (without accepting any panselectionism), because it marks a shift from eliminating decisions from the theory of rational cognition to potentially including them within it, and in particular a shift from excluding aims and method from learning (as respectively conventional (1) and eternal logic) to including them in the dynamic process. The scope of the regulation of decisions now extends from decisions to adopt theories, and presumably to adopt empirical basic sentences as well, all the way up to our decisions to adopt standards or aims themselves. And consistently, both individuals and science are conceived of as systems of such plastic controls. Here then is a unified vision in which reason is implemented as systematic regulation of decisions and in which the regulatory processes superfoliate from sub-individual cognitive state to supraindividual scientific institution, indeed from amoeba to Einstein, as a "growing hierarchical system of plastic controls."[30]

How shall we characterize the rationally regulated decisions? Strikingly, in institutional settings, Popper offers a very different model of decision making than that of logically rigid decisions.

> No action can ever be explained by motive . . . alone; . . . they must be supplemented by a reference to the general situation, and especially to the environment. In the case of human actions, this environment is very largely of a social nature; thus our actions cannot be explained without reference to our social environment, to social institutions and their manner of functioning. (Popper 1966, 2: 90)

Here and elsewhere (1967, 1976), Popper offers a strategic conception of decision. Popper tries to objectify these decisions so that logic might be seen to suffice to deduce the conclusion (Popper 1976, 102-3), leaving a formal characterization of method in scientific institutions untouched. But clearly the appropriate theoretical notion is the decision theoretic one, where that strategy is chosen from among a set of institutionally structured available options, which does sufficiently well at improving the value of some function of a set of relevant utilities under the situational (including institutional) constraints obtaining. Since an agent's accepting each component characterization of this framework as appropriate to their situation is a matter of flexible decision making, this seems clearly a conception of plastic control as Popper intends the term.

But as congenial, indeed essential, as we find it, pursuing this line will certainly lead far away from Popper's official position. A decision theoretic approach to epistemology, in particular to scientific method, departs radically from that of a logically determined approach (chapter 2). Strategic institutional design becomes central to scientific rationality, and we shall then speak of cognitive institutions. Method dynamics are naturally included along with theory and data dynamics, and it places regulated institutional roles at the center of understanding scientific behavior, concluding that "The foundations of science lie in its management philosophy."[31]

Popper is ambivalent about institutions. He acknowledges the ubiquity and importance of institutions and gives many examples of them (1957, 65; 1972, 125). Several social institutions are mentioned as contributing to scientific progress: laboratories for research, scientific periodicals, scientific congresses and conferences, universities and other schools, books, the printing press, writing, and speaking (1957, 154). And he acknowledges the importance of institutional support for science:

> Reason, like science, grows by way of mutual criticism; the only possible way of "planning" its growth is to develop the institutions that safeguard the freedom of this criticism, that is to say, freedom of thought. (Popper 1966, 2: 227)

Despite this, strikingly, the concept of social institution plays no role in Popper's "official" theory of science, which retains its "internalist" character reflecting the empiricist internal/external (normative/descriptive) dichotomy.

Yet there are moments in Popper's discussion of sociopolitical issues, which for him are intimately linked to cognitive ones, where a very different approach appears in embryonic form. At one point, he considers the traditional question of political theory, "Who should rule?," which begs for an authoritarian answer, but he counters:

> This political question is wrongly put and the answers it elicits are paradoxical . . . . It should be replaced by a completely different question such as "*How can we organise our political institutions so that bad or incompetent rulers . . . cannot do too much damage*?" . . . The question about the sources of our knowledge can be replaced in a similar way. It has always been asked in the spirit of: "What are the best sources of our knowledge—the most reliable ones . . ." I propose to assume, instead, that no such ideal sources exist . . . . And I propose to replace, therefore, the question of the sources of our knowledge by the entirely different question: "*How can we hope to detect and eliminate error*?" (Popper 1972, 25).

Here we have the beginnings of a powerful general policy: Treat science as (at least) a specialized social institution, and raise about it all the questions that can be raised about purposive social institutions in general (plus whatever further issues belong distinctively to it). Yet what Popper offers as the "comparable" question is not the comparable one. The analogous question is this: How can we organize our cognitive institutions so that bad or incompetent scientists cannot do too much damage? This kind of question shifts the rationality of science away from sole focus on the qualifications of individuals to include also the design of cognitive institutions. Instead, Popper jumps from the general question of the design of cognitive institutions to an internalist question about abstract method, again revealing his empiricist bias. Nonetheless, the original strategy is a good one, so let us pursue it through his writings. To highlight the implications of Popper's own remarks for the different conception of scientific rationality hidden there, we will insert into the quotes to follow a comparison with cognitive institutions that Popper doesn't make.

> The moral [cognitive] values of a society are clearly bound up with its [cognitive] institutions and traditions, and they cannot survive the destruction of the [cognitive] institutions and traditions of a society. (Popper 1966, 2: 94)

> . . . all long term policy—and especially democratic [cognitive] long-range policy— must be conceived in terms of impersonal institutions. (Popper 1966, 2: 131)

So despite scarcely referring to it in his official (empiricist) account of science, Popper is sensitive to the central roles of institutions in the process of the growth of science. How then are cognitive institutions to be characterized?

> [Cognitive] Institutions are always made by establishing the observance of certain norms, designed with a certain aim in mind. This holds especially for institutions which are consciously created; but even those—the vast majority—which [like science at large] arise as the undesigned results of human actions . . . are the indirect results of purposive actions of some kind or other; and their functioning depends, largely, on the observance of norms. (1966, 1: 67)

Science then is characterized by its institutional norms; it is they that determine the ranges of admissible activities. The question of the origin of science as an institution is replaced by the question of the emergence of the relevant norms.[32] Among these, the norm of valid argument will of course appear, but there will be many others, concerning high content, risky research strategies, technological applicability and security, and so on—even the norm of fighting to improve institutional rationality. Moreover, these norms need to be formulated in context; when the external dependence and imperfections of cognitive institutions are taken into account (see below), the appropriateness of any fixed class of norms seems dubious. Norms are learnable, improvable constructions; their development in interaction with the development of theory and method forms an important component of the overall cognitive dynamics.

Furthermore, the designs of institutions are accessible to us: "[Cognitive] *institutions* can be planned; and are being planned" (Popper 1966, 2: 143) This point seems trivial, but when combined with fallibilism, it opens up the learnability of institutional rationality through improved planning. Indeed, "Human institutions such as the state [including cognitive institutions] are not rational, but we can decide to fight to make them more rational" (Popper 1966, 2: 278). It is important to note that rationality is taken here to be a comparative concept of the kind a strategic conception can be. It is one of the key features that distinguishes Popper's official philosophy of science from what emerges here. "The fight to improve the rationality of scientific institutions is essential because, beside overcoming ignorance [cognitive] institutions are inevitably the result of a compromise with circumstances, interests, etc." (Popper 1966, 1: 159), and furthermore:

> A social [cognitive] institution may, in certain circumstances, function in a way which strikingly contrasts with its prima facie or "proper" function. . . . The ambivalence of social [cognitive] institutions is connected with their character. . . with the fact that they perform certain *prima facie* functions and with the fact that [cognitive] institutions can be controlled only by persons (who are fallible) or by other institutions (which are therefore fallible also). . . . The working of institutions, as of fortresses, depends ultimately upon the persons who man them; and the best that can be done by way of institutional control is to give a superior chance to those persons (if there are any) who intend to use the [cognitive] institutions for their "proper" social [cognitive] purpose. (Popper 1972, 133–4)

Absolute rationality—whatever it means—cannot be realized in human cognitive institutions. First, such distorting factors as human unpredictability and unreliability, ego-fulfillment wishes and group vested interests, financial constraints, and cultural background, among others, have their impact. An institution is therefore a compromise. Even if it were the case that humans were endowed with unerring cognitive capacities, the restrictions on the rationality of human institutions hold. Second, the finitude and error-proneness of humans reinforces this conclusion (Cherniak 1986, Hooker 1994c). That institutional designs are compromises is a profound thesis with far-reaching implications for philosophy of science, and one in sharp contrast to Popper's official philosophy of science.

But then Popper himself moves substantially toward the kind of theory we seek:

> . . . what we call "scientific objectivity" is not a product of the individual scientist's impartiality, but a product of the social or public character of scientific method; and the individual scientist's impartiality is, so far as it exists, not the source but rather the result of this socially or institutionally organised objectivity of science. (Popper 1966, 2: 220; see also ibid. 218; 1957, 155-6)

And with the appropriate design, cognitive institutions can achieve objective science despite the finitude and imperfections of their members (and presumably also of their design):

> . . . there will be always some who come to judgements which are partial, or even cranky. This cannot be helped, and it does not seriously disturb the working of the various *social [and cognitive] institutions* which have been designed to further scientific objectivity and criticism; for instance, the laboratories, the scientific periodicals, the congresses. (Popper 1966, 2: 218)

Here in effect is a proposal to phrase cognitive questions in terms of institutions and the admissible range of activities regulated through their roles, rather than in terms of some formal method. An account along these lines is given in chapter 2. Furthermore, this characterization of institutionalized science provides a route into the rational assessment and redesign of cognitive institutions themselves, in principle to encompass the openness of all decisions to rational assessment through systematic regulated processes. (This is the cognitive functional equivalent of self-reproductive causal closure.) Reason is (imperfectly, improvably) realized in the design of the superfoliating regulatory structure of flexible or plastic controls on decisions. But ultimately, this is not a route that Popper officially permitted himself to travel. The regulating power of unexamined assumptions is very great.

3.III.4. Conclusion

With this, our journey within Popper away from official Popper toward a regulatory systems conception of rationality is completed. We have extracted from Popper the rudiments of a very different conception of science, both individually and socially, than the traditional one he presented and a very different conception of the nature and roles of human decisions within it, yet one that aims to respect the critical search for truth that he enjoined. This entire book is devoted to establishing and elaborating this conception.

Chapter 4

Regulatory Systems and Pragmatism: A Critical Study of Rescher's Evolutionary Epistemology

Introduction and Summary

How is one to understand philosophically the old proverb "Seeing is believing"? Traditional epistemology would construe it along these lines: Each specific occasion of a seeing that p provides sufficient rational warrant, rooted in the visual experiences of that occasion, for affirming the truth of p. The occasions are, as it were, discrete graspings of truths. In his book *Methodological Pragmatism* (1977), Rescher wants to construe it in this way: Perception is a successful practice, incorporating a successful cognitive method and, ceteris paribus, this justifies rational acceptance of each of the individual deliverances of the exercise of this method. The heart of Rescher's proposal is to shift the justificatory emphasis from the truths at issue to the methods through which they are reached.

Theses are like species. The successful adaptation of a species at a given time does not guarantee its continued survival thereafter, nor does its continuing survival guarantee that it is perfectly adapted. In the former case, the environment may change and the species no longer be adapted. In the latter case, there is always the possibility that continuing survival may simply be relatively fortuitous or accidental, but much more important is the fact that adaptation need never be perfect, it only ever needs to be sufficient to allow the species to "get by." Similarly with theses. The truth of a thesis is no guarantee that it will have a scientifically successful career, nor does a scientifically successful career for a thesis guarantee that it is true. A thesis might be true but nonetheless be disbelieved by most or all, whether because of ignorance, social prejudice, or dominant false beliefs inconsistent with it. Conversely, a thesis may be believed by most or all and nonetheless be false, whether this be because of ignorance, social prejudice, or most interestingly, because it nonetheless works within the limits of our current testing capabilities, interests, and imaginative horizons.

The idea then that the theses of science are in some way justified or warranted sheerly by their pragmatic success provides a thoroughly flawed foundation for a theory of scientific knowledge. Moreover, it is not just the specific form of this proposed criterion of thesis truth that presents difficulty, for any attempt to develop an epistemology by providing an epistemically accessible criterion of thesis truth runs up against what Rescher calls the "Wheel Argument": Let C represent any epistemically accessible, hence practically usable, criterion of scientific truth. What validates the employment of C? Only an argument that can independently show that C generates only truths. But any such argument must have true premises and hence itself be in need of validation by the criterion C. (If not, C is irrelevant, and it is the criterion implicit in the justifying argument to which a new version of the Wheel Argument applies once again.)[1]

Notoriously, there are only three general ways out of the Wheel Argument dilemma: dogmatism, rationalism, and pragmatic fallibilism. In the former case, some criterion C is simply dogmatically asserted (e.g., many empiricist accounts of the sensory foundations for knowledge). Rationalism tries to provide self-justifying premises in the argument to justify C (e.g., Leibnizian or Kantian). Once one steps beyond this charmed (if not charming) circle of truth, a regress threatens, a regress that can be stopped only by some further considerations. Thesis Pragmatism (Rescher's phrase), the claim that success is a criterion of thesis truth, is one such pragmatic criterion. Rescher's methodological pragmatism offers us another.[2] Rescher argues that whilst a thesis may not be justified by its success in gaining people's belief in it, a successfully used method does carry rational support for the conclusions drawn with its use (Rescher, 1977, 230–3). The evolutionary epistemology I avow in this book offers yet another, one that is capable of incorporating a (modified) version of Rescher's account within it.

Rescher's methodological pragmatism is, in my view, an important step forward both in the provision of a philosophy of science per se and in the development of an evolutionary epistemology in particular. In respect of philosophy of science, any realistic examination of that cognitive enterprise will reveal its structural and functional complexity (cf. chapter 2), a complexity in which methods play ubiquitous and absolutely central roles in the generation of information, development and testing of theories, and assessment of experimental outcomes and in the conceptual development of science itself. The preoccupation of conventional philosophies of science of all stripes with scientific theses, to the exclusion of conceptual order and methodological organization has been one of its chief weaknesses. In healthy contrast, Rescher recognizes the importance of the systematic structuring of procedures, which moves toward the notion of science as a complex regulatory system (though not, as we shall see, far enough) and provides it with a fundamental relation to the underlying themes of this book.

Similarly, evolutionary epistemology has been confined by the döppelganger approach in which one attempts to identify *bits* of science that are to correspond to bits of biology (e.g., theories as döppelgangers for genes) and by an excessively narrow neo-Darwinian orthodoxy confined just to some process of blind variation, selection, and retention. This cuts biology off from the resources of regulatory systems models of phenotype and ecology and cuts theory of evolutionary epistemology off from these resources along with it (see chapter 1). Under these conditions, the obvious version of evolutionary epistemology is then thesis pragmatism as described—and rejected—above. Rescher sees that there is no future at this level of theorizing, so he proposes to move up one level to escape the problem. Justification shifts from a focus on guaranteeing or at least achieving truth to one on systemization and successful practical action. Not "Should I believe p true?" but "Should I use a systematically supported method M to investigate the world?" becomes the issue. Theses are then derivatively rationally accepted as true if they arise from applying an accepted method. Rescher hopes to show that the problem of justifying practical action is solvable whereas the problem of justifying acceptance-as-true was not. He argues that this expansion suffices to defeat the Wheel Argument.

Certainly one of the standard philosophical objections to the traditional scientific methodologies (empiricism, falsification, etc.) is that they have failed to provide any guarantee that their use will provide truths. This problem, a specific expansion of the Wheel Argument, afflicts all of the traditional approaches across the empiricist-rationalist spectrum. Yet the conceptions of truth from which they severally derive all agree in requiring this of them. (Ironically, it is consideration of common sense and scientific experience that undermines their persuasiveness.) No approach can circumvent this problem. And ultimately, we will see that Rescher does not succeed in doing so either. Rather, the issue needs to be met head on by first distinguishing between offering a guarantee of achieving truth and offering a rationally warranted acceptance of science as making progress toward the truth, and second by affirming systematic fallibilism and so rejecting the demand for a guarantee. The Wheel Argument is replaced, and circularity gives way to a requirement of reflexive consistency that naturalism can meet.[3] And then we will see that Rescher's analysis can make a contribution to the regulatory analysis of cognitive progress, one nicely complementary to that of chapter 2, which confines rational acceptance criteria to cognitively accessible features while not trying to insist that truth criteria be so confined.

Trying to define truth criteria in terms of cognitively accessible features is a mistake, made, for example, by those pragmatists who seek to define truth in terms of successful practice. While doing so is able to satisfy the demands of conventional philosophy for a proof of method as a truth achiever, it does so by robbing truth of its central roles (see chapter 1). In particular, it robs human knowing of its open-system properties, yet emphasis on these is otherwise one of the characteristic strengths of pragmatism. It also robs pragmatism of its natural connection to a naturalist understanding of intelligence. If an approach of Rescher's kind can avoid the problem while still delivering a rationally persuasive account of method, it can retain these strengths of pragmatism and so hold a particular interest in the present context.

The essence of the evolutionary naturalist realism I avow lies in taking as a cardinal virtue a regress of steps of Rescher's kind. However, I shall argue that Rescher's moves, while in the right direction, are incomplete, partly confused, and flawed by the presence of an antinaturalist rationalism. I aim here to expose the key ideas, merits, and deficits of Rescher's position and show how it points the way beyond itself—points, in fact, toward integration into the evolutionary epistemology and naturalist realism presented herein. Because of Rescher's valuable investigation of the systemic structure of rational yet fallibilist methodology, integration (with modification) enriches both sides.

4.I. Thesis Pragmatism and Its Critique

Rescher construes thesis pragmatism as follows:

> (RQ1). . .what determines the truth-acceptability of propositions is the factor of their *utility*: a proposition is to count as true if the practical consequences of its acceptance outweigh those of its non-acceptance or perhaps rather—and better—those of its rejection (i.e., the acceptance of its contradictory denial). The rationale of this pragmatic theory seems to reside in the (surely ill-advised) view that one cannot profit by error, and fare better by rejecting a true proposition than by accepting it (or by accepting a false proposition than by rejecting it.) Perhaps by contraposing the plausible thesis that being right is the most advantageous circumstance, such an approach construes maximal utility as a safe indicator of truth. Accordingly, we are to determine whether or not a proposition qualifies as true by assessing the *utility* of endorsing this thesis as compared with its possible alternatives. The truth, to put it crudely, is that whose acceptance is maximally utile and "works out for the best." As William James puts it, "ideas become true in so far as they help us to get into satisfactory relation with other parts of our experience." (37)

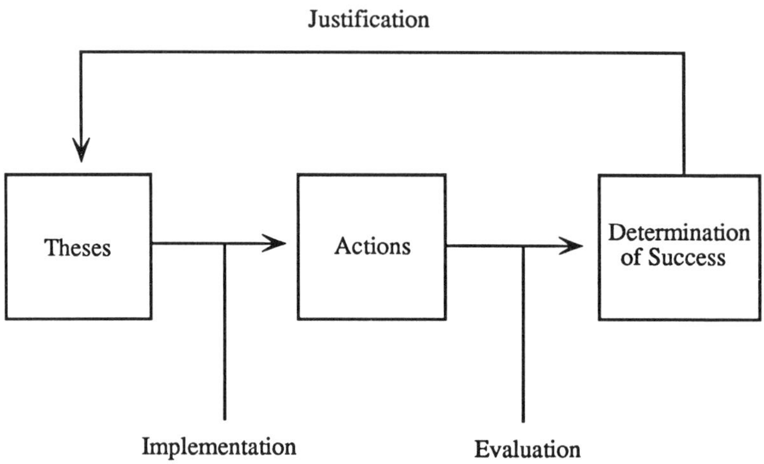

Diagram 4.1 The Pragmatic Justification of a Thesis
Arrows indicate the sequence of schematic cognitive stages which together make up the cycle of theory testing with consequent rejection or entrenchment.

(This quote is from Rescher 1977 as will be all unsourced quotes hereafter, and the emphases here are Rescher's, as will be all those in quotes to follow unless explicitly stated otherwise.) The drift is clear enough, even if the position is put somewhat vaguely and contains at least three distinct versions of a pragmatist acceptance criterion.[4] Anticipating Rescher's later use of block diagrams, his position is schematized in diagram 4.1.

Though Rescher couches the matter in the technical jargon of utilities and decision theory, his essential criticism of thesis pragmatism is that given in the introduction: On the basis of pragmatic success alone, a thesis has neither a logically nor even a practically warranted connection with truth.[5] As Rescher notes, this is true equally of success measured in the manners of James or Dewey or even of success measured by survival or staying power after the manner of Peirce (cf. chapter 1, section 1.I).

4.II. Methodological Pragmatism

In response to these well-known difficulties, Rescher introduces an extra step in the justificatory cycle given in diagram 4.2. The focus now shifts to the rational acceptance and rejection of methods,

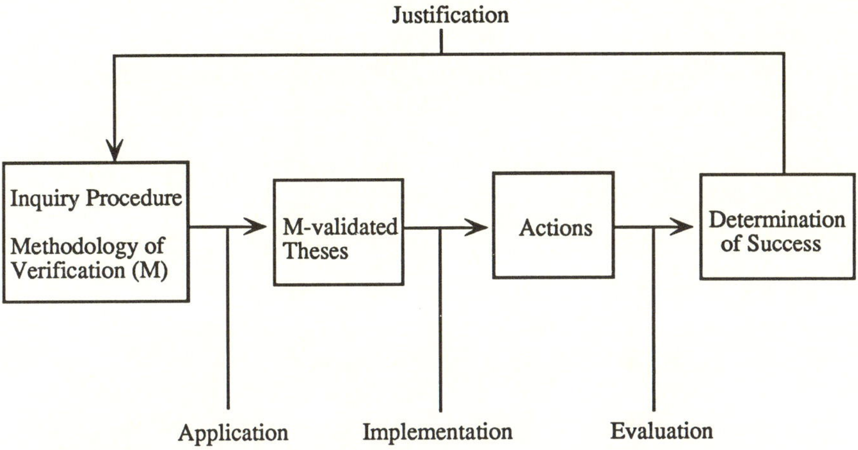

Diagram 4.2 The Pragmatic Justification of an Inquiry-Methodology
Arrows indicate the sequence of schematic cognitive stages which together make up the cycle of theory testing with consequent rejection or entrenchment. (Taken from Rescher 1977, 66.)

rather than of theses. As already noted, for Rescher the importance of this shift is that justification also shifts, from a focus on achieving truth to one on systematic coherence and practical action.

> (RQ2) . . . it becomes possible to break the regress of justifying theses by theses: a thesis can be justified by application of a method, and the adoption of this method is justified by reference to certain *practical* criteria (preeminently, success in prediction and efficacy in control). This two-stage division of labor represents the characteristic idea of a specifically *methodological* pragmatism. (67)

> (RQ3) By focusing upon *cognitive methods*—methods of inference, testing, checking, and other ways and means of task-accomplishment in the area of cognitive substantiation—we regard how-to knowledge as more fundamental than knowledge-that. The latter is held to be ultimately rooted in the former. (70)

> (RQ4) This line of thought drives us towards a Kant-reminiscent "Copernican Inversion." Later findings do not rest on a superior methodological basis because they are "truer"; rather they must count as truer because they rest on a superior basis. In effect this Copernican Inversion proposes that we not judge

> a method of inquiry by the truth of its results, but rather judge the claims to truth of the results in terms of the merit of the method that produces them (assessing this merit by both internal [coherentist] and external [pragmatist] standards). (179)

Since visual search is a successful practice, at least in the vast majority of cases, we accept "seeing is believing" as a general rule, and so we rationally accept the products of this rule, visual beliefs, as true.

But now the Wheel Argument applies once again:

> (RQ5) It remains a *rationally warranted presumption* that the truth-claims based on a superior inquiry procedure are themselves superior in their rational legitimation. Accordingly, the warrant at issue in the "warranted assertability" of the truth-claims validated by a duly legitimated inquiry procedure resides in a *regulative principle of rationality*.
>
> But by just what justificatory rationale can such a regulative presumption be legitimated? Clearly, the key question remains: How is one to validate the linkage between the factor of methodological success on the one hand and that of thesis-truthfulness on the other. (80)

Bluntly, why not a successful method that nonetheless yields (some) falsehood? Surely every practical method we can aspire to will be imperfect; in at least some range of circumstances, it will generate at most partial truths (i.e., falsehoods if taken as true simplicitur). And it is not hard to imagine circumstances in which our best particular methods go systematically wrong in part (e.g., simply failing to resolve differences below some discrimination threshold).

Rescher's immediate response is to develop a regulative theory of acceptable inquiry methods. Such methods, he asserts, must be inherently systematic, general, and public (inter-subjective). Roughly, systematic methods are explainable, codifiable, and teachable (74–6). A general method is one that applies across many areas of inquiry, that is open-ended in its range of application, a method that is "generic," versatile, and non-ad hoc (82). A method is public if it satisfies the systematic and general criteria in a way which is accessible to all and that naturally generates agreement among all in these respects. These features are designed to rule out *local* ignorance, bias, and errors and to mitigate finitude, as sources of falsehood in a pragmatically successful method. It is already clear that Rescher conceives of science as characterized by a regulatory system, although their components are picked out entirely in terms of their semantic contents (see further below). His account is beginning to flesh out the orders of systematic feedback and feedforward structures that are needed to rationally regulate scientific acceptance.

But beyond a regulative theory of acceptable inquiry methods, Rescher adds a further regulatory component: metaphysics. Rescher's justificatory use of metaphysics is direct:

> (RQ6) Given a suitable framework of metaphysical assumptions, it is effectively impossible that success should crown the products of *systematically* error-producing cognitive procedures. (90)
>
> (RQ7) Here all of the safeguards built into the statistical theory of the "design of experiments" come into play with respect to the probative significance of the number and variety of instances. It is inconceivable that a *systematic* success across so broad a range should be gratuitous. (82)

Roughly, (1) our beliefs find their coherent place within a metaphysical framework, the lineaments of a world view, and the result of collecting accepted beliefs should go on confirming, "fleshing out", that framework, and (2) that framework should underpin, or provide a "justificatory rationale" for, moving from pragmatically justified methods to rationally accepting as true the theses to which they systematically lead on their application.

What are the general features of the metaphysics Rescher believes methodological pragmatism requires? He summarizes these in diagram 4.3. Rescher's justificatory structure is now that of

INQUIRERS	WORLD
Individual	responsiveness of nature (to human intervention)
reasonableness activism	
interactionism (with the "external world") sensitivity (to feedback)	nonconspiratoriality of nature (neither on the angelic nor the demonic side)
Communal	
methodological continuity purposive constancy	uniformity of nature

Diagram 4.3 Metaphysical Presuppositions of Knowledge
The characteristics of inquirers and their investigatory environment which are jointly required in order for a coherent empirical knowledge process to be practically feasible. (From Rescher 1977, 89.)

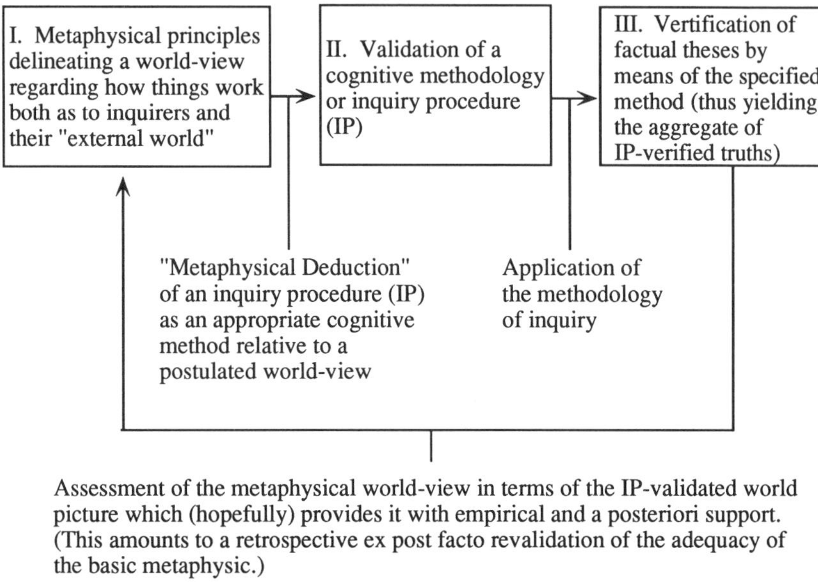

Diagram 4.4 The "Great Circle" Legitimation of an Inquiry Procedure
Arrows indicate the sequences of shematic cognitive stages constituting the cycle of specification and testing of an inquiry procedure, with consequent entrenchment or rejection. (Taken from Rescher 1977, 99.)

diagram 4.4. These metaphysical features are Rescher's response to a question along the lines (cf. 94): Given that knowledge (in particular, science) is possible, what must we and the world be like? Stated unqualifiedly, Rescher's responses here are controversial. Surely, for example, nature doesn't have to be uniform for us to be able to know it, and in fact it isn't. And what is appropriate purposive constancy? Should not our purposes change as we learn more? As will emerge, my own critique of these entries has to do with the failure to read them in thoroughly regulatory systems terms. It is reasonable to suppose, for example, that a precondition of a knowable world is only that there must be enough interactiveness, uniformity, and purposive constancy at each level for us to be able to discover and represent change relative to uniformities at other levels.

4.III. Presumptions, Regulative Principles, Constitutive Theses, and Justification

But now we must face an expository difficulty with Rescher's text. Recall that Rescher is of the opinion that once metaphysics is included along with requirements on what can count as a cognitive rule, it is "effectively impossible" or even "inconceivable" for our rational beliefs to go systematically wrong. The problem is to reconstruct Rescher's argument clearly, and it is at this point that I find his text frustratingly vague. So far as I can judge, the entries of diagram 4.3 simultaneously have two distinct statuses or play two distinct roles: factual presumption and regulative principle. We are told:

> (RQ8) . . . a suitable justificatory rationale links the success engendered by use of an inquiry procedure and the presumptive truthfulness of its deliverances. This linkage . . . resides in the operation of . . . certain metaphysical principles and is underwritten by a series of complex (and largely factual) assumptions about the nature of the world and of man's place in it—assumptions we have signalised with such labels as *rationality, activism, interactivism, feedback and sensitivity*. (106–7)

But later we are told:

> (RQ9) Regulative principles generally contain implicit correlative presumptions in just the way the *procedural injunction* "Treat people as innocent until proven guilty" embodies the *presumption* of innocence-in-the-absence-of-proven-guilt. Just this is the case with the various principles comprising the basis for the "metaphysical deduction" of our inquiry-methodology. To be sure, these were set out in the preceding chapter as theses rather than overtly regulative principles, but this should not disguise their true nature beyond recognition. They are at bottom of regulative and presumptive standing. (For example: "activism"—"Conceive of people actively involved in dealings within the setting of their environment!," "uniformity of nature"—"Extrapolate uniformly from accessible data!", etc.) The *theses* correlative with such regulative principles are not (necessarily) certified truths that hold without exception "across the board"; they are mere presumptions: propositions provisionally accepted in the context of a purposive venture (viz., the validation of an inquiry method), and conceivably subject to correction or abandonment in the light of further considerations. (116–7)

The relationship between these two roles is further clouded by Rescher's speaking also of the metaphysical "box" split into two kinds of propositions, regulative principles *and* their "correlative factual presumptions" (116); cf. "an apparatus of systematic coherence at the theoretical level (a coherence in which factual presumptions *and* metaphysical presuppositions both play a crucial part)" (110; my italics). Moreover, Rescher quite generally introduces presumptions as "low grade data" and as "concepts having a tentative impetus toward truth" (114). This intuitively fits neither the cognitive status assigned factual presumptions in RQ8 and RQ9 nor what we are inclined to assign the elements of diagram 4.3 (which tends, I suggest, to be either rather high or rather low).

Rescher represents the relations among these components in the manner of diagram 4.5, and the structure of diagram 4.4 is then rerepresented as in diagram 4.6.

To complete the (my) confusion, at some points, Rescher speaks as if regulatory and factual status are utterly distinct, for example:

> (RQ10) Methods, as we have insisted time and time again, are inherently general. Their function is *regulative* and their role is thus quite different from that of cognitively *constitutive* general

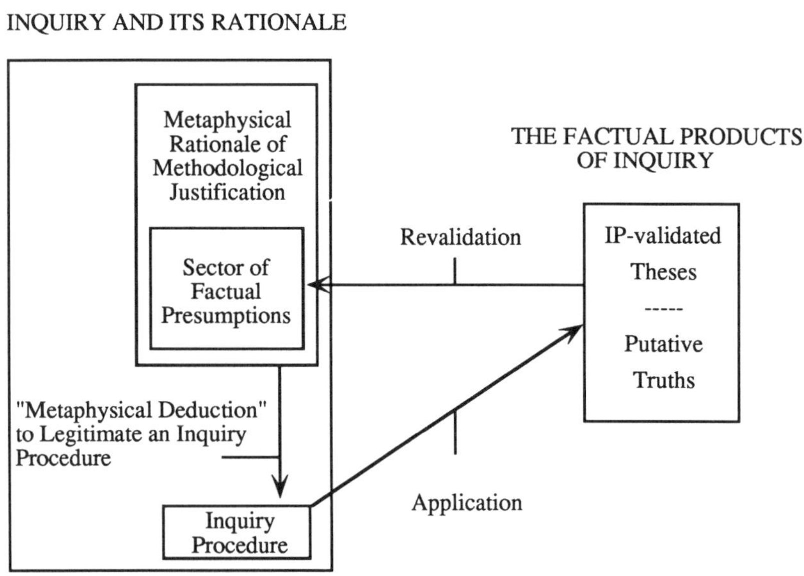

Diagram 4.5 The Interaction of Factual and Metaphysical Elements in Inquiry
Arrows indicate the sequence of cognitive stages constituting the cycle of specification and testing of an inquiry procedure, with consequent entrenchment or rejection. (Taken from Rescher 1977, 182.)

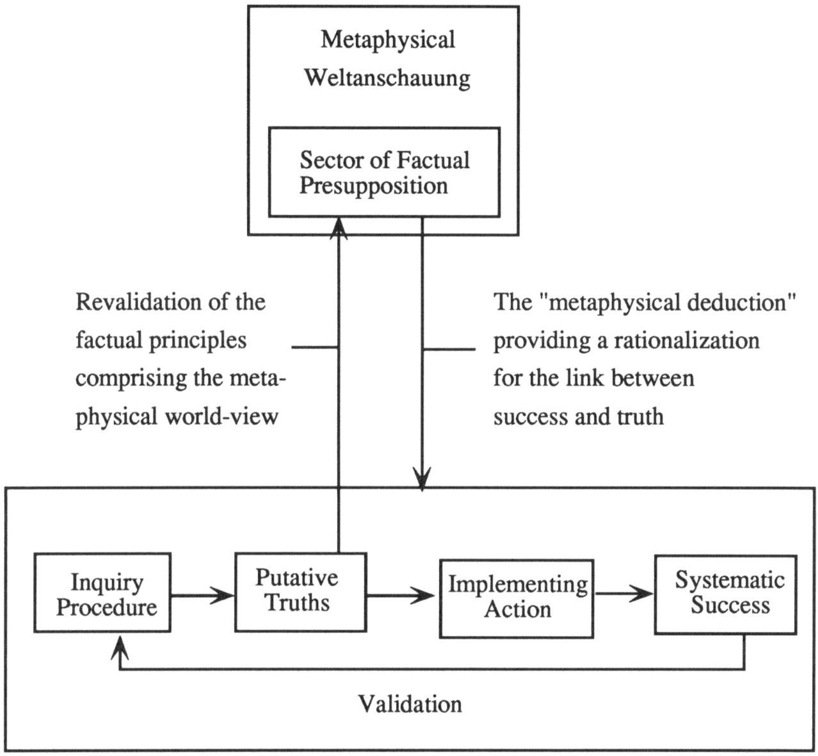

Diagram 4.6 The Double Circle of Pragmatic Justification
Arrows indicate the sequence of cognitive stages jointly constituting the cycles of theory testing and inquiry procedure with consequent entrenchment or rejection. (Taken from Rescher 1977, 187)

> theses. Moreover, they have an entirely different (and theoretically more tractable) manner of justification since their legitimation lies in the *practical* sector, where ground-rules different from those of the *theoretical* sector come into play. (p.158)

But then here he is speaking of *constitutive* theses and not mere presumptions. If this is to make a difference, we need an account of constitutiveness. A relationship between presumption and constitutiveness is given in the following passage, but there the regulatory/factual boundary is blurred and crossable, and essentially so:

> (RQ11) One simply cannot validate a methodology of inquiry in terms of its yielding "the real truth." The point of the Wheel Argument (*diallelus*) is precisely that this would be viciously

> circular, since we have no independent access to "the truth" as such. What we must do is pull ourselves up by our own bootstraps. We begin by provisionally accepting certain theses whose initial status is not that of certified truths at all, but merely that of plausible postulations, whose role in inquiry is (at this stage) one of regulative facilitation. Eventually these are retrospectively revalidated (*ex post facto*) by the results of that inquiry. At *that* stage their epistemic status—though not their content—changes. *In the first instance* these presumptions have a merely provisional and regulative standing, though in the final instance they attain a suitable degree of factual/constitutive substantiation. (119–20)

This gives us the needed presumption/constitutive relationship: A thesis changes from presumptive to constitutive as it becomes well enough confirmed in the cycle of inquiry. But it doesn't help at all in understanding the regulatory/factual relationship.

Presumptions have sources:

> (RQ12) Presumptive truths generally do not stand alone. They come in clusters or families. The ancient theorists of knowledge usually construed these groupings in terms of the epistemic *sources* from which the theses at issue have their provenience. The "data of memory" and the "deliverances of the senses" were the paradigmatic cases then envisaged. Such an approach seems perfectly reasonable—presumptions should be understood as forthcoming systematically and diffusedly in terms of some general policy, method, or procedure. (116)

What kinds of sources and presumptions go into Rescher's metaphysical box? Their characterization as low-grade data combines with the foregoing discussion to suggest specific features of the world (e.g., the color of my dog) as the content of typical presumptions and their sources as the traditional ones Rescher mentions—hardly suitable for metaphysical generalizations.

So this picture of presumptions fits neither diagram 4.3 nor talk of regulative principles. Yet this last quote goes on immediately: "This systematic status [of presumptions] is correlative with a generalised role in the regulative guidance of inquiry." And Rescher then continues: "The appropriateness of such a basis is ultimately subject to the controls of revalidation operative in the cyclic validating process in which its correlative presumptions are utilised" (116).

That there is an intended regulatory/factual relationship is clear. What precisely that relationship is and how it changes (if it does) along the presumption- constitutive spectrum is much less clear.

The position is further complicated by Rescher's later explicit introduction of a metaphysical subsystem of regulative principles in addition to presumptions, and a new relation of legitimation, represented in diagram 4.7. Certainly there must be regulative principles that are presupposed by the system of science. First, there is the wide range of principles covered by implication in RQ9 and diagram 4.3, from ones concerned with metaphysical adequacy (e.g., in representing agency) to those narrowly concerned with formal inductive procedure.[6] Further, Rescher needs some regulatory principle governing the relative entrenchment of presumptions as a result of their being confirmed in theoretical development, testing and appli-

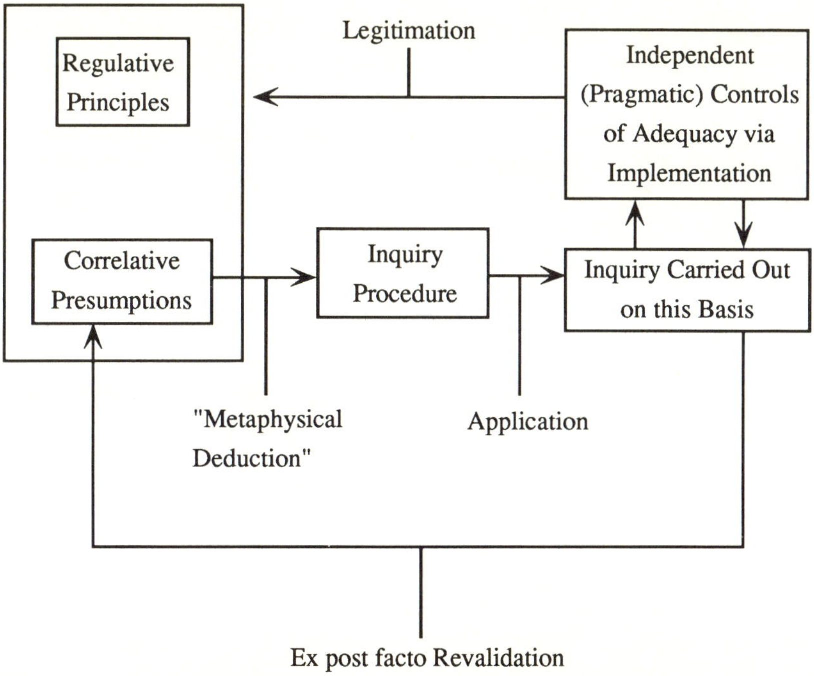

Diagram 4.7 The Revalidation of Presumptions
Arrows indicate the sequence of cognitive stages constituting the cycle of specification and validation of presumptions with consequent entrenchment or rejection. (Taken from Rescher 1977, 177)

cation—that is the essence of Rescher's notion of "epistemic stratification" (113–4), where presumptiveness gives way to constitutiveness.[7] Similarly, there is a need for regulatory principles to cover the case where input and output clash. The "metaphysical deduction" may work fine when the factual presumptions that emerge from the pursuit of science reinforce those that "undergird" (Rescher's term) the metaphysics and the methodology, but what happens when they don't? Presumably, we then need further "regulative principles of rationality" to guide rational resolution of these clashes; for example, how to choose new metaphysics (cf. the ongoing debate between atomic and field metaphysics in physics). And what of principles employing the standard notions of intertheory comparison, simplicity, explanatory unification, and the like that Rescher tantalisingly mentions (see RQ13 below) but never discusses explicitly: What role do they play in the scheme of things? Though the existence of all these kinds of regulatory principles is clearly implied (e.g., by the quote RQ9 above), Rescher is silent on the issue. Evidently, these regulatory principles form a (largely unaddressed) core to Rescher's metaphysical/Weltanschauung box.

Furthermore, we stand in need of an account of the legitimation relation. How does this relation stand to that of ex post facto revalidation? And what are the "independent (pragmatic) controls of adequacy" that serve to differentiate the two? How are they related to the regulatory principles?

Rescher sums up his position as follows:

> (RQ13) The present theory of knowledge may thus be characterised as a "dialectical idealism." It is *dialectical* because of the nature of the feedback mechanism operative in the justificatory process that provides its fundamental *modus operandi*. And it is an *idealism* because of the crucial role of mind-contributed presumptions which are not only mind-contributed but (like simplicity, regularity, reasonableness, etc.) are also mind-patterned. (123)

This passage quite strikingly gives the mind as source of the presumptions. One is tempted immediately by a Kantian rationalist reading, which would certainly give presumptions regulatory status by turning them into a priori truths. But then they would lose their relationship to the factual spectrum.

And it is hard to see how Rescher's idealism can be a rationalist one, given the assessment in the quote RQ11 above. And in fact, Rescher invokes a principle of fallibilism (Rescher, 1977, chapter XI) for all factual propositions, quite properly given his avowed pragmatism, and this evidently extends to metaphysics:

> (RQ14) On the present approach the metaphysical rationale is a potentially changeable theoretical structure whose justification lies not so much in whatever deep roots it may have in the course of human habituation, but in its functional capacity to accomplish its tasks in the framework of legitimation. (Our "metaphysic" is thus vastly more amenable to rational readjustment than that envisaged by Peirce.) (124–5)

Later, Rescher extends this fallibilism even to formal propositions (principally logic) as well (Rescher, 1977, chapter XIV), an extension of considerable interest to a fallibilist naturalist account but hardly supportive of rationalism.

We are left then with adequacy controls, regulatory and metaphysical principles (if different), presumptions (of varying degrees of constitutiveness), and sources of presumption without any adequate account of their natures, status, and roles in the dynamics of knowledge validation.

4.IV. Reprise and Prospect

There are important similarities between RQ13 and Piaget's approach. An active constructivism is placed at the center of understanding cognition. This constructivism can be given a rationalist reading, and we find clear rationalist tendencies in both Piaget and Rescher. In Piaget's case, this tendency proved damaging, leading to an overemphasis on formalism and static structures, as opposed to process, and to a Lamarckist tendency in biology (chapter 5). I shall shortly expose the same kinds of difficulties in Rescher (see respectively, sections 4.V through 4.VIII below). But in Piaget's case, there is a stronger naturalist anti-rationalist fallibilism underpinning a basic regulatory systems approach. And this supports his unified model of constructive self-organization, a model that incorporates an evolutionary epistemology. I shall argue that Rescher's position, which already includes an evolutionary epistemology, can be construed in a closely similar fashion, albeit with severer correction than for Piaget, and that, thus construed, his general account then makes a valuable contribution to a theory of cognitive dynamics exactly where Piaget does not elaborate, namely, to a self-organizing constructivist model of normative learning in science.

What then are we to make of Rescher's account in section 4.III? Pursuing the approach just outlined suggests the following treatment. Every declarative sentence within a body of knowledge, in particular within science, is a presumption. The scope of presumptions thus includes everything from detailed, particular factual assertions through to more general empirical and theoretical claims, including factual presuppositions of methods, and extending finally to the general assertions of metaphysics. Each of these presumptions has some kind of weighting attached to it representing its current degree of entrenchment as a consequence of the operation of the regulatory process of scientific development sketched in diagrams 4.4-6. As the weighting increases, one advances from presumptive toward constitutive status. As part of this, there are (undiscussed) basic or most general regulative rationality principles in metaphysics that select from these presumptions the further regulative principles that are to operate. On this basis, and presumably using yet further presumptions (undiscussed), methodology is developed. Methods too, being modifiable, must have some weighting attached, but here too Rescher is silent; perhaps their weighting is that of their weakest presupposed factual presumption. Putting these lacunae aside, the full regulative apparatus is now at hand to turn on those presumptions operating in the more directly testable scientific domain. Success and failure of theories is fed back to weighting changes in the methods through which they were developed; changes in data and theory (and methods?) are fed back to the metaphysics box, and the whole self-correcting regulative process is off and running.

From this point of view, the account looks remarkably like a version of Popper translated into elementary control system presentation. And Rescher exhibits the Popperian tendency to reduce methodology to logical tests of coherence and contradiction (cf. section 4.V below). Rescher does criticize Popper's use of evolutionary analogies, and I shall examine this criticism shortly (section 4.VIII below), but this is compatible with their sharing a basic formalist and empiricist metaphilosophy (see below, and cf. Popper's critique of empiricism, chapter 3). Just as Popper's non-naturalist empiricist metaphilosophical commitments proved damaging and distorting (chapter 3), and the same will prove true of Piaget's non-naturalist and formalist rationalist component (chapter 5), so the damage done by the formalist rationalist and empiricist components in Rescher must be now assessed. Since methods and method dynamics stand at the heart of cognitive dynamics, I turn first to a critical examination of Rescher's account of methodology.

4.V. Methodological Dynamics

Certainly the idea that there corresponds to every methodology a set of substantive assumptions is a sound one. Rescher himself agrees:

> (RQ15) To be sure, the methodology itself rests on a basis of presumptions which themselves represent a generalised "knowledge" of the world, a knowledge arrived at on the basis of the "lessons" of past experience in whose make-up trial and error has played a prominent part. (160)

The use of visual search for empirical information acquisition is a basic epistemic methodology. But its use requires the assumptions that across the narrow visual band, the causal interaction of electromagnetic radiation with the environment carries information concerning the structure and properties of that environment, that visual information processing in our nervous systems extracts at least some of that information over at least a range of recognizable circumstances at some level of reliability sufficient to function, and that its representation in our nervous systems preserves that reliability.

More generally, there is an intricate and intimate correspondence between theories (theses) and methods; see chapter 2. As emphasized there, theory and method are thoroughly intertwined with one another; each order of the one may in principle affect each order of the other. It is better then to think of theory and method across the orders of science as but two dimensions of a single complex structure idiosyncratic to each domain.[8]

There is no serious attention given to this complex methodological structure by Rescher, and hence one finds no attention to methodological pluralism, to the notion of a variety of methods, complementary, competing, and independent, a variety that leads to the development of methodology itself. Rescher considers only "methodological schizophrenia," the alleged existence of inconsistent but equally successful methods, and quite rightly argues against it (111–3), but at this point mentions pluralism only once in a passing footnote (113, note 6). Yet methodological pluralism is of the essence for a realistic account of scientific practice. Our most reasonable construction of what exists is what is invariant across methods (e.g., present for both sight and touch). The scientifically decisive moment in the acceptance of atoms, for example, came when several different methods for determining Avogadro's number all gave closely related answers.[9]

Moreover, methodological pluralism is essential to a fully developed relationship between biological and cognitive evolutionary processes (see below). Rescher himself sees this:

> (RQ16) It is not difficult to give examples of the operation of Darwinian processes in the cognitive area. . . . Examples of such defunct methods for the acquisition and explanatory utilisation of information include astrology, numerology, oracles, dream-interpretation, the reading of tea leaves or the entrails of birds, animism, the teleological physics of the Presocratics, and so on. There is nothing intrinsically absurd or contemptible about such unorthodox cognitive programs, even the most occult of them have a long and not wholly unsuccessful history. (Think, for example, of the prominent role of numerological explanation from Pythagoreanism, through Platonism, to the medieval Arabs, down to Kepler in the Renaissance.) Distinctly different scientific methodologies and programs have been mooted: Ptolemaic "saving the phenomena" vs. the hypothetico-deductive method, or again, Baconian collectionism vs. the post-Newtonian theory of experimental science, etc. The development of the means of inquiry and explanation invites a Darwinian account.
>
> There can be little question that the history of science encompasses the development of scientific methods and explanatory mechanisms as well as the development of theories and explanatory models. (138–9)

One then waits for a spelling out of a theory of method dynamics, but the wait is in vain. At that point in the discussion, Rescher merely mentions that the historical study of methodology is recent and incomplete, and he passes on to other issues. Later on Rescher remarks:

> (RQ17) Once an even modestly satisfactory inquiry method is at hand, progress at the thesis level can be very swift because of the inherent power and generality of such a method. Moreover, thanks to the cyclically self-corrective aspect of such a method, further substantial progress in the methodological side itself becomes a real prospect. (162)

Here we have what is evidently the only explicit acknowledgment of methodological development in the book. A note closing this remark reads as follows:

> (RQ18) Although stated flatly and without much development here, this point is actually one of great importance. (Presumably the reader can work out its ramifications for himself, given the structure of the cyclic models of methodological reappraisal set out in Chapter VII above.) (162)

However, when we return to chapter VII, which is focused on presenting various versions of diagram 4.4 above and explaining their operation, we find no substantial discussion of the *methodology* of methodological development. What is offered instead is only a single feature: coherence. The system checks for coherence, or lack of it, between initial presumptions and the results of theory development, testing, and application. Presumably, this is a purely formal logical test. For most philosophers, the a priori status conventionally attributed to logic then undermines any need to refer to a methodological regulatory structure at this point, and this may explain Rescher's neglect of the matter. If so, it provides another instance of the naturalist/rationalist schizophrenia running through his account, for this latter requires setting aside Rescher's own approach to logic; see section 4.VI below.

But in fact, Rescher cannot rest content with a purely logical notion of coherence even for the case where there is agreement between input and output in diagram 4.4. As noted in section 4.III, Rescher needs some notion of the relative entrenchment of presumptions to yield constitutiveness. The need for something more than logical coherence is equally urgent in the case where input and output clash. Indeed, none of the regulatory principles uncovered in section 4.III can be adequately expressed in simple formal notions, a fortiori by logical coherence alone. Thus despite Rescher's passing mention of methodological development, no substantial account of that development is provided.

In a later chapter, when discussing evolutionary epistemology, Rescher asserts that methodological change occurs solely through trial and error, where the trials are blind to the utility of the outcomes to which they eventually lead:

> (RQ19) The combination of a model of method-learning based on the blind groping of trial and error and of thesis-learning based on the use of methods makes it possible to have the best of both worlds. (159)

The adequacy of this two-tier conception for evolutionary epistemology will be challenged in section 4.VIII below, but for the moment, the point is that while Rescher may have thought that the occurrence of blind trial and error at the methodological level obviates the need for any further theory of methodological dynamics, this is not so. Methods, as we have just had cause to note, carry with them factual presuppositions or presumptions. Hence one cannot apply a method in isolation from the rest of the scientific system to see whether it generates theories that are supported through test and application. Rather, in order to trial it, a methodological variant will have to be combined with, and possibly have to modify, the system of metaphysics, methods and theories into which it is asserted—or be modified itself. And as is the case with theories themselves, we can expect no simple outcome of these trials; rather, they can be expected to lend weight to either incorporating or rejecting that variant. Both of these actions, initial incorporation of the methodological variant and the fed-back decision as to its retention or rejection, themselves require methods. Tests for simple logical coherence are not enough. The lesson remains: Rescher has not provided the substantive account of methodological development that his own theory demands.

But note here that the methodology/rationality components with which Rescher is tacitly working has a rich internal regulatory structure to it. The regulative principles used to derive methods for testing are of higher order than the methods themselves (which in turn regulate theory), and the regulative principles governing method entrenchment are evidently higher order again, since presumably we may learn from the success and failure of our method development procedures. Rescher himself fleetingly recognizes this at one point but fails to grasp its general significance:

> (RQ20) Moreover, the above concession that methodological progress might proceed purely and solely by blind trial and error (i.e., by wholly random and rationally unguided variation) may well go too far. After all, even methodological innovation is not wholly haphazard. In any field, say woodworking or chess—and so even perhaps in the cognitive sphere of fact-substantiation—methodological innovation is itself to some extent guided by methodological considerations. It makes sense to conceive of the issue of methods for making methods on analogy with the enterprise of engineering machine-tools, i.e., machines for making machines. (165)

This capacity to ascend the regulatory order is in fact the single most powerful and distinctive feature of science, as the dynamics of technological development testify, and of central importance to scientific progress (chapter 2). Meanwhile, note that it is evidently Rescher's rationalism that cuts him off from considering the full regulatory structure. So I turn next to examine Rescher's account of rationality.

4.VI. Rationality

> (RQ21) Rationality emerges as a key element in this picture. The survival of effective methods is not inevitable and automatic ("natural") and assured by some inexorable agency of nature. Rather, the crucial link between "success" and "survival"obtains because the rational man places his bets in theory *and* practice in consonance with those methods that prove themselves successful, tending to adopt those that succeed and to abandon—or readjust—those that fail. (135.)

It is "regulative principles of rationality" that guarantee that the system hangs together in the correct manner. (That a complex system is self-reinforcing is of no help in itself, for many incompatible and erroneous systems display that feature.) Moreover, it is evidently some such principles that lie behind Rescher's Kantian-idealist assurance that in this system, it is "impossible" (or even "inconceivable") to go systematically astray.

What then is the status of the principles that enter this evidently complex theory of rationality? The discussion of sections 4.III and 4.V (cf. the quotes RQ11, 14) make it clear that at least with respect to all other statements entering the system, from metaphysics to empirical observations, fallibilism reigns—they are all in principle open to change. Does this apply as well to the regulative principles of rationality and whatever other statements lie within rationality theory? Evidently the regulative principles that enter the "metaphysical deductions" are of the form: "In a world of kind K, with creatures of kind C, cognitive methods of kind M cannot deliver systematically erroneous theses." In the standard trichotomy, statements of these sorts might be held to be vacuous but analytically true (translatable into theorems of logic), to be strongly true yet informative (something like mathematical "deep truths" or a Kantian synthetic a priori), or be just contingently true. While Rescher's claims to justification may suggest one of the first two possibilities, these are evidently ruled out:

> (RQ22) One thing very much needs to be said regarding this proposed metaphysical "deduction": It is *not* in fact to be

> regarded as providing anything approaching a demonstrative proof. For were this so, then, deductive logic being what it is, we would simply have to load into the premises (in duly disguised form) the very conclusion we are trying to extract. Rather, the inference at issue is essentially a *plausibility*-argument—one that builds a good case for its conclusion, providing it with a solid rational warrant which (admittedly) stops short of giving a logically airtight guarantee. The person who, in the face of such an argument, refuses to accept its conclusion (while yet granting the premises) is not being *inconsistent* but simply *unreasonable*. (96–7)

But perhaps the last remark here could be construed so as to support the Kantian synthetic a priori alternative. This position is at least consonant with the quotes RQ6, 7, 9, and 13. It is also consonant with Rescher's later appeal to "rational selection" in his discussion of evolutionary epistemology:

> (RQ23) [With respect to the prima facie gap between Kantian transcendental justification and pragmatic justification] . . . the gap between these two approaches is drastically narrowed by an evolutionary theory that envisages the development of our cognitive tools through rational selection. For if those cognitive instrumentalities we men in fact have—the only ones with which *we* can work, since they alone are at our disposal—are the products of a selection process itself based on considerations of instrumental efficiency and effectiveness, then the convergence and concordance of these two seemingly opposed approaches is assured. (But, of course, our approach mitigates the element of absolutism of a "transcendental" argument by relativising its operation to a contingent methodological basis.) (131–2.)

And while, as noted in Section 4.III, Rescher later introduces a principle of fallibilism (1977, chapter XI), and even applies it to logic (chapter XIV), it is restricted to "knowledge claims," and these evidently do not include the regulative principles of rationality themselves. (But this again breaks their connection to their factual presumptions.) In sum, there is strong evidence that the third, mere contingency, alternative for the status of the regulative principles of rationality is not an accurate reading of Rescher either. Some much stronger knowledge status is required.

On balance then, Rescher is best understood as adopting a quasi-Kantian idealist conception of rationality and our knowledge of it. All this sits ill with a thoroughgoing fallibilism of the sort a thoroughgoing evolutionary naturalism requires (see chapter 1). Most importantly from the dynamic systems perspective, it places an absolute "cap" on introducing any higher-order regulatory components to Rescher's two-component system, a conclusion reached in section 4.V above. (Since the strongest possible justification is already achieved with this first two-component structure, there is no point to looking further.) In this way, Rescher's position cuts itself off from being a serious evolutionary epistemology; we are left with a schizophrenic account in which a naturalist selection process is set in a non-naturalist framework of non-naturalist rational selection (see also section 4.VII below).

But even on Rescher's own terms, this account is evidently still too simple. A clue is provided with Rescher's courageous and insightful exploration of the notion that use of logics too is pragmatically justified (1977, chapter XIV). But L-valid reasoning (= the statement corresponding to the argument form is valid in logic L) is a key component of rationality. By what rationality judgments do we then choose logics? And may we not learn to improve these choices? (Surely just this is what the contemporary explosion in logics implies; see chapter 1.) Then how can the statement that rationality lies in using a particular set of L-valid reasonings be itself strongly (Kantianly) true?

As we have seen (see section 4.V above), Rescher's non-naturalist concept of rationality is also characterized by a failure to consider methodological development fully. Moreover, consider the whole diagram 4.4, rationality principles and all; what kind of judgment is it that it all hangs together in the right way? Presumably one that also belongs to rationality theory. But then is there no room to learn about alternative systems schemes here? What of a metaphysics where empirical states were significantly mind dependent, perhaps functions of our values and/or our beliefs; would the same justificatory considerations apply in the same way? Here we change the metaphysics and so use a different principle for justification (cf. Hooker 1992a and the telling investigation by Churchman 1968, 1972). Is there to be no opportunities of these kinds to learn what it is to be rational?

These considerations point to another difficulty with Rescher's account: his evident assumption that rationality is not only divorced from fact, both in nature and status, but similarly from value as well. Reasonableness is defined as a complex coherence among beliefs, methods, and actions (85); values are not mentioned. One has the sense that lurking behind Rescher's otherwise anti-empiricist writing are the old empiricist metaphilosophical dichotomies between fact and value, cognitive and pragmatic (cf. the closing remarks to section 4.V and notes 2, 4). These are precisely what a thoroughgoing evolutionary epistemology needs to reject.

In sum, despite the beginning acknowledgment of theory and method dynamics in Rescher's account, his conception of rationality remains schizophrenically locked into an eternally fixed formalism by a quasi-Kantian idealism and empiricist fact/value, cognitive/pragmatic dichotomies. This shuts it off from interaction with all the other components of a dynamically developing cognitive system (including values). These are interactions that Rescher himself tantalizingly acknowledges but does not address and that are central to understanding cognitive dynamics. By contrast, a thoroughgoing naturalist evolutionary epistemology suggests that the concept of rationality or reasonableness picks out the distinctive regulatory structure characterizing self-organizing, developing cognitive processes (see chapters 2, 6). In this case, it operates across the entire mental landscape, in respect of beliefs, methods, values, and reasonableness itself and must itself be expressed as a design to a complex regulatory system.

4.VII. Regulation

Rescher clearly gives regulation a central role. Rescher's cognitive system structure for the generation of knowledge, sampled in the diagrams 4.2–7, are clearly and explicitly intended to represent regulatory systems. Moreover, their sequential operation is intended to generate a history for science, similar in significant respects to the history of life itself. (One enters initial knowledge presumptions of various kinds and follows their evolution as there is subjected to methodical testing, modified, brought into coherence with the underlying metaphysics, re-presented for application and testing, etc.) Indeed, Rescher likens his cognitive regulatory system (diagram 4.4) to a quality control system in industrial production, a self-regulating system aimed at quality output.

> (RQ24) This circular-structured process of validation conforms to the performance- monitoring *modus operandi* of a self-evaluating servomechanism, since it provides for a quality-control feedback loop that leads from the *product* ("factual theses") back to the *process* (the deployment of a method) which generated it. (103)

He represents it as in diagram 4.8. As the titles to Rescher's diagrams imply, he wishes to represent the cognitive development sys-

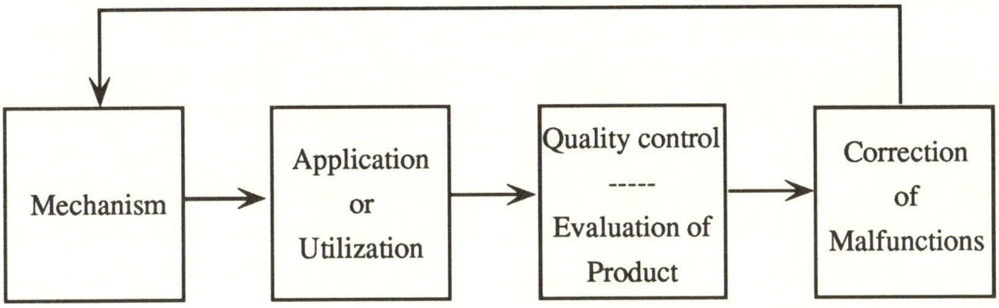

Diagram 4.8 The Dynamics of Performance Monitoring
Arrows indicate the sequence of stages which comprise the cycle of monitoring and subsequently improving the performance of some mechanism or process. (Taken from Rescher 1977, 7)

tem as a set of interlocking quality control loops of this kind. Later on (chapter X) Rescher stresses the conception of science as an autonomously self-correcting (i.e., self-regulating) system. Reason here is the cognitive quality control process and knowledge is quality controlled (i.e., rationally warranted) acceptance. This seems to me to be the essentially correct insight into both reason and knowledge.[10]

But Rescher's notion of regulation does not run deep enough. In essence, what goes wrong is that the regulatory system that Rescher describes cannot itself evolve. The net effect of Rescher's rationalist idealism, already noted, has been to freeze his proposed regulatory structure in the ghostly realm of the a priori, allowing only contents to develop. Its staticness and transcendence lies in uneasy tension with the dynamicalness and fallibilism that characterize the lower levels of Rescher's system. Moreover, we have seen that the evolution of regulation is of the essence of both biological and epistemic development, since in actual systems there is no structure/content dichotomy because content is stored as active structure (chapter 2).

Consider again the regulatory methodological structure of science. For reasons ranging from theoretical dissensus to different laboratory conditions, there is a distribution of methodological structures spread across the population of scientists (chapter 2). Understanding the evolution of that distributed methodological web involves understanding both the internal regulatory structures that scientists possess and the web's regulated expression in their institutionalized interactions with one another and with their environments. The importance of this structure Rescher clearly grasps (see the institutional features incorporated in the metaphysical structure, diagram 4.3 above), but he fails to develop this feature of evolutionary epistemology. Similarly, recall that the discussion of sections 4.V and 4.VI has also made clear the importance of at least three orders of methodological structure to the regulatory framework and plausibly of orders beyond that. However, the same discussion revealed that Rescher had failed to develop any substantive theory of this kind. Thus while the notion of method as a regulatory structure exists for Rescher, this regulatory structure does not penetrate very deeply. The significance of this failure will become more apparent as we proceed.

But first, consider its parallel appearance in Rescher's treatment of values. Now for methodological pragmatism, values or goals are at the foundation of knowledge.

> (RQ25) A merely instrumental justification could in principle aim at *any* sort of goals; a pragmatic justification is committed to the specific family of welfare-oriented goals. (85.)

> (RQ26) But, of course, *real* success lies in realising the reasonable and rational goals of people. Objectives that are *pathological*—the courting of pain, privation, anxiety, harm, etc.—must be excluded. In the sense operative with respect to pragmatic justification, "success" must be taken to lie not in the realisation of objectives as such, but in the realisation of our orthodox affective objectives, that is, those that have a positive bearing on human welfare. In its pragmatically relevant sense, "success" is oriented towards realising the *real interests* of people in satisfying their recognised welfare needs. (86)[11]

Here Rescher has values playing the central role of shaping all purposive action, that of science in particular. But from whence come our goals or values, and how do we learn about them and modify them? Rescher is silent on the issue. Nowhere in the book does he discuss their interaction with any other component of the system.

Surely we may learn about valuing. Is there not a rational feedback from experience to the choice of values? Or is all value change merely causal and noncognitive? Anyone who holds this latter, empiricist view defeats the purpose of a Rescher-style program by undermining the reasonableness of science. For an empiricist, science is based on fact and logic, and the subjectivity and rational opaqueness of values make no impact on the reasonableness of science. But for Rescher, the reasonableness of science hangs on the rational design of the multifeedback system he lays out (or begins to lay out). If this design reflects our purposes, which in turn reflect our values or goals, and if these are opaque to reason, then after all, the whole scientific enterprise is founded on a non-cognitive commitment, a foundation that may shift, and so shift the design of science, without any cognitive controls on direction.

Cognitive values (epistemic utilities) certainly play a fundamental regulatory role in their structuring of proxies for truth, that is, of what counts as the valuable kinds of information that structure the kinds of quality control we pursue. These cognitive values appear in all the myriad individual acceptance decisions constituting the science process. And in their roles as regulators of actions (internal and external), they form an ordered regulatory structure, since they themselves form a complex ordered structure; for example, values may take priority over others, be interdependent on others, and so on. Moreover, the inclusion of values becomes unavoidable in a full regulatory setting because of the intimate role in the cognitive process played by the disturbance and technological transformation of the environment into artifacts (see chapter 2). For an artifact is an art-i(n)-fact, a factually realized human design, and designs express values because they are based on choices, strategic selections from among the possibilities, and thus always involve the values expressed in their selection criteria. This is true specifically of the methodological and instrumental designs in science. But then the development of this world, and of every science within it, becomes a complex dynamic interaction between fact and value through design choice and artifact production, experience in one artifactual condition helping to determine both how our knowledge system changes and how we next transform our artifacts.[12] Within this dynamics, the conditions for the epistemic enterprise can be distinguished in a principled, but not apriori, way vis-à-vis those for larger human flourishing; they include, for example, sufficient variety in the artifact pool to maintain the processes of regulatory superfoliation and support for scientific institutions of sufficient regulatory complexity (see chapter 2). The role of values is as central as that of facts in this regulatory process.

But values and value dynamics are really excluded from the realm of rational development in Rescher's discussion (section 4.VII above), and the regulatory apparatus is not extended to them. No concept of environmental transformation emerges, for example, and there is no discussion of its relevance for the relation of science to human values. There is equally no attempt to relate the nature of technology to the conceptual development of science and thence to its methodological development as well. This is a particularly striking failing in a theory that gives a central role to successful technological practice in both the justification of science and the process of scientific advance (cf. IX below).

In sum then, Rescher has provided a valuable and stimulating initial development of epistemology as a regulatory system. However, the concept of regulatory structure remains both too limited and too superficial. Both of these deficiencies are evidently traceable to Rescher's idealism and/or lingering empiricism, which blocks taking a thoroughgoing regulatory systems approach.[13] Now let us examine the consequences for Rescher's evolutionary epistemology.

4.VIII. Rescher on Evolutionary Epistemology and Method Darwinism

Rescher's review of evolutionary epistemology has two foci: Thesis versus Method Darwinism and natural versus rational selection. First, Method Darwinism. As noted, Rescher views the failure of Thesis Darwinism as correlative with the absence of any theoretical justification for progress in adaptation within biology (see Introduction above). By contrast, for Method Darwinism the ultimate objects of selection are methods, not theses, but these are selected only indirectly via the proximate selection of the theses to which they give rise and on grounds of success in prediction, control, etc., not truth per se. Rescher views defense of Method Darwinism as a part of the very nature of rationality in a world where science is possible at all (see sections 4.II, 4.VII above).

My own evaluation of these moves will perhaps now be somewhat clearer. I believe that Rescher's move to Method Darwinism is a move in the right direction, because it enlarges the scope of regulatory process, a move that provides for increased adaptability, not just increased adaptation (cf. chapter 2). But I cannot accept the attempt at an idealist quasi guarantee of the connection of successful methods to the truth of the theses they generate, nor the cut-off in regulation it engenders. Rather, when Rescher's regulatory structures are extended to the full, open-ended regulatory structure of cognition, Rescher's idealist guarantee for methods is replaced by rational commitment conditionalized on the appropriate regulatory assumptions. The practical success of a thesis is grounds for its warranted assertion, conditional upon its having been evaluated using the most adequate methodology available and (so) its corroborative consistency with the most warranted metaphysics. A method's success in generating successful theses is grounds for our rational commitment to it, conditional upon its having been evaluated by our most adequate method methodology, for example, by its metaphysically systematic generation of successful theses. And so on. (Cf. the nested fallible commitments in Hooker 1987, chapter 8.)

For evolutionary naturalism, all levels are fallible, and nothing is guaranteed. Even when Rescher's cognitive system generates a coherent metaphysics supporting a science whose successful methods in turn generate theses coherent with the metaphysics, it is still possible that the structure of the world is sufficiently complex and subtle that our system has locked us into a self-reinforcing illusion (at least with respect to sufficiently subtle regulatory features). Indeed, religious mystics of various persuasions insist on just that. But without resorting to issues at this scale, every level of intelligence in the evolutionary sequence yields its own examples. Simple conditioned creatures living in a regular, predictable environment develop simple sensory-motor methodologies for flourishing in that environment, methods that are successful, generating a variety of "beliefs" concerning the forms and locations of predators and of food, etc. Yet the little cycle of self-reinforcement that these creatures exhibit can be easily thwarted by any process that perturbs the regularity of their environment, not only the chance external catastrophe, but a long term geo-ecological environmental shift or even an ecological regulatory cycle of many of their generations in length or the preselective activity of a creature with greater cognitive capacities. We differ from them only in degree.

The essence of intelligence is to be able to develop a flexible superfoliating cognitive regulatory structure. An intelligent scientist will, within the context of an experimental situation, accept a variety of theories and methods, but in a larger regulatory setting, he may hold any one or more of these open to question. And in fact, it is of the essence of scientific rationality to be able to create new higher regulatory orders and to suspend lower-order regulatory commitments in this way (chapter 2). Though Rescher insists that he is not attempting a transcendental deduction of the success of science, there is also none of this kind of regulatory flexibility, none of this more powerful kind of adaptability, in Rescher's account.

Second, selection. Here we find the same problem of inadequate regulatory conception. Rescher employs a generalized variation-selection-retention (VSR) model of evolutionary process, in which selection is given the generic functional meaning of elimination, with two subspecies, natural and rational selection:

> (RQ27) As regards *selection*, the crucial factor is that of *critical rationality* in adopting, from among competing alternatives, that method which proves in the course of applications to be more successful in point of goal-realisation—and correspondingly in abandoning those methods that have shown themselves less successful. Thus in our methodological case, where

> overtly purposive instrumentalities rather than biological organisms are at issue, the operative factor in the developmental process is not that of *natural selection*, but that of *rational selection* in the light of explicitly purpose-oriented considerations. This, of course, is a significant point of difference from evolution in its classical Darwinian form where survival alone, rather than any other more elaborately rational purpose is the operative factor. In the present case, where methods are overtly purpose-correlative, an explicitly rational teleology is called for. This point of difference is, however, quite irrelevant to the basic evolutionary pattern of the present model of the historical process. It has the classic form of an evolutionary pattern based on variation and selection. (9)

> (RQ28) Rational selection is a process of fundamentally the same *sort* as natural selection—both are simply devices for elimination from transmission. But their actual workings differ, since elimination by rational selection is not telically blind and bio-physical, but rather preferential/teleological and overtly rational. In this sense, rational selection is essentially Lamarckian: survival favours certain "preferred" conditions, with preferability determined in an explicitly purposive way. Standard neo-Darwinism is in effect a way of *removing* teleology; it provides a way of accounting for *seeming* purposiveness in purpose-free terms, by deploying the mechanism of a *blindly eliminative* annihilation of certain forms in place of any recourse to preferential considerations. But our present rationally-oriented neo-Darwinism can only operate with respect to beings endowed with intelligence and action, with reasoning and purposes—its mechanism being the *deliberately rejective* annihilation of forms that are not purpose-serving. (133–4.)

These passages should be read in conjunction with RQ21 and RQ23; see also Rescher 1977, 134, note 11, where rational selection of methods is linked to cultural selection and transmission, Rescher 1977, 136, note 13, and Rescher 1990, 8 and throughout.

Rescher's Lamarckism here is evidently rooted in his conception of reason; it is because methods are rationally selected in the light of our goals, and so of our values, that there is hereditary transmission of acquired characteristics. Although Rescher nowhere discusses the matter explicitly, one assumes that this Lamarckism is what lies behind his conception of methodological progress in science, which is in turn central to Rescher's defense of scientific progress (see below). It would be understandable in these circumstances to link natural selection with its standard genetic model of blind variation and rational selection with guided variation, and this is the linkage suggested in the above quotes.

The general model suggested by Rescher is that while both evolution and science are governed by a generic VSR process, every part of this process in the case of science is teleological, with a resulting feedback from an intentional or purposive S (via R) to V, so that V is not blind. But the very genericness of the model distracts from precision: We can see how theory variation might be considered directed, namely directed by methods; but method evolution, Rescher's distinctive idea, is not discussed. However, these passages certainly seem to be primarily about the rational selection of methods. This view is corroborated by the following:

> (RQ29) Orthodox Darwinian evolutionary theory envisages no "coupling" between the sources of variation and selection; mutational variation is seen as a fundamentally blind process untouched by any "precognition" of ultimate selective outcomes. In this sense, the evolution of methods by a process of rational rather than natural selection departs from the orthodox picture. . . . (165)

We are then scarcely prepared to read:

> (RQ30) The present theory provides a natural basis for combining a natural selection process at the level of *theories* with an epistemology of blind-variation-and-selective-retention at the level of *methods*. (160)

> (RQ31) [There is] . . . a crucial disanalogy between biological and cognitive evolution indicative of the quasi-"vitalistic" character of the latter. In biological evolution the mutations that actually arise fall across the entire spectrum of possible alternatives with equal probability, and so the direction of evolution is not determined by the direction of mutation. . . . In the case of cognitive evolution *viewed from the standpoint of thesis-acceptance*, the case is exactly opposite: the actualisation of possible mutant alternatives is probabilistically skewed, favourable mutations predominate, and the direction of evolution is governed as much by the inherent selectivity of mutation as by selection proper. But, as we shall see, in the cognitive case—unlike the biological—there is nothing occult about any of this, because one can embed the "vitalistic" features of epistemological evolution at the thesis level within an orthodoxly

> randomised and blindly unguided evolutionary model at the methodological level. (157)

This two-tier view reverses our expectations. Perhaps we should have been prepared for it by RQ19 and RQ20 above.

Evidently, there is an internal discrepancy in Rescher's account. While RQ21, RQ23, RQ27 and RQ29 take method to be rationally selected, so evidently with non-blind V and intentional S, RQ19, RQ30 and RQ31 hold that method V is blind. (They are vague about method S.) To further complicate matters RQ20 holds that method V is "to some extent" non-blind, and RQ29 continues:

> (RQ32) . . . to be sure it [the evolution of methods] calls for uncoupling [between the sources of variation and selection] in the *final* analysis, but in the first and second (and third, etc.) analyses it does not see a blind stumbling in the dark, but a groping in the dim half-light of the preestablished and *in some degree* prevalidated methodologies for handling analogous cases. (165–6)

Meanwhile, RQ30 has theory naturally selected while RQ31 asserts non-blind theory V and, supported by RQ19, presumably intentional S, that is, rational selection.

Rescher seems to have assumed that blind variation and intentional selection yield a mutually exclusive, exhaustive classification of VSR processes. But as we see, this is not so. Nor, given the developments in cognitive science (e.g., Holland et al. 1986 on the role of genetic algorithms) and a naturalist account of teleological capacity in species (see below), is there any reason to accept that natural and rational selection can be distinguished on this basis. (This will become clearer in the sequel.) Finally, there is the issue of making sense of RQ20 and RQ32 with their notion of internal self-guidance and only asymptotic disconnection between variation and selection. I suggest that these latter are the key and that the roots of the present confusion lie in Rescher's inadequate conception of regulatory systems and processes, already exposed.

Let us investigate further, beginning with some preliminary remarks on Lamarckism. Lamarckism refers to the inheritance of acquired characteristics, and its controversial essence is the postulate of some causal feedback to genetic structure from phenotypic structure causally acquired in interaction with the environment, feedback of a kind that causally results in transmission of a suitably modified genetic structure. So it would be the causal history of scientific ideas that is relevant, but Rescher offers no account of this.

Of course, Rescher intends content to be inherited—successful theses and methods are preserved—but in itself, this may just correspond to the persistence of the genotypes of survivors, part of the usual Darwinian process (cf. Hull 1988b, chapter 12). Nor does he offer any account of what either theses or methods "acquiring characteristics" might be. Rescher works entirely within the information representation (theses etc. as propositions), but modified theses and methods are logically distinct from their originals; how then to tell, by logical criteria, which are modifications and which are simply new theses and methods resulting from independent invention ("mutation")? This problem is, I suggest, insoluble as posed, because it is misposed; it arises from conflating a causal process (Lamarckism) with a formal, functional one.

However one is to clarify Rescher's account of Lamarckism, I see no reason to leap to Rescher's Lamarckist conclusions about selection. Evidently he moves directly from the intentionality (purposiveness) of belief modifications in scientists to the conclusion that the scientific process is Lamarckian. This move is too fast, even on the analogy account. For a Darwinian random (blind) variation process with ontogenetic constraints shows a similar teleonomy, since the only mutations that are presented for environmental selection are those that have survived the internal test of viable ontogenesis and so have in effect been screened for compatibility with the extant genetically incorporated environmental information. This is functionally similar to the scientific method of screening new proposals against accepted theories (as noted by the orthodox Darwinian Campbell 1974; cf. chapter 2). Of course, this does not touch any causal differences in the generation of the variations, but here too caution is needed. The division between Lamarckian and Darwinian processes is not wide and obvious; a causal regulatory systems analysis shows that between the usual extreme versions of each, there are many subtle intermediate positions possible (see chapter 5 and Hooker 1994d on Piaget). Certainly, *if* phenotypic characteristics are acquired intentionally *and* Lamarckian inheritance applies, *then* the overall process has a teleological character. But it is invalid to infer Lamarckian inheritance from purposive modification alone, or even from that together with a teleonomic character to the overall process.

I shall return to Rescher shortly, but first I want to enter a cautionary comment. On this doctrinal sea, there is a rationalist Scylla and a naturalist Charybdis between which one must steer. Scylla: If a non-naturalist account of reason is presumed, then any distinction between rational and natural selection will involve commitment to a rational/causal dichotomy at the basis of one's evolutionary epistemology. Popper presents a classic case of the impossible struggle to maintain continuity across evolution in the face of this dichotomy (chapter 3). A defensible naturalism must avoid this position. Charybdis: If an analogy account of evolutionary epistemology is presumed, then a naturalist must show that selection is the same in both biology and cognition. But this is difficult, since the regulatory contexts for each—indeed, for selection processes at differing places within each—are different. A defensible naturalism must also avoid this position. Mice have greater teleological capacity than caterpillars, and it shows, for example, in the complexity of environment-altering behavior, which in turn modifies selection pressures. The difference is greater still for the scientist purposively testing a new theory. I defend a naturalist account of reason (chapter 6) and replace the analogy account of evolutionary epistemology with an account of cognition as a continuous extension of biological evolution (no döppelganger genotype or selection). In this way, I hope to steer between these two unacceptable views. In particular, I see no reason to defend an exact parallel between biological and cognitive selection, so long as the latter is a natural(izable) extension of the former.

Returning to Rescher, an important part of the issue here concerns the regulatory loci at which purposiveness and blindness of variation appear and their significance for the nature of selection (see especially RQ32). But Rescher's position is obscure. He introduces his discussion of blindness of mutation with a critique of Popper's PEE (chapter 3). Since Popper rejects induction of any sort, search among alternatives must be both among all logically possible alternatives and be random. Rescher raises a cognitive version of the old Darwinian problem of the rate of evolution:

> (RQ33) But genetic mutations are of limited variety—unlike the possibilities for hypothesis formation where literally endless variations are available. In the cognitive case the timespan is just too short to account for the phenomenon of progress in terms of a blind trial-and-error groping amidst infinite possibilities. (151)

> (RQ34) Popper's theory thus faces a vitiating dilemma: he must choose between having the Darwinian selection process operate between all *conceivable* (i.e., theoretically available) theories or between all *proposed* (i.e., actually espoused) theories. If he opts for the second course—taking the (intrinsically surely more attractive and plausible) line that Darwinian selection operates

> with respect to the *actually* proposed and genuinely espoused alternatives—then the difficulty of accounting for substantial progress within a limited timespan becomes pressing. And it can be solved only if one grants man a capacity for efficiency in hypothesis-conjecture—a kind of inductive skill—so that those hypotheses actually conjectured are in fact likely to prove among the intrinsically superior alternatives. (154)

(Of course, with the first option the difficulty of accounting for substantial progress within a limited time span is equally pressing and a solution equally unavailable, and not just to Popper.) But "a capacity for efficiency in hypothesis-conjecture—a kind of inductive skill" is evidently what the historical accumulation of internal regulatory filters or selection bestows. This was Campbell's intention when he spoke of the "inductive achievements" brought about by blind variation and selection processes (Campbell 1974, 421). Speaking of a nested hierarchy of vicarious selectors, Campbell discusses as an example Salamander leg length:

> [The] . . . encompassing selection system is the organism-environment interaction. Nested in a hierarchical way within it is the selective system directly operating on leg length, the "settings" or criteria for which are themselves subject to change by natural selection. What are criteria at one level are but "trials" of the criteria at the next higher, more fundamental, more encompassing, less frequently invoked level. (Ibid.)

Indeed, Rescher himself is moved at one point to speak of methodological regulatory structure as a "'nested hierarchy' of heuristic processes" (160). But all of Rescher's discussion is in terms of the methodology of testing and application, methods that cannot be used to prescreen hypotheses in order to decide which of them are plausible candidates for testing and application. In fact, Rescher offers no clear account of this aspect of methodology. Yet without it, the primary locus for discussion of rates of evolution here has been removed. Further, it is reasonable to suppose that Rescher would adopt a fairly straightforward account along the following lines: All methods presuppose background assumptions or theses (section 4.VII above); candidate "mutant" hypotheses are simply assessed against general methodological requirements (e.g., non-ad-hocness and consistency) and for consistency/plausibility against the remaining background content of science, including all of those presuppositions of methods in use—and especially including the metaphysical component, which Rescher so rightly locates and stresses—as well as accepted theories and data. But then it is unclear why he draws so sharp a distinction between biological and cognitive selection.

Now let us consider blindness itself. Rescher considers rational selection to be only asymptotically blind (see RQ32). Whatever this is to mean precisely, the definition must take into account that biological variation is also screened for viable ontogeneses before they are presented for environmental selection. Darwinian theory requires blindness of variation, roughly that the range of variants presented for selection is not chosen in relation to some overall goal or purpose that determines success in advance but arises "blind" to the nature of the successful (selected) outcomes. But this is compatible with nested selection of selectors with the consequence that, so long as the environment is sufficiently stable, the probability that those mutations that can lead to viable phenotypic variations should in fact succeed and be entrenched will be greater than for mutations at large and the probability difference will increase as more information about the character of the environment is built into the genotype, and expressed in embryogenesis. Where then is the blindness?

The problem, however, is less absence of blindness than multiple ambiguity in the specification of blindness. Does blindness lie in the fact that the generator of variants produces variants that are random with respect to the current embryogenesis? Or does the generator produce variants randomly with respect to success in the current populational social structure, or in the current environment? (To the degree that the current embryogenesis, social structure, and reality are mismatched, it is not obvious that all three randomizations can be satisfied simultaneously.) Or is it that those variants favored by the current embryogenesis must be so favored independently of whether or not they will be entrenched in future environments? And why must this be so? Cannot a regulatory structure succeed in representing a long term environmental pattern (e.g., expressing a fundamental law)? Or does blindness require an assumption of a sufficiently dynamic environment, sufficiently rich that all regulatory structures are in effect ignorant of it?

Whatever the proper resolution of this complex and subtle issue might be, and one intuits that its obscurity may reflect the shortcomings of hitherto neo-Darwinian mathematical models that lack nonlinear self-organization, it suffices here to note the parallel with the case of science. For brevity's sake, and recalling my rejection of 'döppelganger' evolutionary epistemology (Introduction), take cognitive variations to be any regulatory change in the cognitive regulatory structure. Among all the possible cognitive variations few will successfully combine with remaining science to yield viable alternatives and be subject to selection through testing and application in the environment.[14] Blindness of variation for scientific development requires roughly that the range of cognitive variants presented for selection is not chosen in relation to some overall goal or purpose that determines cognitive success in advance (though variants are "locally" intended to achieve truth), but arises blind to the nature of the cognitively successful outcomes. However, given a learnable world, the probability that those variants that can combine with the existing structure (i.e., with existing method, theory and data) to produce viable cognitive variations should in fact succeed and be entrenched will be greater than that for variations at large, and the probability difference will increase as more information is built into the scientific regulatory structure.

Scientific examples of this process abound. Within physics, we have learned a great deal about spatiotemporal symmetry structures and conservation laws. Barring dramatic new evidence (cf. major environmental shift or radiation to a new environment), all new detailed physical theories will respect these symmetry constraints; otherwise, they would not be seriously investigated in the first place, physical theory being sufficiently entrenched in this respect to reject them as nonviable. The only exceptions would be those theories that challenge the foundations of physics itself, but as time goes on and data and workable lower-level theories superfoliate, viable foundational challenges become harder and harder to invent. (Nonetheless, this does happen through our increasing capacity to explore new environments, as the discovery of parity violation in high-energy particle interactions shows.) Similar remarks apply to methodology. A proposed new astronomical methodology based on the assumption that virtual images generated within optical telescopes give the true spatial locations of the objects viewed would be quite reasonably rejected without the kind of serious evaluation that would be given to other proposals (since, for example, it would require also defending a new optical theory for which there was no supporting evidence). And similar remarks apply across all of the sciences. Where then is the blindness?

Indeed, it seems that the problem is much more acute for cognitive processes. For surely the individual scientist acts rationally and with foresight in the pursuit of goals, so that every variation that is seriously introduced has been introduced with an overall goal or purpose in mind, namely to achieve truer theories and data, or at least to achieve theories and data that exhibit increasing amounts of the proxies for truth (explanatory power, technological applicability and reliability, etc.). Furthermore, surely these aims apply as well to the institutionalized processes through which science passes in selecting which theories and data will be accepted across the scientific community. An intentional process of scientific selection is surely better labeled artificial and rational rather than natural, and artificial rational selection processes are evidently anything but blind. Yet the best of intentions can be defeated; as the discussion of the optical proposal above reminds us, Galileo was in exactly the same position—he lacked a supporting optics, and even a dynamics (Feyerabend 1978a)—yet succeeded. The defeat of Newtonian mechanics after 200 years of success similarly attests: In classic biological fashion, new instruments and mathematical tools, largely based on its own insights, plus other "competing species" (electromagnetic theory) led to its demise.

Indeed, the parallel set of questions that appeared in the biological case can be regenerated for cognitive processes also, since it is just the introduction of a regulatory framework that leads to ambiguity in the question of blindness. Does blindness lie in the fact that the generator of variants produces variants that are random with respect to the current scientific regulatory structure? Or does the generator produce variants randomly with respect to practical success in the current institutional structure of science, or in the wider socioeconomic environment? (To the degree that the current regulatory structure, socially structured scientific processes and reality are mismatched, it is not obvious that these three randomizations can be satisfied simultaneously.) Or is it that those variants favored by the current scientific regulatory structure must be so favored independently of whether or not they will be entrenched in the future? And why must this be so? Cannot a cognitive regulatory structure succeed in representing a long-term environmental pattern (e.g., expressing a fundamental law)? Or does blindness require an assumption of a sufficiently dynamic environment, sufficiently rich that all cognitive structures must in effect remain ignorant of it? The scientific situation is evidently on a par with the biological one; there is no more reason to think blindness fails in the one case than the other.

This ambiguity concerning purposiveness and blindness is a feature of any teleonomic search device that utilizes nested selection processes; it does not require Lamarckism, contra Rescher, and it does not require any döppelganger cognitive genes or phenotypes. Nested nonblindness is quite compatible with blindness overall.

(This case has also been argued by Stein and Lipton 1989, who review the blindness debate.) A scientist may, acting purposefully, set out to develop a new theory to explain some phenomena, conduct a new experiment to confirm a theory, develop a new technology from accepted theory, and so on. But whether the resulting changes to the cognitive commitment structure will prove successful, will propagate across the population of scientists and become entrenched, is not under that scientist's control or indeed under the control of contemporary scientists collectively. Balmer did not set out to refute classical mechanics when he recorded the hydrogen emission spectrum, nor Lobachevsky or Riemann to invent General Relativity, nor de Broglie to develop the electron microscope, nor Rutherford to develop nuclear weapons, though in each case their work had a specific purpose in mind within the context or sub-system within which they were working. The same applies to phlogiston chemists and caloric engineers, though we no longer follow either theoretical notion.

Thus there is no case for deducing a rationalist Lamarckism from purposive selection per se in either biology or science. Lamarckism is an empirical matter and probably occurs rarely if at all (but see chapter 5). Stein and Lipton 1989 take the view that nested selection processes suffice to explain biological phenomena without violating Darwinian blindness and without appeal to Lamarckism. Moreover, recall that there would in any case be no difficulty in acknowledging differences between biological and cognitive selection processes within an overall naturalist regulatory framework.

This discussion has thus far left open the problem of how to theorize teleology, especially intentional pursuit of truth, naturalistically. There is space here to present only a brief sketch of a positive research program for resolving this large and complex problem. There is no claim to have a complete resolution to hand. First, naturalistic models of goal-directed behavior, or teleonomy, can be introduced in terms of well-known cybernetic models of "homing" feedback/forward systems (see Christensen 1992, Sommerhof 1974; cf. Piaget at PQ31, chapter 5). Second, we can understand naturalistically the capacity for logical reasoning in terms of the occurrence of logical models as the projective structure of visual and kinaesthetic geometry and similarly for mathematics generally (cf. Hooker 1988). The actions or operations from which logico-mathematical structures may be extracted can be quite elementary (see the discussions by Finkelstein and Giles respectively in Hooker 1979) and have been studied by psychologists, by Piaget and his followers especially. And Piaget himself provides the beginnings of a promising naturalistic model of how non-natural goals such as truth may come to be naturalistically formulated (see the discussion of ideal formation in chapter 5 below).

Taken together, these elements form the outline of a general theory of naturalistic cognitive teleonomy. Based on this, a plausible model of teleology can be given that distinguishes it *within* teleonomy, rather than *from* teleonomy metaphysically, as one with a particular complexity to regulation, namely the construction and use of internal representations of goals and actions (Christensen 1992). When this theory is wedded to the dynamical development notions discussed in chapter 2, we obtain the outlines of a theory of teleonomic development as increasing individual and social teleonomic complexity, a process that includes teleological development as a special case. But carrying this out in detail is a complex and subtle task (cf. Dyke 1988, Hooker 1991a, and chapter 2), which I do not pursue here. The preceding discussion has already shown how this naturalistic conception of teleology can be embedded within and support (phenotype-extended) Darwinian biological dynamics.[15]

There remains, contrary to Stein/Lipton 1989, the closely related question of the causal nature of the process of generating hypotheses (variations). This is essentially the problem of causally understanding creativity. Here I currently admit defeat, I have no satisfying naturalist model to offer. It seems clear, as Popper says, that neither a rigidly determinist nor an utterly random process does justice to our sense of the phenomenon. Popper coined the descriptor *plastic*, but did not solve the problem (chapter 3). Piaget too worried the problem, trying on the one hand to make a model of biological mutation that would make it more environmentally responsive and so more creative-like, and trying on the other hand to understand choice as sufficiently complex causally-realizable regulation (teleonomy); but no clear resolution emerges (chapter 5). Pursuant to the treatment of teleology, I believe that the latter route (choice = regulatory design) is the most promising one. But since I cannot at this time refer readers to a better model than Piaget offers, I confine myself to arguing that this issue should not prevent a naturalist account of reason from proceeding (chapter 6). Especially since the standard logicist/AI models of mind make creativity equally mysterious. So let us return to Rescher's overall account.

Rescher's position is that the key locus of variation and selection is that of methodology, whereas theories are methodologically screened before being put to test and application. But we have already seen that methods themselves constitute a complex regulatory structure—indeed, one that Rescher is forced to tacitly acknowledge (see RQ15). And that beyond what Rescher recognizes as methods lie the more general methods of pure mathematics and philosophy. From this point of view, therefore, the essential locus of progress lies in superfoliating the open-ended regulatory system of cognition, rather than any one move within it. Then just as a viable mutation is one that survives all the regulatory loci of screening, from those imposed by the genetic system and the stages of embryogenesis to those—called selection—imposed by the ecosystem, so a viable hypothesis or theory is one that survives all the regulatory loci of screening, from those imposed by the various levels of background metaphysical and methodological screening plus background facts and theories to the consequences of being tested against reality at large. The ground is now prepared to examine Rescher's account of scientific progress.

4.IX. Scientific Progress

Contrary to Rescher's view, the basic problem for Popperian methodology is not that it cannot account for the speed of progress, but that it cannot account for progress on its own terms at all (see chapter 3). The point is not that Popper cannot guarantee progress—any fallibilism must reject that requirement—but that within the constraints of his own account, Popper has no basis for rendering a claim to progress itself (fallibly) rationally acceptable. Rescher for his part surveys and rejects four alternative conceptions of scientific progress in terms of some notion of getting closer to the truth: (1) the Peircean attempt at a guarantee by defining truth just as that which science ultimately agrees upon, (2) the successive approximation idea that successor theories are better approximations to the truth than their predecessors (e.g., Popper's verisimilitude notion—see chapter 3), (3) a probabilistic enhancement theory, according to which later stages of science are more probably true than earlier stages (Carnapian inductive logic), and (4) an error-elimination view, under which successor theories reduce the distance to the truth by eliminating earlier errors. As Rescher rightly remarks, all of these conceptions tacitly presume some objective way to judge what is really true—but just this is what is at issue (cf. the Wheel Argument). If, following an evolutionary naturalism, fallibilism is systematically adopted, then there is no privileged access to truth. What then, Rescher asks, might be an appropriate account of scientific progress? Is any such account available at all?

Rescher's own answer is already clear from the foregoing: There is progress in increasing systematic internal coherence (among the components in diagram 4.4 above) and in increasing external pragmatic adequacy (i.e., technological applicability, predictive accuracy, etc.). But the special character of scientific progress lies in the increased rational warrant that the contents of science have when these two forms of systemic progress occur together in a mutually coordinated fashion.

> (RQ35) We thus envisage a two-fold criteriology of cognitive adequacy (and ergo of scientific progress), namely the internal factor of systematisation and the external factor of pragmatic efficacy. Yet while the theoretical factors of systematicity and comprehensiveness do figure indispensably as prime criteria of progress, it must be said that the whole structure of our discussion underpins the claim that this second factor is no less crucial to rational legitimation. . . . It is here that we find the ultimate guarantors for the cognitive procedure we indeed employ in the factual domain—the scientific method.
>
> On this dualistic approach to cognitive legitimation, it appears that the superiority of the scientific framework does not revolve only—or even primarily—about theoretical considerations (such as that of explanatory adequacy, or—more broadly—considerations of coherence in general), but also and particularly about the pragmatic issues of *problem-solving* and *control*. . . . In the final analysis, the credentials of science derive from a suitable coordination of strictly practical with purely theoretical considerations. And any advance in adequacy in these regards on the part of methodology places the theses validated by these latter-day methods on a cognitively firmer basis of enhanced warrant. (177)
>
> (RQ36) Improvement in the warrant for claims to "scientific knowledge" is always possible, and it is in terms of such improvements that the idea of a "scientific progress" which leads "nearer to the truth" must be understood. Enhanced adequacy in the grounding of an inquiry procedure certainly does not *guarantee* a "closer approximation to the truth" for its deliverances. All one can say is that it is reasonable to *presume* the truth of the more adequately grounded alternative. This

> greater reasonableness of presumption neither guarantees nor requires actual correctness. The very fact of more adequate grounding is of itself enough for its establishment. (181)

Notice that, despite all the quasi-rationalist talk (e.g., RQ23) this position is consistent with fallibilism.

> (RQ37) There is a vast gulf fixed between the sum-total of the informational *content* of the thesis that "P is true" and the set of checks and balances within our reach of operative control in entering upon a *rationally warranted presumption* of P's truth. An adequate basis of rational warrant stops well short of an unqualified guarantee. (182)

It is for the following conclusion that all Rescher's earlier and problematic work on regulative/substantive presumption has been prepared:

> (RQ38) Thus later applications do indeed—all our earlier counter-argumentation notwithstanding—provide "truer" results than the earlier ones, but on the basis of *regulative* rather than *substantive* considerations (to use Kant's very useful distinction once more). That is, we adopt the stance that one does not here augment the stock of available truth through *assured possession*, but only through *reasonable presumption*. (181)

Thus here Rescher comes down on the fallibilist side of the uneasy fallibilism/rationalism division running through his work. This does nothing to explain the mind-contributed, idealist component or the ultimate regulatory rational principles involved, but it does sit well with a naturalist approach to evolutionary epistemology.

And in my view, Rescher is roughly on the right track in his account of progress, but once again fails to appreciate the profound connection with biological evolutionary processes and, correlatively, mislocates the locus of progress. Specifically, Rescher is on the right track concerning an account of progress in three respects: (1) that it is the coordination of some kind of internal systematicity with external pragmatic success that is important to understanding progress; (2) that the ascent from theses to methods is important for developing a conception of regulatory structure to science within which this insight can be properly formulated; and (3) that the ground for warrantedly claiming progress does not lie in some privileged access to the real truth, but in an internal assessment within the historical human community itself.

Nonetheless, Rescher's own account needs modification because of the defects in his conception of regulatory structure and because of his underlying rationalist idealism. With respect to regulatory structure, a simple ascent from theses to method is not sufficient; rather, it is the superfoliation of the regulatory system itself that is the primary foundation for progress, rather than the achievement of increased capacity at any one or more specific loci (chapter 2). However, as in evolutionary circumstances, it is coordinated superfoliation that is required (cf. diagram 2.4 and text); research programs or sciences that are not successful at lower regulatory levels will scarcely survive long enough and/or well enough to open up the higher levels. But these modifications represent more an enrichment of Rescher's conception rather than a rejection of it. What does need to be rejected is the rationalist idealism mold in which Rescher casts his ideas and that truncates the recognition of the open-ended dynamics of science. But where Rescher seeks to distance his theory of science from biological processes (e.g., RQ28–31), I have been at pains to establish that a naturalist regulatory systems version is available. I hold that the same applies with Rescher's general theory of progress.

The key to a unified understanding is a quite general regulatory systems formulation of the scientific case. (1) Increased internal systematicity combined with increased access to higher regulatory levels yields increased adaptability, for example, an increased power to adapt science to an ever wider variety of circumstances. (2) This increased adaptability is then coordinated with an increased capacity for successful pragmatic control, for example, for successful specific adaptations within specific circumstances. According to Rescher's more limited version, it is the coordination of increased methodological adaptation with increased thesis adaptation that provides the secret to progress. According to the more general version, it is increasing adaptability, which requires increasing access to higher order regulatory capacities as well as increasingly well adapted specific methods and theses coordinated with these, that is the secret to cognitive progress.

4.X. Conclusion

I have tried to show in what sense Rescher's *Methodological Pragmatism* represents an important advance over earlier philosophies of science, at least from an evolutionary naturalist point of view. I have also tried to uncover its deficits and remove them by showing how it may be embedded in a more thoroughgoing naturalist evolutionary epistemology couched in the terms of regulatory systems theory. By adding Rescher's thus-renovated insights to those of Popper and Piaget, where it complements the absence in both of them of a theory of methodological dynamics, I have sought also to enrich the account presented herein.

Chapter 5

Regulatory Constructivism: Jean Piaget

Introduction

An immediate aim of this chapter is to elucidate the relationship between Piaget's genetic epistemology and the evolutionary epistemology espoused herein. This is a natural task, considering not only the influence of Piaget's ideas, but the centrality in Piaget's thought of a unified (self-)regulatory systems model encompassing biological and cognitive processes. The latter is expressed, for example, in his extensive use of the biological regulatory ideas of C. H. Waddington, which also have an important connection to the evolutionary epistemology espoused in this book. This leads to a second aim: to understand more deeply the nature of genetic epistemology by situating it within the framework of Piaget's unified regulatory systems model. But this leads into a third and more important aim: to relate Piaget's unified regulatory systems approach to that presented in this book and to see how his research helps to enrich the regulatory systems paradigm for cognition and epistemology in particular.

For many, these aims will not seem promising. Piaget's genetic epistemology seems to be clearly about individual development, while evolutionary epistemology seems equally clearly about public or population-level evolution; so the two are completely separate studies. Any attempt to relate them must conflate development with evolution and hence be committed to discredited Lamarckism. That is the standard line or received view (cf. Gillièron 1986). But this view quickly runs into difficulty. On the one hand, it is now widely acknowledged that evolution and development must indirectly interact via the environment (viz., interaction between embryo and environment to produce developed phenotypic capacities and interaction between thus-equipped phenotypes and environment to modify selection pressures), and this is compatible with orthodox neo-Darwinism (chapter 2). On the other hand, the received conception of genetic epistemology is forced to ignore most of Piaget's work on biology, to which he devoted many years, as wrong-headed, irrelevant, or both. I believe that we can do much better than this, indeed, that the received view is itself wrong-headed. The key to a better understanding lies in taking a dynamic regulatory systems approach. What comes out of the reexamination, I shall attempt to show herein, is not only a much improved understanding of Piaget's position, especially his biology, but of both genetic and evolutionary epistemology and a naturalistic approach to cognition.

Whether or not some version of genetic or evolutionary epistemology proves supportable, the aim of understanding Piaget is a worthwhile one for anyone developing a regulatory systems naturalism. There are, indeed, difficulties with both understanding what Piaget says and agreeing with all that one does understand. Not all that Piaget says belongs to the basic naturalist framework I shall delineate; there is a Lamarckian-formalist-rationalist strand as well, which I shall reject. Nonetheless, I contend that there is a unifying naturalist regulatory systems framework basic to Piaget's thought that provides a key to understanding his doctrines, by comparison to which the Lamarckian-formalist-rationalist strand is secondary and can be set to one side without losing what is central to Piaget's position.

It is important to be clear about what is being claimed here. The idea of connecting development to evolution, though suppressed until recently by a narrow and aggressive neo-Darwinism, is not new and is enjoying a deserved reaffirmation with the new nonlinear systems models as support (see Bonner 1982; Goodwin et al. 1983, 1989; Raff/Kaufman 1983). But there is a continuing tendency to create a dichotomy between neo-Darwinians as advocates of simple gene-centered stochastic selection models and a neo-rationalism that attempts to replace these factors with universal, indeed necessary, biological forms as the key to both development and evolution (see Smith 1992). Certainly Piaget can be read in these terms as belonging to the neorationalist side (see Kitchener 1986, 1987; cf. below). But first, there is no need to accept this dichotomy; without going to rationalist extremes, one can resituate neo-Darwinist processes within the complex systems framework, affirming still the development-evolution connection and the active role of the organism or phenotype in selection and criticizing genetic "atomism" and determinism (see chapter 2 and below). And second, I contend that just this kind of naturalist position captures Piaget's core position more faithfully and illuminatingly than does the neo-rationalist reading. Moreover, that basic naturalist framework turns out to provide an appropriate embedding for a regulatory systems evolutionary epistemology. And it also provides important initial bases for an appropriate self-organizing systems account of knowledge and reason.[1]

This chapter makes no pretence to offer an authoritative assessment of the whole of Piaget's work; to the contrary, it is selective and concentrates just on the aims stated. (The key works here are those in which Piaget explicitly discusses the notion of a genetic epistemology and its general theoretical foundations, and the references are confined to these but will guide the reader to the larger literature.) But since the issues are fundamental to both understanding Piaget and epistemology, I have some hope that this discussion will contribute something of value. An account of Piaget's genetic epistemology in regulatory systems terms appears elsewhere (Hooker 1994d); recounting the basic ideas is necessary if the reader is to follow this chapter's exposition, but the reader is referred there for most of the details pertinent to section 5.I.

Piaget describes his own position as *genetic epistemology*, a term coined by a prominent predecessor in the general tradition of a biology of knowledge: J. M. Baldwin.[2] Its orientation is dynamic:

> (PQ1) . . . if we wish to set up a truly scientific epistemology . . . we have only to ask, not "what is definitely scientific knowledge"—envisaged as a static, unchanging whole—but "how do the various forms of knowledge grow" . . . . (Piaget 1970/71, 69.)

(In what follows, all but some brief quotations from Piaget are numbered consecutively to facilitate later cross-reference.) Of particular interest here is Piaget's lifelong focus on a conception of living organisms as dynamic, constructive, self-regulating systems. He conceived of the whole of life as a multilevel interacting complex of such systems bound together, both within each level and across levels, by positive and negative feedback and feedforward processes. In a manner to be explained, each system in this grand cybernetic device strives for the stability of its own viability conditions (homeostases) and its own processes (homeorheses), importantly through developing improved ones when the former fail. In this way, each system develops increasing endogenous functional completeness, thus further propelling its own development/evolution and autonomy. In the process, environmental information is incorporated into regulatory design. For Piaget the design of the sensorimotor schema in the young child codifies the elementary group of spatial displacement operations, and so on through successively more complex structures to ultimately include all science, logic, and mathematics. The resulting functional changes, coupled with their grounding structural changes, form the system's adaptations to its environment. Thus cognition emerges out of biology. Piaget conceives of cognition as an extension of biological development, speaking of early sensorimotor intelligence as an embryology of reflexes and thought as the embryology of intelligence (see PQ15 below and Piaget 1970/71, 17–9). As noted in chapter 1, this phylogenetic approach to the origins of epistemic structure carries with it epistemic fallibilism.

From this perspective, the project of genetic epistemology is nothing less than to understand the whole of cognition, from the reflexes of the protozoan to the esoterica of philosophy, as an expression of fundamental regulatory processes.[3] Piaget pursued the development of this conception from his earliest biological and philosophical researches, which predate his psychological research, through his well-known research in developmental psychology to philosophy and returning full circle to biology in his later years.[4] Though some evidence for the fundamentalness of this regulatory conception will be cited below, even a casual reading of Piaget's reflective works will drive the point home. These conceptions structured Piaget's work throughout his life and account for his particular choice of the label *genetic epistemology*.

5.I. Piaget's Regulatory Systems Framework

5.I.1. Piaget's Developmental Psychology and Biology

For Piaget, the fundamental manifestation of regulatory activity is endogenous (internal) construction. Construction takes place in the context of organism-environment interactions. Through seeking to regulate themselves (always themselves—autoregulation) in relation to their environment, regulatory systems construct themselves and typically the environment around them as well.

> (PQ2) . . . there seems to be a common postulate of accepted epistemologies, viz. the assumption that there exists at all levels a subject aware of its powers in various degrees (even if these are reduced to the mere perception of objects); that there are objects existing as such for the subject . . . and above all intermediaries (perceptions or concepts) which mediate between the subject and objects . . . Now the first results of psychogenetic analysis seem to contradict these assumptions. . . . knowledge arises neither from a self-conscious subject, nor from objects already constituted (from the point of view of the subject) . . . it arises from interactions that take place mid-way between the two . . .but by reason of their complete undifferentiation . . . if there is at the start neither a subject in the epistemological sense of the word, nor objects conceived as such, nor invariant intermediaries, the initial problem of knowledge will therefore be the construction of such intermediaries: . . . they will develop in the two complementary directions given by the external and the internal, and it is on this twofold progressive construction that any sound elaboration of subject and objects depends. (Piaget 1970/72, 19–20.)

This epistemological conception is grounded deeply in Piaget's approach to biology as a whole. Piaget accepts the general Darwinian process of evolution, by variation, natural selection, and genetic retention but regards it as having a limited scope (Bickhard 1988, Chapman 1988). What he wishes to do is extend and modify this skeletal framework by attention to the capacities of the active organism, for example, in internally preselecting genetic mutations for viable embryogeneses and externally preselecting its environment through various modifications (see chapter 2 and Piaget 1976/78). Piaget conceives of this activity as "attenuating" the role of chance and amplifying organism construction, both internal and external (see below). He speaks approvingly of Waddington's similar emphasis (see Piaget 1970/72, 56–7, and Piaget 1976/78). Witness the centrality for Piaget of the notion of the phenotype as a self-regulating system:

> (PQ3) . . . the self-regulations exhibit all of the three following characteristics: they constitute the antecedent condition of hereditary transmissions; they are more general than these latter; and lead eventually to higher-order necessity. After all, regulations (with their feedbacks, etc.) are found at all organic levels, from the genome onwards. The latter includes regulatory genes as operants, and functions, as Dobzhansky has said, in the manner of an orchestra and not as a group of soloists (cf. polygeny and pleiotropism, that is, the many-one or one-many correspondences between genes and transmitted characteristics) . . . . It is therefore clear that certain regulations already condition hereditary transmission, and that they do this without transmitting themselves in the strict sense, since they continue to function. Now whereas the transmitted characteristics vary from species to species, and in cases from individual to individual, the regulations exhibit a much more general form. (Piaget 1970/72, 57)

Self-regulation is universal:

> (PQ4) These regulatory systems are to be found at all levels of the organism's functioning, from the genome up to the field of behaviour itself, and therefore appear to reflect the most general characteristics of the organisation of life. . . . [There are] regulations and equilibrium states observable at all levels of cognitive behaviour; self-regulation seems to constitute one of the most universal characteristics of life as well as the most

> general mechanism common to organic and cognitive reactions. (Ibid., 60–61).

Here cognition is certainly to be understood as an extension of biological process, not a separate realm (pace Popper): (PQ5) "To conclude, we wish to remind the reader that the constant aim of our genetic epistemology has been to show that the spontaneous development of knowledge has its source in biological organisations and tends toward the construction of logico-mathematical structures" (Piaget and Garcia 1983/89, 274.) But in these quotes, there is also again a disturbing mention of "higher order necessity" = logic + mathematics, which seems to clash with the basic naturalism. Piaget speaks of (PQ6) ". . . the two-fold construction of logico-mathematical and physical knowledge and above all the intrinsic necessity attained by the former . . ." (Piaget 1970/72, 52; see also later at pp. 68–76 and Piaget 1976). Section 5.II argues that what there is here of a rationalist strain in Piaget—and there is wider scope for a naturalist account than might be supposed—can safely be set aside without seriously altering his basic self-regulatory systems position.

For Piaget, the process of regulatory development meant that neither system nor environment, subject or object, could be thought of as the sole locus (cf. PQ2). He employs the regulatory model to distance himself from two great dichotomies. In biology, he rejects a simplistic dichotomy between extreme Darwinism and extreme Lamarckism (see Piaget 1970/72, 59–60), both sides of which, he argues, assign too passive a role to organisms. Indeed, so strong is his emphasis on active endogenous construction by organisms that in his phenocopy theory, he not only emphasized the idea that the regulatory requirements of viable ontogenesis act as a first selection environment for genotypes (which is compatible with, though an extension of, neo-Darwinian orthodoxy, see also chapter 4, section 4.VIII), but toyed with the quasi-Lamarckian idea that this same internal environment could apply some kind of directing force to genetic mutation. Hooker 1994d examines his phenocopy doctrine in detail and argues that the quasi-Lamarckian element is a minor, empirically testable, and if appropriate, excisable addition to a basically acceptable regulatory conception of selection dynamics. (It is an empirical, not an ideological, issue; see Cairns et al. 1988, Sarkar 1990, and in Tauber 1991.) In psychology, Piaget similarly rejected both pure externalism (behaviorism) and pure internalism (maturationism or preformationism) on the ground that the endogenous constructions are new, a product of organism-environment interaction and not preexisting anywhere (see Piaget 1970/72, 55, 61).

The positive content of Piaget's developmental psychology is too well known (and too vast) to warrant repeating here (see bibliography). Its outline structural form is that of a series of stages. At any given time or stage of development, input stimuli receive characteristic processings for their features currently held cognitively salient at that stage—a process that Piaget labels assimilation—and they result in a characteristic set of responses, both internal and motor. Each stage is marked by a temporary, one might say metastable, equilibrium in which the sensorimotor coordinations achieved suffice for the organism to cope with its environment, within the confines set by its biological maturation and the environmental information thus far incorporated.

But inevitably, there arise inputs that cannot be assimilated. Indeed, it is intrinsic to the nature of equilibria that this is so. The organism's capacities at a given equilibrium typically will make possible new explorations. A baby, having learned to crawl, can now crawl into new places. (PQ7) "Any scheme of assimilation tends to feed itself, that is, to incorporate outside elements compatible with its nature into itself" (Piaget 1975/77, 7; cf. p. 83). The extension of schemes is actively pursued by organisms using whatever powers of generalization and anticipation they currently possess (Piaget 1967/71, 206–11). Inevitably, these explorations ultimately reveal the limits of that equilibrium (see also section 5.II.3). The baby can crawl into tight corners or holes and so on from which crawling alone won't rescue it. Instead, the organism must accommodate to the new inputs, that is, must alter its current internal sensorimotor coordinations in order subsequently to be able to assimilate them. The process of bringing assimilation and accommodation into a new harmony or equilibrium that supports the organism's life is equilibration.[5] The passage to each new stage of equilibrated internal organization is thus initiated by organisms and marked by a breakdown in the current equilibrium, and the new stage is marked by a new complexity to, and extension of, sensorimotor activity.

> (PQ8) The central concept in our explanation of cognitive development (whether we speak of the history of science or of psychogenesis) is therefore that of successive improvements of the forms of equilibrium; in other words, of an "increasing equilibration" (Piaget 1975/77, 178.)

For convenience, I shall henceforth use *transequilibration* to refer to the whole process of increasing equilibration, hence to incorporate not only constructing a new equilibrium following breakdown of the old (equilibration), but also the active explorations that lead to breakdown of equilibria.

Sensorimotor coordinations require an intervening operational structure in the central nervous system. Characterized in the information representation (chapter 2), these structures represent contents, environmental information stored as operational structure.

> (PQ9) . . . knowing an object does not mean copying it—it means acting upon it. It means constructing systems of transformations that can be carried out on or with this object. . . . Knowledge, then, is a system of transformations that become progressively adequate. (Piaget 1971, 15.)

See also Piaget 1970. We may speak of concepts and representations (cf. Bickhard 1993, Hooker et al. 1992b). And so the framework of cognitive agency is manifested as the organism develops.

One of his latest works, Piaget and Garcia 1983/89, provides possibly the clearest summary of these developmental dynamics as they have been elaborated in psychological terms over the years. (Unsourced page references in the remainder of this section refer to this book.) Piaget and Garcia distinguish (A) instruments of, (B) mechanisms for, and (C) processes within, development (see 268 ff.). These notions will be reviewed *very* briefly.

A. The "instruments" of cognitive development are the assimilation, accommodation, and equilibration process already noted (cf. 234) and its cognitive equivalent: empirical abstraction, reflective abstraction, and generalization (empirical, completive, and constructive). Empirical abstraction (cf. 205–6, 270) consists in selecting some features of objects, situations, etc. as salient and ignoring others (cf. assimilation as signal from noise discrimination on stimuli).[6] Reflective abstraction (cf. 12, 172, 212, 270) comprises two activities: (1) relocating a set of operations, relations, etc. one level of abstraction "higher" (e.g., extracting a set of relations among sensorimotor schema and "reflecting" it as a set of relations among cognitive operations), and (2) reorganizing the reflected structure in the light of the richer set of possibilities available on the higher level of abstraction (reflection) (e.g., reorganizing Newtonian dynamical theory once the framework of LaGrangians and Hamiltonians was made available by reflection of action and energy relations at the level of specific dynamical equations to the level of general theory of differential equations).[7] These operations are complemented by not only mere empirical or numerical generalization but the constructive generalisation that completes the internal representation of the new abstract objects (see 172, 212, 270 on simple completion and 169, 270 on constructive completion).

B. There are two mechanisms by which cognitive development occurs. (1) The replacement of old epistemic frameworks by new ones, expressed in the rejection of old pseudonecessities in favor of more empirically and formally adequate ones (85–7), cf. the Newtonian rejection of the Aristotelian requirement that the natural dynamical state must be one of rest. (2) The transition from intra-entity features (qualities, relations) to inter-entity comparative relations and thence to trans-entity structures (133–40, 169–73, 182–4), for example, from internal properties of geometrical figures, to comparative transformation relations among figures (e.g., displacement and rotation), to abstract algebraic (group theoretic) and differential manifold characterizations of geometrical structures. Or again, the transition from Aristotelian motion as quality to relations among spatiotemporal features of motions (Galileo), to particular motions as transforms of initial conditions along state-space trajectories (Newton), to Hamiltonian characterization of generalized motions.

C. Finally, Piaget and Garcia identify a number of processes as characteristic of cognitive development by these instruments, exhibiting these mechanisms. (1) Thematization (65–6, 113, 273) makes an internal operation acquired in use the explicit cognitive object of other operations under reflective abstraction. (2) Differentiation and integration result from the cognitive ascent of the intra-inter-trans sequence, the former as richer detail becomes accessible, the latter because a minimal level of organismic coherence is essential (see 130–4). Internal differentiation, for example, eventually separates spatial from logical relations, unseparated in sensorimotor schemata (131) and, through increasingly rich (differentiated) cognitive frameworks for characterizing physical entities, also leads to a search for scientific objectivity (130–1). The third process characterizes the kind of differentiation involved: (3) constructing representational frameworks of increasingly differentiated dual necessities and possibilities. Again, the passage to Hamiltonian dynamics provides a good illustration, for there we have a powerful abstract structure—flows on a symplectic manifold—which characterizes the dynamics of any energy-conservative system; and so its theorems provide the structure of (nomological) necessities for such systems (e.g., that all state trajectories are unique, i.e., nonintersecting), and the selection of a flow (i.e., of an interaction Hamiltonian), and within a flow, of a trajectory (i.e., of initial conditions), characterizes precisely the structure of available possibilities. The most important process is

(4) the "seeking of reasons," which "means relating the results obtained to a 'structure' or coordinated schemata" (27).

> (PQ10) It now becomes possible to explain the general process, which generates the various forms of necessity as they develop and whose acquisition proceeds only gradually. The relations between elements proper to the intra level either have no necessity or, if they do, they achieve only very limited forms that are still very close to simple generality. Understanding states as resulting from transformations and, to this end, performing local transformations of elements—as is the case at the inter level—provides a first access to the necessary connections which intrinsically determine their own reasons. But transformations, in turn, require an explanatory rationale, and the search characteristic of the trans level—which leads to structures— is the response to this new need, since a total system of transformations generates new transformations and provides the reasons for their systemic composition. But it is clear that this "total" character itself remains relative and that the movement continues, as we have seen, among other things, with the transition from structures to categories of structures. In a word, the sequence from the intra to the inter to the trans levels is merely an expression of an identical process. On the subjective side it is the search for explanation; on the objective side it is the achievement of necessity, which always remains relative, but which increases constantly from one stage to the next. (Piaget and Garcia 1983/89, 169–70.)

This notion of necessity will be reexamined later (section 5.II); for the moment, note that this passage makes clear Piaget's emphasis on abstract structures and on their rational necessity in particular. Indeed, according to Piaget, each stage is so structured as to realize an operational representation of a member of some recognizable class of abstract structures—Piaget favors mathematical groups—such that the logical sequence of structures generated by successive formal enrichments of the most primitive structure generates the (abstract representations of) stages. It is in this formalist sense that Piaget explicates the requirement that each stage must suitably incorporate the previous stage and must also be a preparation for a successor stage in this same precise sense. This process gives a special status to logic and mathematics. The development of cognition is characterized by the construction of an internal operational structure adequate to experience in general, but the crowning achievements of that construction are the formal systems of logic, algebra, and geometry. These are held to describe the relations among operations themselves, operations of any kind whatsoever, operation on objects of any kind. They are therefore always strongly true of our operations (Piaget 1970/72, 64, 72, 74). Here we find Piaget's formalist, structuralist, rationalist side, in typically European Kantian mode. But the deemphasis of this aspect implicit in Piaget and Garcia's emphasis on the *process* of cognitive development and its "functional" determinants is more faithful to the naturalist fallibilist reading of Piaget's basic position defended here.

What this conception of cognitive development emphasizes is the constructive activity of the organism, and this is universal:

> (PQ11) Now, this interpretation of the three levels, in terms of exogenous, exo-endogenous (if one might use that term), and finally more and more endogenous truths enables us to give an acceptable sense to our efforts to uncover the common transition mechanisms between one level and the next (let us insist on this point), both in psychological development and the history of the sciences. . . . Since we are not interested in the contents of the developmental levels, but in their mode of construction, it does not seem to be more far fetched to compare the mechanisms involved in the sequence of stages in history to that found in psychological development than it is to look for common evolutionary mechanisms at vastly different levels of zoological phyla. (Piaget and Garcia 1983/89, 139–40)

This passage concludes by emphasizing the biological ubiquity of endogenous construction, typified by the phenocopy process Piaget investigated (see PQ14 and Hooker 1994d).

This is a search for endogenous completeness, the possession of a suitable group of responses for every environmental contingency. Self-reproducing completeness or closure of biochemical processes is a prerequisite of life; through its metabolitic processes, an organism must constantly reproduce its own form and processes, even while its specific substance is constantly changing (see Piaget 1967/71, 149 and passim). Closure conditions in themselves, while still a challenge to empirical understanding, are not Piaget's focus and I assume them in what follows. Rather, it is the capacity of such self-reproducing systems to develop that is important for Piaget, since only then can they adapt so as to maintain their self-reproduction. Indeed, development through transequilibration is always aimed at improving endogenous completeness, it is a search for more functionally adequate closures. This remains true from cellular immune system response to the esoteric reaches of cognitive abstraction.[8]

Transequilibration is associated not only with increasing universality (completeness), but also with increasing "necessity" and objectivity and the construction of normative ideals (section 5.II.3 below). And organism autonomy thereby increases. Each successive equilibrium is an improvement over its predecessors in autonomy, since input is regulated in relation to output such that the organism's homeostases and homeorheses are maintained across a wider range of environments. The organism becomes more independent of its environment, more autonomous. And the richer its internal operational structure, the more it has learned about the systematic way to improve equilibria, and so the more its subsequent accommodatory efforts are directed by this learning, that is, internally directed (Piaget 1967/71, 206 ff.). In the framework provided by chapter 2, section 2.II, the organism has made epistemic progress, through regulatory refinement (assimilation, accommodation) and ascent (reflective abstraction and generalization) Roughly, superfoliation = transequilibration (within science).[9] As well as increased effectiveness, this represents a further reinforcement of autonomy. Thus completeness/universality, normativity (here, truth seeking), objectivity, necessity, and autonomy all codevelop. The overall effect is of a flourishing life, the flourishing life of a developing regulatory system.

Over all, Piaget's general conception of psychogenesis has proven a very fruitful stimulus to psychological research and one supported *in a general way* by a large range of empirical research. It is possible to challenge the details that Piaget suggests for each of his stages, or the rigidity of their sequencing, or even whether there are true stages rather than a more complex, nonlinear pattern of development. Boden (1979) holds that although Piagetian doctrine is usually roughly correct, Piaget's psychological theorizing and experimentation is too vague and nonspecific and that sharp Piagetian stages often evaporate under careful scrutiny (see also Brainerd 1978; Siegel and Brainerd 1978; but cf. Chapman 1988). All of this may be accurate, but it doesn't touch the overall regulatory process conception, which is the heart of Piaget's position. These criticisms can be regarded as aimed at the first approximate, simplified version of this approach. (See further below.) Fortunately, there is no need to pursue these issues to a detailed resolution here in order to investigate the lessons that Piaget's doctrines might hold for the development of a fruitful biologically based theory of knowledge and reason, although it is essential that the overall or general regulatory conception to be delineated in the following sections be empirically correct.

5.I.2. The Structure and Scope of Genetic Epistemology

5.I.2.1. Genetic Epistemology as Process. The basic conceptual framework for Piaget's genetic epistemology is the regulatory conception derived from his biology (section 5.I.1). Piaget defines his genetic epistemology as follows:

> (PQ12) Genetic epistemology attempts to explain knowledge, and in particular scientific knowledge, on the basis of its history, its sociogenesis, and especially the psychological origins of the notions and operations upon which it is based. . . .
>
> The fundamental hypothesis of genetic epistemology is that there is a parallelism between the progress made in the logical and rational organisation of knowledge and the corresponding formative psychological processes. Well, now, if that is our hypothesis, what will be our field of study? Of course the most fruitful, the most obvious field of study would be reconstituting human history—the history of human thinking in prehistoric man. Unfortunately, we are not very well informed about the psychology of Neanderthal man or about the psychology of *Homo siniensis* of Teilhard de Chardin. (Piaget 1971, 1, 13)

Genetic epistemology concerns the origin and dynamics of knowledge formation and includes public scientific knowledge as well as individual knowledge. It draws some connection between psychological development and the historical development of public knowledge, a cognitive version of a connection between ontogenesis and evolutionary history. But we are given little idea of this connection; there is only baffling talk of a "parallelism." (*Whose* "logical and rational organisation of knowledge"? That of the individual? Our species? Science?) And the later remarks are also puzzling, considering that Piaget's lifetime work was not on phylogenesis and the evolution of mental function, but on the developmental psychology of contemporary children, and to some extent on scientific development. They evidently point to an extension of the parallelism between psychological and historical development; Piaget evidently expected the evolutionary prehistory of *Homo sapiens* to be reflected in each individual ontogenesis in some way.[10]

However we are to understand this passage—and I shall return to make specific sense of it in section 5.I.2.2—this is still too narrow a conception of the relationship Piaget evidently has in mind, for it ignores his lifetime emphasis on biology. Writing in 1976, he says:

> (PQ13) . . . my efforts directed toward the psychogenesis of knowledge were for me only a link between two dominant preoccupations: the search for the mechanisms of biological adaptation and the analysis of the higher form of adaptation which is scientific thought, the epistemological interpretation of which has always been my central aim. (Piaget, Foreword to Gruber/Voneche 1977)

In many places, Piaget writes as if he aimed at uncovering quite general diachronic regulatory processes and laws, for example, that the transequilibration process applies from evolutionary development to scientific development and to both individuals and communities (populations considered as regulatory systems); see PQ3, PQ4 and:

> (PQ14) If our interpretation of the phenocopy were to prove valid, it would allow us to furnish a common answer to the classic doctrines of both neo-Darwinism and behaviourism. This answer would be that the environment in fact plays a fundamental part at every level, but as something to be overcome, not as a causal agent of formation. . . . On the one hand this would simply mean that conquest of the environment, besides being considered an extension of the basic assimilatory tendency of life, usually begins with simple trials by phenotypic accommodation or by empirical knowledge. On the other hand it also means that, by virtue of the internal requirements of equilibration, these trials will subsequently give rise to more secure forms of assimilation. These in turn would be arranged in ascending degrees over every level of development, beginning with that of "genetic assimilation" . . . and ultimately attaining the various levels of cognitive assimilation, including those of scientific thought. (Piaget 1974/80, 79)

See also Piaget 1950, 1970, 1971. This aim is reinforced by Piaget and Garcia, 1983/89; cf. PQ11 above, which explicitly emphasizes the importance of the search for universal developmental *processes*, even when the developmental *products* vary.[11]

So we should look for a wider relationship, one that encompasses biology as well, relating historical change to individual development. And the primary relationships will be those among the regulatory processes involved (rather than among the products of those processes). The place to begin is with the connection between the biological and the cognitive. For Piaget, psychogenesis is an extension of embryogenesis; for example,

> (PQ15) . . . child psychology certainly constitutes a kind of embryology of the mind, both in describing developmental stages in the individual, and especially in studying the actual mechanism of this development itself. Developmental psychology [psychogenesis], moreover, represents an integral part of developmental embryology [embryogenesis]. . . .(Piaget 1970/71, 17–8.)

The initial idea here may in turn be expressed by affirming a mapping, E_{op}, from ontogenesis to psychogenesis, such that the latter is in some proper regulatory sense an extension of the former. A front end loader extends the shoveling process to new domains by scaling it up; a symbolic labeling and division process for economic property extends the literal physical division process by abstraction. What is intended here is that, at some sufficiently general level of specification, the regulatory processes that account for the embryological developmental modification of the organism continue to account for its psychological modification. The general form of those processes is to incorporate environmental information into regulatory structure at key junctures so as to achieve transequilibration, that is, homeorhesis of widening equilibria (cf. section 5.I.1 above). The mapping E_{op} can be given both a causal and a complementary informational (functional) reading (see chapter 2, section 2.I.1); for brevity I shall continue to refer to both readings ambiguously throughout the discussion of maps to follow.

Note that while the continuity of the process is primary, there is no requirement that the detailed processes be exactly the same at every developmental stage. What continues unchanged, presumably, is some generically specified process, functionally explained by invoking sufficiently high-level regulatory processes that account for transequilibration. Within the operation of these, the specific regulatory dynamics can be expected to be continually changing as the regulatory structure of the organism is built up (or with aging, degrades). This is characteristic of self-organizing nonlinear dynamic systems. A map from earlier to later states in such a genetic sequence would leave continuing basic regulations invariant while mapping simple initial relations and structures on to later more complex relations and structures (cf. leaving basic economic rules invariant while mapping simple initial products and interaction rules on to later more complex products and interaction rules).

Correspondingly, there is no requirement that the products of a process remain the same across its extension. There must be certain kinds of continuity of features of products; for example, the modifications of neural interconnections, which are the causal instantiations of adult belief formation, are not different in generic kind from those occurring during embryogenesis of the nervous system. But there are many equally important differences of detail between the two kinds of products, befitting their differing positions in the regulatory sequence. Piaget, however, wanted to discuss both processes and products, which can cause unnecessary difficulty (see 5.I.2.2 below).

Piaget holds that phylogenesis maps environmental information into genetic structure and expresses it (indirectly) as regulatory phenotypic structures. Suppose we add to this the idea that the essence of the process of science is also to provide a (systematic) mapping of environmental information into a regulatory structure. From this perspective, the development of public knowledge—let us call it cognogenesis—is a regulatory extension of phylogenesis. (PQ16) ". . . consider epistemology . . . as a theory of intellectual evolution or of the adaptation of the mind to reality. . ." (Piaget 1970/71, 18). It seems quite clear from passages such as this, PQ5, PQ11, PQ13, and many others that Piaget holds a view of this kind, in fact a conception similar to that already stated concerning the ontogenesis-psychogenesis mapping: At some sufficiently general level of specification, the regulatory processes that account for the differentiation of communities or populations (including speciation) continue to account for their socio-cognitive modification. Compare, for example, the adaptive radiation and specialization of species with that of scientific specialists. The general form of those processes is to incorporate environmental information into regulatory structure at key junctures so as to achieve expanded equilibria (cf. section 5.I.1 above). This idea may be expressed by affirming a mapping, E_{pc}, from phylogenesis to cognogenesis, such that the latter is in some proper regulatory sense an extension of the former.

Evolutionary epistemology expresses one form of just this view, since it postulates some interesting regulatory commonality between phylogenesis and cognogenesis (see chapter 1, section 1.II, and section 5.I.3 below). In the present terms, an evolutionary epistemology postulates that E_{pc} preserves some processes in common, that is, these processes should be invariant under E_{pc}, most commonly, variation, selection, and retention. Piaget's regulatory systems position can be summarized in the doctrine that the regulatory extension primarily preserves the process of transequilibration through the incorporation of environmental information into regulatory structure. This applies to both the ontogenesis-psychogenesis process and the phylogenesis-cognogenesis process giving them, and hence also E_{pc} and E_{op}, same general regulatory form.[12] The E maps make single processes out of phylogenesis - cognogenesis and ontogenesis-psychogenesis, whence label them respectively phylocognogenesis and ontopsychogenesis. Like evolutionary epistemology, the E maps are primarily concerned with the regulatory processes themselves, not their products (though there must again be some suitable relations among the products), and are compatible with variations in the regulatory details as the phylogenetic or cognogenetic system is built up. That is, as life and science evolve, their regulatory character may change at various levels compatibly with the overall process proceeding.

Two mappings have now been introduced. Piaget himself introduces a relationship between cognogenesis and psychogenesis that represents a third map, H_{cp}. It is central to Piaget's specification of genetic epistemology (PQ12 above) and is clearly crucial to his entire approach (see PQ8). These maps suggest, what Piaget's remarks confirm (PQ12, PQ13), that we consider a fourth and final mapping, H_{po}, which relates phylogenesis to ontogenesis.[13] The completed regulatory structure is given in diagram 5.1. Each of these four maps is subtly unique. The E maps are synsystemic (operate at the same system levels) but express diachronic relations, while the H maps are synchronic but express diasystemic relations (relations across system/subsystem levels). Though the E maps operate at different system levels to each other, they both express extension of regulatory process, because there is direct continuity of causal modification processes.[14] Though the H maps operate at different regulatory levels, they both specify similarity of regulatory process between individual and community or population system levels.

My proposal is that *diagram 5.1, read in terms of relations among regulatory processes, represents the basic regulatory conception with which Piaget works*. Various passages from diverse parts of Piaget's corpus can, in isolation, be taken to support narrower readings, some fragment of diagram 5.1, but it is the structure as a whole that I suggest makes best overall sense of his position.

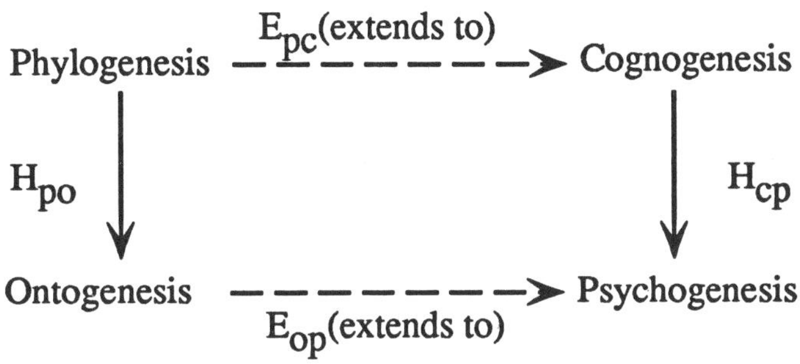

Diagram 5.1 Piagetian Process Homologies
Arrows indicate postulated dynamical relationships among the processes constituting biological evolution/development (phylogenesis, ontogenesis) and cognitive evolution/development (cognogenesis, psychogenesis). The E relations extend a biological process into the cognitive domain; the H relations extract dynamical similarity between processes at different system levels.

Against this background, a natural construal of "the fundamental hypothesis of genetic epistemology" (see PQ12) is that "there is a parallelism between" the phylocognogenesis and ontopsychogenesis processes, where parallelism is now explicated as the claim that H_{po} and H_{cp} exist, are at least homomorphisms (specifying commonality of regulatory processes) and are essentially the same homomorphism. (To express this latter idea, let us add to diagram 5.1 a mapping $H_{po:cp}$, from H_{po} to H_{cp}. The simplest form for this map is an isomorphism, which is also the current proposal.) In effect, *genetic epistemology affirms the entire regulatory structure of diagram 5.1*. This seems to me the right reading of Piaget, and it gives genetic epistemology a universal scope, it is concerned to uncover the regulatory processes beneath the generation of all forms of correctable information storage wherever it occurs. This explains (cf. section 5.I.1 above) why it can neither be reduced to developmental psychology nor to philosophy. Indeed, if there is a difficulty with this position, it concerns its defense rather than its attribution to Piaget. In PQ14 above, see how quickly Piaget moves from the phenocopy phenomenon to all regulatory processes; the obvious problem is not whether Piaget aims to develop a unified regulatory conception, but whether the evidence supports its existence.[15]

To repeat, the basic framework proposed here is that of system(at)ic relations among regulatory processes themselves. Piaget's whole emphasis is on this, for example, the importance of the transequilibration process and the intra-inter-trans succession (section

5.I.1 above). Consider again the quote PQ13 above, the one that sent us looking for a biological extension of the cognitive relations. It is concerned with the *mechanism* of adaptation, that is, with the regulatory processes involved in adaptation; it is not focused on the products of that process. This is true too of Piaget's work on phenocopy; it is not the particular biological instances of phenocopy that are important to him—these are relatively rare—but the regulatory process itself, because it may be generalizable to other situations.[16]

The regulatory structure presented in diagram 5.1 provides a satisfyingly unified and comprehensive initial framework for understanding both natural cognition and Piaget, but it is to be given a carefully naturalist reading. For this, two matters need further comment, it must be read in terms of processes, not products, and it must be read as a first-order approximation, not as specifying a set of necessarily universal forms. I turn to these in order.

5.I.2.2. Piaget and Products of Processes. But now what of the products of regulatory processes, for example, the anatomies of organisms or the cognitive structures that make up Piagetian stages? Are there not relationships among these to be read off diagram 5.1 as well? Did not Piaget often read the relationships in exactly this way? Did he not place great store by generating the cognitive sequence of stages from a consideration of the logical relations between their contents?

It certainly is possible to consider the products of the regulatory processes, and the relations among those products, in light of the relationships asserted in diagram 5.1, both at the biological and cognitive levels. As noted, these relationships are likely to be very complex. Any attempt to read diagram 5.1 as directly specifying a simple set of relationships among the products themselves leads to immediate difficulties.

One kind of problem stems from the argument that any H relations among products (not processes) should lead to some version of the claim that ontogeny recapitulates phylogeny (cf. H_{po}): because of accumulating ontopsychogenetic constraints the products at the individual level should recapitulate the history of products at the population level, and vice versa wherever learning and social interaction are important. Piaget sometimes seems to accept some such "biogenetic law," but mostly rejects it (see Kitchener 1986, 156–7). In his now classic study of the issue, Gould 1977 makes out a strong case for the view that, while a rigid recapitulationist thesis cannot survive scientific examination, a weaker more generalized version emerges from a new appreciation of the role played by heterochrony in the evolutionary process. (Heterochrony is the relative developmental acceleration or retardation of sexual against nonsexual features in ontogenesis.) What emerges from Gould's study is the importance of the detailed interconnections between phylogenesis and embryogenesis (more generally, ontogenesis) for the understanding of evolutionary processes. In any event, we may consider that Piaget is committed to at most some broad reading of the relationship consistent with Gould and that genetic epistemology as characterized in diagram 5.1 does not require any further product relationship.

Another kind of problem stems from Piaget's tendency to characterize cognitive products solely in terms of formal structure and contents, with the consequence that their inter-relationships become specified independently of the regulatory processes that generated them (indeed specified timelessly). Piaget especially employed this latter approach to characterize stages, with many attendant difficulties. (And he did so even while he inveighed against empiricist and rationalist epistemologies for obscuring the processes of construction that lay behind their contents!) But why should Piaget adopt a position of this kind, one apparently so different from his basic approach?

Various answers are possible. Two of the more plausible are: (1) He may simply have unquestioningly adopted the mode of analysis of the historico-critical school with which he grew up (note 10); and (2) he may really be a rationalist for whom naturalism was a sometime convenient strategy. Whatever the roles of such factors, a different basic answer is worth considering: (3) because it is as yet too empirically difficult to study the regulatory processes directly.[17] When looking at carefully, a reader of Piaget's works will find that there is much more attention given to the nature of the regulatory processes than concentration on formal stages would lead one to suppose (cf. PQ2, PQ5, PQ15, even PQ8). If the primacy of process is the right general reading of Piaget, then this last methodological explanation makes best sense of Piaget's approach.

A principled (as opposed to pragmatic) reliance on formal relations instead of dynamic ones seems both empirically and formally dubious (see Apostol 1982; Brainerd 1978; Kitchener and Kitchener 1981; Vuyk 1981). It also invites criticisms of the sort Fodor 1980 and Haroutunian 1983 make, namely, that Piaget doesn't offer, and can't offer, any serious learning theory. He certainly doesn't offer the kind of formal programming theory Fodor and Haroutunian have in mind, but from a naturalist regulatory perspective, the primary place to locate a learning theory is in the basic regulatory processes. Learning corresponds to regulatory superfoliation (chapter 2), especially ascent, opening the way to develop a naturalist regulatory learning theory.[18] This is evidently what Piaget also wanted to do when he sought to generalize biological regulatory processes (cf. Piaget 1980). Transequilibration is certainly a useful beginning on a regulatory account (cf. Kitchener 1987), though it only provides a bare skeletal form for thinking about models of learning processes. But then his commitment to formal relations hamstrung him. These are complex issues for which the reader must be directed elsewhere (see e.g., Bickhard 1988 and note 36 below and Hooker 1994d, e for further discussion).

Thus a purely regulatory process reading of diagram 5.1 forms the primary framework of genetic epistemology. With the jettisoning of any simple requirement of correspondences among the products of these processes (i.e., among stages as contents), especially any necessary formal one, all the worst difficulties facing Piaget are avoided. Each sequence of states (we may even refrain from a sharp notion of stage) at some given time and regulatory level may need its own detailed explanation in terms of preceding history, current environment, and regulatory features. Insofar as commonalities emerge, well and good, but the real commonalities will be at the level of "mechanisms" or regulatory processes, namely the set of similarity relationships among regulatory processes presented in diagram 5.1 together with whatever more determinate detail is to be added following detailed investigation. We have then a conception of life as a complex of dynamic, interrelated regulatory processes that over time map environmental information into regulatory processes and subserving structures in specific ways, and of cognition as an extension and leading edge of that overall process. This is the fundamental conception that characterizes Piaget's work.

From this perspective, we can now return to make some sense of Piaget's otherwise rather obscure remarks on researching genetic epistemology contained in quote PQ12 above. For as Piaget implies, the investigation of genetic epistemology, thus understood, would require first examining independently the phylocognogenesis and ontopsychogenesis processes and then comparing the results to examine the H relations. Unfortunately, as Piaget notes, only a fragment of this task is really empirically accessible to any substantial degree. Piaget then opts to work on the accessible portions, chiefly on ontopsychogenesis (E_{op}) and on H_{cp}. By understanding the regulatory dynamics underlying E_{op} Piaget evidently hopes both to render this unified approach to biology and cognition more plausible and, by uncovering the specific regulatory processes involved, to gain insight into the character of E_{pc} via H_{cp}, and perhaps also render phylocognogenesis more amenable to (at least indirect) investigation. On the other hand, to the extent that $H_{po:cp}$ is either assumed to be an isomorphism or can be otherwise known, then H_{cp} can be studied as the accessible but representative fragment of the whole relationship. We have empirical access to both psychogenesis and the development of scientific knowledge, and evidently we have access to them moderately independently of one another. Note that the specification of the aim of genetic epistemology with which PQ12 opens puts particular emphasis on scientific knowledge (i.e., on developments well within historical time. This approach then makes methodological good sense out of Piaget's approach and lifetime's work. (Hooker 1994d shows how it also makes sense out of other of Piaget's discussions, for example, why a study of embryogenesis would be of value to epistemology.) It is essentially only by beginning with this kind of regulatory framework that one can fully understand Piaget's position.[19]

5.1.2.3. Processes and Universal Form. Diagram 5.1 and text provides only an approximate or first-order theory of the maps. Ontogenesis, for example, creates increasing systemic levels as it builds individuals; in some sense then, it operates across systemic levels. Once it is recognized that individuals and communities or populations are separated only by regulatory degree and design (chapter 2), and once the complex feedback and feedforward relations between communities/populations and their individual members via their environment is fully represented (see chapter 2), there is no clear sense in which the H maps could be strictly synchronic or the E maps neatly separated. But to work out the details is beyond the scope of this discussion and really beyond the scope of current theory, both because the fundamental concepts of self-organizing systems theory are not yet agreed and because the empirical information required far exceeds our present knowledge.

Further, as it stands, part of the reading of diagram 5.1 includes $H_{po:cp}$ as an isomorphism. But since there is every reason to believe that the regulatory processes may change in detail as regulatory structure is built up, this cannot be assumed as more than a first approximation. (This prospect has already been alluded to.) Piaget himself acknowledges this when he acknowledges important differences, though not of a fundamental kind, between cognition and all other organs (see Piaget 1974/80, 1976/78) and specifically when he introduces the distinctive role of reflective abstraction. A simple thermostat, for example, does not exhibit the same detailed regulatory processes as one equipped with a control system that corrects for oscillatory overshoot (and that fact shows in its sequences of states and its performance). Similarly, a cognitive system equipped with a rich methodological structure as a result of its history will not employ the same detailed assessment procedures as one equipped with only primitive random trial generators and all-or-nothing refutation/acceptance gates. So we should be prepared to allow that the extension relations E_{pc} and E_{op} are not identities and therefore that H_{po} is not the same as H_{cp} (i.e., $H_{po:cp}$ is not an isomorphism). Rather, it too will presumably be a general homomorphism (and perhaps not that if, for example, we find that advances in public scientific knowledge and its allied technological products permit individuals to skip cognitive stages in their intellectual development). It will be for future science to specify these regulatory processes in detail.

In contradistinction to this analysis, Piaget sometimes speaks as if he were aiming to specify universal, perhaps necessary, forms for these processes. Even the recent Piaget/Garcia 1983/89, which emphasizes process over product and empirical inquiry over rationalist structure, still speaks about the intra-inter-trans progression in ways reminiscent of this approach. Nevertheless, it is clear that a naturalist approach to dynamic processes requires its excision, and as noted, there is also support for that in Piaget's writings.

5.I.2.4. Non-naturalist Interpretations of Piaget. The naturalist regulatory process construal of Piaget's position is certainly not the only possible one, the rationalist strands in Piaget's thought may be emphasized instead. By rejecting the search for necessary product structures in favor of processes and then rejecting necessarily universal processes, these rationalist strands have been stripped off to reveal the core Piaget. (This removes him from the simplistic neo-Darwinist/neo-rationalist dichotomy Smith 1992 sets up, see introduction above.) While rationalist readings can find textual support in Piaget, the basic issue for rival interpretations is not merely to point to rationalist strands in Piaget, but to be able to argue a convincing *core* rationalist interpretation. We have seen that this is not so easily done, and this conclusion will be further argued in section 5.II below. In preparation, two other initially plausible interpretations of diagram 5.1 will be briefly considered here.

(A) The fundamental thesis of genetic epistemology is just H_{cp}, in keeping with Piaget's emphasis on knowledge.[20] But then Piaget's emphasis on E_{op} would remain methodologically irrelevant to establishing a genetic epistemology. His concern with finding processes that encompass biology as well as cognition would be difficult to understand, as would his looking to prehuman psychology as his first choice of focus. Finally, his less intensive and more qualitative and philosophical study of science becomes a problem to explain. No attempt is made, for example, to analyze any specific theoretical revolution, the most obvious cases of complex regulatory transitions.[21]

(2) Suppose, picking up on Piaget's talk of "levels" of regulatory structure and "analogies" among them (see PQ4), we attribute to Piaget the (at least) quasi-dualist idea that thought is epiphenomenal to, or supervenient on, brain operation. Elsewhere (1963/68), Piaget speaks of a parallelism between causal chains at the physiological level and logical ("implicative") chains at the conscious level,.and this parallelism is related to a parallelism between mathematics and physics. Taken at face value, such talk presents a straightforward dualism, in fact a double dualism, of mind and body and of natural and formal-symbolic worlds. In this setting, the fundamental thesis of genetic epistemology might be given an entirely individualist reading: the progress made in the logical and rational organization of knowledge of each individual parallels (dualistically) their embryogenesis.

This reading makes sense of the emphasis on E_{op} and the relative neglect of science. Better; it makes one kind of sense of that tendency in Piaget to be interested in modern science only to the extent that he can connect it to (universals in) individual psychology. But this alternative construal does no better at understanding the initial selection of prehuman psychology. And of course it must brush aside, or collapse, Piaget's many discussions of the evolutionary process and his evidently quite naturalist approach to regulatory processes (see almost any quote and reference above). And though we could avoid it by adding an objectionable third dualism between normative and descriptive and by confining the normative to an abstract symbolic world (cf. Popper), this interpretation also neglects Piaget's equal concern with science as a system of public contents.

Neither of these two otherwise unattractive rivals to the interpretation I suggest seem to me to do overall justice to Piaget's written commitments. I can find no other plausible readings of Piaget's doctrine of genetic epistemology.

5.I.3. Genetic Epistemology and Evolutionary Epistemology

If section 5.I.2 presented the right understanding of what genetic epistemology is, then its relationship to evolutionary epistemology can be clarified. Evolutionary epistemology postulates some interesting regulatory commonality between phylogenesis and cognogenesis (i.e., some version of E_{pc}). (This already rules out what Bradie 1986 calls evolutionary epistemology of mechanisms, which accepts evolution of the brain but prescinds from any account of brain function, as too weak to form an evolutionary *epistemology*, and it equally rules out what Hahlweg/Hooker 1989a call the bioepistemology of the Lorenzians, who seek selectionist defense of a quasi-Kantian epistemology, as not offering a truly *evolutionary* epistemological dynamics.) There is widespread disagreement about the commonality (Chapter 1, section 1.II): recall that evolutionary epistemologies can be conveniently grouped according to the degree or strength of the commonality affirmed and that, in order of increasing strength, Selective Analogy, Extended Analogy, Phenotype Extended Analogy, and Embedded theories were specified.

We are now in a position to discuss the relation of evolutionary to genetic epistemology. Against a background assertion of E_{pc} and E_{op}, genetic epistemology asserts the existence and dynamical relatedness of H_{po} and H_{cp}, all relations read in terms of regulatory processes rather than products (refer to diagram 5.1). Thus in general, evolutionary epistemology offers a version of one of the component relations characterized in genetic epistemology. But whether E_{pc} is the only relation involved and whether evolutionary epistemology can be embedded within genetic epistemology depends upon the precise versions of E_{pc} and evolutionary epistemology at issue.

Consider a genetic epistemology where there is a basic variation-selection-retention process invariant under E_{pc}; call this VSR genetic epistemology. Then the defining relation of the minimal evolutionary epistemology, the Selective Analogy account, can be directly embedded in this E_{pc}. So the Selective Analogy account of evolutionary epistemology as it stands above can be embedded in VSR genetic epistemology. On the other hand, were we to add to the Selective Analogy account the denial that there is any more invariant content to E_{pc} than the variation-selection-retention process, and Campbell 1974, for example, comes close to this, then that account could no longer be embedded in even VSR genetic epistemology, since Piaget certainly has a richer version of E_{pc} than that (see section 5.I.2).

There is a further consideration. For Piaget, E_{pc} defines an *extension* of phylogenetic processes to cognogenetic processes. A simple version of extension is invariance, as discussed above. But as pointed out in sections 5.I.1 and 5.I.2.3, an extension of a process will in general involve at least its superfoliation and may involve its modification as well. This may not prove a difficulty for the Selective Analogy account, which asserts only the variation-selection-retention process to be in common, since these can all likely be accommodated in a Piagetian extension relation. (Although Piaget himself has little to say about the matter, tending to ignore or play down appeals to selectionist explanation, even while not denying their basic applicability.) But even this compatibility is not obvious (it will need detailed empirical investigation of which we are only now becoming capable), and it is potentially a serious problem for stronger versions of evolutionary epistemology that may assert too strict and detailed versions of E_{pc} to be compatible with the modifications to the process required by the Piagetian notion.

Similar considerations apply for the Extended and Phenotype Extended Analogy accounts, insofar as E_{pc} is concerned. Suppose a suitably extended version of genetic epistemology is adopted that includes a genotype/phenotype structure invariant under E_{pc}, and call the result ExtVSR genetic epistemology. Then the defining relation of the Extended Analogy account can be directly embedded in the E_{pc} of ExtVSR genetic epistemology; but whether the Extended Analogy account as a whole could be embedded depends on what else is included under the E_{pc} relation and how much regulatory modification E_{pc} encompasses and whether the Extended Analogy account includes a denial of anything else in E_{pc}. Equally, whether the Extended Analogy account represents a compatible enrichment of the E_{pc} of VSR genetic epistemology depends on these same two considerations.

The situation with the Phenotype Extended Analogy account is exactly parallel, except that more is involved than just consideration of E_{pc}, for it is committed to the active intervention of the phenotype in shaping selection and so involves some account of the other relations in a genetic epistemology as well and also some account of superfoliation under the E relations. This will increase its general similarity with Piagetian genetic epistemology, but the two accounts may still differ in structural detail. If its account is structurally compatible with Piaget's then, insofar, the Phenotype Extended Analogy account can be embedded in genetic epistemology. Since Hahlweg's version of the Phenotype Extended Analogy account, for example, incorporates Waddingtonian regulatory processes, as does Piaget's version, it seems a likely candidate for embedding in Piagetian genetic epistemology. (So long as, like Waddington, it eschews Lamarckian processes.) Parallel considerations apply to the relation of the Embedded account of evolutionary epistemology to Piagetian genetic epistemology. But now the embedding problems are further eased because the necessity to find exact cognitive counterparts to genotype and phenotype is eliminated. In particular, I take the version of the (biologically) Embedded account I advocate to be also directly embeddable in Piagetian genetic epistemology.

Whatever the precise relations between a genetic and an evolutionary epistemology are at issue, I hope to have shown that the general framework of regulatory processes represented by diagram 5.1 provides a useful way to articulate both kinds of theories.

5.II. Piaget's Normative Epistemology

5.II.1. Introduction

The discussion in Part 5.I has sought to elucidate the conception of regulatory systems dynamics that provides the framework for Piaget's thought. This same framework, pruned of rationalist tendencies, also provides an appropriate embedding for the kind of evolutionary epistemology espoused herein. But that discussion has not addressed any normative issues.

In particular, nothing has been said about what gives normative force to genetic epistemology. Why isn't genetic epistemology just a component or variant of descriptive developmental epistemology? What makes it normative? Since epistemology is the application of reason to information flow, this is equivalent to asking: What is Piaget's concept of reason and how is it realized in his genetic epistemology? In particular, what is his conception of scientific development and how is it argued to be rational? What is its relation to genetic epistemology? How do Piaget's notions of reason, norms, and rational development compare to those provided by the evolutionary epistemology espoused herein? Part 5.II addresses these questions.

5.II.2. The Status of Philosophical Construction

Philosophically, Piaget's conception of genetic epistemology formed in antipathetic reaction to the major components of the Continental European philosophy with which he was surrounded. Where they offered "fixed points" and pure natures Piaget emphasized the dynamics of regulatory constructivism. Just as he saw the active constructions of the organism as the basis for rejecting both behaviorism and preformationism in psychology, so also he saw it as the basis for rejecting both positivism (empiricism) and rationalism. On the one side, Piaget argued against the logical positivism of the

Vienna Circle and its adherents that the notion of a transparent, wholly reliable and theoretically uncommitted perceptual or sensory foundation for science was an illusion (Piaget 1970/71, chapter 4). And as against the positivists' conception of logic and mathematics as expressing merely a set of empirically empty conventions abstracted from language, Piaget argued that the empirical evidence points to the reverse conclusion, that logic precedes language and that it needs to be constructed by the cognitive agents, its structure reflecting a profound operational exploration of the external world (Piaget 1971, chapter 1).

On the other side, Piaget was equally critical of the Continental phenomenological and rationalist traditions. Against these traditions, he argued forcefully that the phenomenological world as it appears to us, far from being a finished product at any stage, is the result of an elaborate, limited, and fallible construction process. It should be seen as at most in dynamic equilibrium, each new experience at least a potential modifier of its construction. Introspection reveals but the surface finish of this edifice, accessible to us in certain limited respects. To endow the deliverances of introspection with either completeness or indubitability is both to wholly mistake their proper status and to obscure our understanding of their origin and dynamics. In similar vein, Piaget rejected the judgments of rationalists as reflecting, not an infallible access to a realm of eternal certainties, but the reportage of those intuitive structures thus far equilibrated on human experience. He delighted in pointing out how time and again science constructed counterexamples to rationalist claims. (See e.g., Piaget 1965/71.)

Naturalist realism endorses these twin criticisms. Indeed, it is of particular importance to the evolutionary epistemology espoused here that "reason" itself be seen as an evolving structure. Reason expresses itself, for example, in the structure of scientific method, and the dynamics of methodological evolution are central to understanding the nature of scientific knowledge. As information and resources accumulate, humans are able to superfoliate methodology, achieving increasingly open and adaptable systems as they go (see chapters 2 and 6). Philosophical theory is also included within the scope of this dynamic. A fortiori any historical stage of science-philosophy is a thoroughly fallible, criticizable, and revizable one.

Though one might expect to receive Piaget's unambiguous support for this naturalist fallibilist open-endedness to cognitive development, unfortunately Piaget's pronouncements on the philosophical enterprise are rather more ambiguous. One can in fact find textual support in Piaget's writings for no less than three views concerning the nature and status of philosophy generally and of genetic epistemology in particular. (1) When he is reacting most strongly to the positivism, rationalism, and phenomenology around him, Piaget is inclined to want to divorce science from philosophy altogether. He sees this as the only way to free scientific method from the arbitrary imposition of philosophical pronouncements announced as a priori true, which thereby ignores their constructivist origins. Philosophy is said to be a "wisdom" that is "essential for coordinating . . . values" but "is not knowledge properly so called," because it does not possess "methods of verification" or other "safeguards" (Piaget 1965/71, xiii and passim, but especially the postscript to the second edition). If we accept this position, then genetic epistemology must be construed as a purely empirical science, but naturalist realism can accept rejection of the a priori, anti-constructivist line without relinquishing philosophy.

(2) Piaget often writes as if favoring some kind of naturalist position in which epistemology is itself a fallible theory but clearly distinguished by its level of generality and its critical role vis-à-vis scientific theories themselves. He hints that genetic epistemology is perhaps best regarded as a theory to be constructed using the same methodological principles as are employed for science, but he speaks of epistemology as "expressing fallible norms" as well as "containing factual elements." Referring to those who, relying on a sharp normative/descriptive dichotomy, wish to impose values on science, Piaget speaks of physicists and logicians "knowing best" what a good theory is in their domains (Piaget 1971, 13) or of "cooperating with" psychologists (Piaget 1970/71, 8); he affirms that (PQ17) ". . . only in the real development of the sciences can we discover the implicit values and norms that guide, inspire, and regulate them" (Piaget 1971, 4; cf. 8–10.) And speaking of the search for truth, he says:

> (PQ18) . . . the whole trouble has been, from Plato to Husserl, that this transcendental subject [Truth] has been changing its appearance all the time but with no improvements other than those due to the progress of science—the progress of the real model rather than the transcendental one. . . . we are attempting to interpret knowledge in terms of its construction, which is no longer an absurd method since knowledge is *essentially construction*. (Piaget 1967/71, 362; Piaget's emphasis)

See also PQ26.

(3) But Piaget also asserts a certain autonomy and necessity for at least logic and mathematics as the characterization of rational cognitive structures (see PQ3, PQ6, and section 5.II.3). As such, they

are manifested only in "coordination with" empirical cognitive structures. He speaks of logistic as "complementing" genetic analysis and on occasion even of having a Kantian-like status superior to mere empirical detail, yet it has to be constructed genetically by us. In this case, genetic epistemology is a curiously hybrid activity; though working with contingent experience, it constructs the necessary. (It might seem as though there were Platonic-like rational structures that the developing mind explores experimentally and fallibly through its natural embodiment or that there was a Kantian transcendental structure to uncover diachronically, except that Piaget explicitly rejects these alternatives; Piaget 1970/72, chapter 3, IV.) On this account, empirical inquiry can illuminate these processes and help to select the most appropriate representation of the uncovered rational structures involved, but epistemology lies distinctively on the side of the rational structures themselves.

The fact is that Piaget never tackled the nature and status of philosophy head-on in an unambiguous manner. Kitchener (1986) makes plausible cases from Piaget's writings that he shares fundamental principles with Cartesianism, Kantianism, dialectics, and pragmatic naturalism. Piaget's total *oeuvre* certainly shows at least a complex eclecticism (and perhaps worse), but prodigal borrowing alone does not explain his pronouncements. Rather, there is something more and distinctively Piagetian involved, deriving from his fundamental regulatory systems constructivism. For just as Piaget wished to see the regulatory structures that are constructed by the organism as deriving neither from a purely environmental origin (neo-Darwinism and behaviorism) nor being purely innate (Lamarckism and preformationism), so also philosophy is to be understood as neither purely empiricist nor purely rationalist. Indeed, there are hints that Piaget does not consider philosophy to be purely anything else either (e.g., neither purely realist nor purely idealist). The realist and idealist positions come, Piaget suggests, from emphasizing the object of knowledge to the exclusion of the activity of the knowing subject (realism), and from the reverse fault (idealism). The truth lies in the constructive interaction between object and knowing subject. According to Piaget, the true philosophy, like the adequate psychology and biology, has the form of an interactive construction that is a dynamic synthesis of all of the foregoing philosophical positions, correcting each in the light of the active constructions of the organism. In this sense then, the true philosophy transcends each of the philosophical positions that standard philosophy knows, and this has only been possible because of its interaction with empirical science. (See Piaget 1970/71, 83–4; 1967; Gillièron 1986.)

These views cannot be properly understood except within the framework of a regulatory systems constructivism. If one standardly takes philosophical positions as finished and static, then it is difficult to accept as more than metaphorical (and misleadingly so) an attempt to combine idealism and realism in some higher synthesis. But if philosophy itself is a fallible and/or partial construction, then it can be improved in the light of experience. A first step in this direction is to recognize the virtues of a *critical* realism or idealism, where criticisms coming from the other point of view are appreciated and taken into account. (But without relinquishing the arguments for realism given in chapter 1.) A second step is to recognize the fallibilist status of philosophy and its dynamic interaction with science, generating a mutually open-ended feedback loop (cf. chapters 1 and 4). It is the character of the resultant evolving system that matters. This is a leading theme throughout this book. From this point of view, reading 1 is set aside as incompatible with a thoroughgoing constructivism, and reading 2 is adopted as basic, enriched by the addition of a basic naturalist constructivist version of reading 3. This is in agreement with Kitchener 1986 who selects pragmatic naturalism as most basic to Piaget's approach.

It is difficult to obtain a clear idea of this constructive status for philosophy in Piaget because of important lacunae in his work. Most of his career saw little attention to the history of science, and what there was focused on the contents of concepts. Piaget neither attempts any explicit interactional analysis of the kind provided in chapter 4 nor any related methodological analysis employing method in a dynamic systems setting (cf. note 9). These lacunae leave Piaget with a damagingly truncated framework; he has a dynamic constructivist orientation but insufficient means to express it. The result is undue reliance on a formalist structuralism and attendant rationalism.[22] Not until Piaget and Garcia 1983/89 was process emphasized over formal product in the analysis of science, and even then it still tends to be modeled as very general universal formal process and still focuses on individual psychological processes. And we have yet to deal with Piaget's views on the necessity of the ultimate products of construction (see section 5.II.4 below), which sits ill with naturalist fallibilism. Although there are indeed rationalist tendencies that find their expression here, the core of what Piaget has to say is again best read within a thoroughly naturalist, constructivist, fallibilist framework. The absence of a fuller dynamic systems perspective is perhaps understandable when we recall the recentness of these ideas. Once we add to Piaget's remarks the missing components noted above, we can begin to construct an account of philosophy as itself an evolving construction and so gain some clearer conception of this notion, which Piaget was struggling to express so many years before his time. There are further insights that Piaget's position has to offer philosophy, especially for conceptions of reason and norms.

5.II.3. Piaget's Conception of Reason

According to Piaget, the cardinal feature of living regulatory systems is that they are constructively self-organizing, that they manifest an intrinsic striving for development through transequilibration (see section 5.I.1), that is, toward equilibria resulting from internal operational adequacies sustainable across an ever wider range of environments and organism activities. See PQ22 and:

> (PQ19) We are forced to conclude that the ultimate aim of behaviour is nothing less than the expansion of the habitable—and, later, of the knowable—environment. This expansion begins as "exploration" in animals of various degrees of complexity, but it extends far beyond the needs of immediate utility, and of precautions, until we find it operating on levels where a part is played by simple curiosity about objects or events, as well as by the subject's pursuit of every possible activity. (Piaget 1976/78, xix–xx.)

To remind: An equilibrium involves an organism (system) being able to maintain within viable limits the collection of interacting homeostases and homeorheses essential to its life through its autoregulation, that is, its internal control over its input and output and internal input-output processing (including those of internal subsystems). Then the organism (system) has an organized internal regulatory state that realizes these autoregulatory capacities (causal representation), one whose semantic contents (information representation) arise from their operational or regulatory roles (not vice versa; PQ9). Successive equilibria represent the successive incorporation of environmental information into the organism. The system of concepts or information processing structures then in use suffices to comprehend the input so as to produce outputs that satisfy its needs (as they currently appear). As a first crude approximation, needs correspond to homeostases and homeorheses (= process homeostases); whenever one of either is disrupted, a need arises. (The disruption of blood sugar homeostasis corresponds to hunger, the need to eat, etc.)[23] The activity satisfying the need has as its causal consequence the reestablishing of the homeostasis or homeorhesis. We arrive at fully fledged cognitive systems when distinctively cognitive needs are recognized: desire for consistency or completeness, satisfaction of curiosity, etc. But cognitive needs introduce a special complication of this simple account, as we shall see, giving a special force to an intrinsic drive for development.

According to Piaget, the transequilibration process itself is the fundamental autoregulatory homeorhesis, a stabilized drive to development that is ubiquitous. In contrast to neo-Darwinism, which suppresses the process, Piaget insists that neither phenotypic behavior, ecological dynamics, nor ultimately evolution itself can be really understood without acknowledging its centrality. (See Piaget 1967/71, 1974/80, 1976/78, and section 5.I.1 above.) Cognitive development, whether in individuals or in scientific communities, is but the most explicitly information-oriented facet of this universal drive.

Piaget identifies reason with this universal homeorhetic drive, so that it is reason that drives us individually and collectively on to more informed states. Piaget even regards logic as a structure emerging from a homeorhetic drive; for example, noncontradiction is a "functional invariant" and expresses a drive for "coherence and unity of thought, organisation" (Kitchener 1986, 194, cf. Bickhard 1988, 1991b). Here is Kitchener's helpful summary:

> To say that an epistemic subject acted rationally is to say it made the epistemic transitions it did because a later stage was more equilibrated than an earlier one. Hence to act rationally means for Piaget to act in such a manner as to become more equilibrated. The concept of rationality, therefore, is not one involving merely the formal and evidential relations between propositions, nor is it a pragmatic means-end rationality. Piaget's concept of rationality, however, includes both of these aspects. . . . Thus the concept of rationality is that of autoregulation, adaptation or, in general, equilibration. . . . (Kitchener 1987, 361–2)

This conception of reason is thoroughly within the Western tradition, where reason is that capacity that both enables and drives us to transcend our defects: ignorance, prejudice, bias, egocentrism, anthropocentrism, and anthropomorphism. These defects, along with unavoidable finitude itself, constitute the current limits of our epistemic condition. (See the discussion at chapter 6, section 6.I.3.1.) But in Piaget, as opposed to the core Western tradition, reason grows out of natural biological activity. *Reason is the (self-)organizing component of the cognitive face (i.e., information representation) of normal or healthy regulatory systems development.*

However, by including adaptation here, Kitchener conflates the regulatory process itself with the products of that process. It is necessary to distinguish reason as an autoregulatory capacity from the content that results from exercising that capacity, even though the actual dynamics of the system involve an interaction between the two (cf. below and chapter 2). Meanwhile, observe that there is no commitment in this formulation to identifying reason with logic or indeed to reason being a formal structure at all.[24]

There is here a tension between a universal naturalism, because of the roots in biology, and an equally universal rationalism, because of the extension of reason to all developmental processes; this issue will be examined in section 5.II.4, where again a fundamentally naturalist reading of Piaget will be defended.

To develop this Piagetian notion of reason, consider first an organism or regulatory system of organisms (population group, social institution) that is temporarily equilibrated. Even here, assimilation must include some accommodatory activity. Not every surprise for a system precipitates sufficient operational organization to deserve being labeled a new stage in development. Though we are dealing with degrees of change, let us call these small intra-equilibrium adjustments microaccommodations and those inter-equilibrium adjustments precipitating a new stage macroaccommodations. A child whose concept of a flower is that of a four-petaled display at a branch end, for example, will be surprised when asters are first discovered and more surprised still to discover grevilleas from Australia; but while these discoveries will necessitate some minor conceptual revision (microaccommodation), they are unlikely in themselves to precipitate a new developmental stage. Piaget assimilates microaccommodation to assimilation; the asters and grevilleas will have been successfully assimilated to the existing category of flower.[25]

There is already creative autoregulatory activity at work here; microaccommodation is far more than memory update, say the assimilation of the location of one more example of a tulip (the one that grandma says must not be trampled). The autoregulatory capacities associated with each equilibrium will have to include not only memory update, but simple expansion and differentiation of categories, formation of new connections among categories (e.g., that dogs don't eat flowers but squirrels and some birds might), reorganization of goals in the light of these changes, and so on (cf. PQ7). In short, all of the "housekeeping" activities that keep the system well organized, perhaps we could say efficiently autoregulatory, are involved. The cognitive face of this microprocess is certainly a manifestation of reason.

The exercise of microaccommodatory reason at equilibrium yields a fairly standard conception of science viewed as an autoregulatory learning system. A system in equilibrium exhibits a dual mastery, a mastery over its own internal processes and a mastery over its relations with its environment. The former includes the ability to generate explanations and predictions by assimilation to nomic patterns (stored as operational structures), together with the ability to update and modify concepts, beliefs, and goals, check for consistency, etc. in microaccommodatory ways so as to better satisfy needs, including cognitive needs. The latter includes the ability to position itself via prediction so as to acquire relevant information, and it typically also includes an ability to alter the environment to a condition more favorable to need satisfaction (subcase: alterations that produce new relevant information). Environmental abilities are manifested through bodily behavior and so are extended by technologies, which are themselves typically the product of previous environmental abilities. These are all the capacities one expects to find in what Kuhn would call normal science. A science in equilibrium exhibits a dual mastery, a mastery over its own internal processes and a mastery over its relations with its environment, where now the positioning involved in the latter includes the ability to organize and improve experimentation and information-collecting instrumentation. (See the discussion of the systems nature of science-technology in chapter 2.) It then becomes reasonable that normal science should exhibit some amount of all the usual virtues: consistency (otherwise preconditions for mastery, like uniqueness of connection, fail), empirical adequacy (otherwise responses will not be successful in maintaining homeostases and homeorheses), explanatory power (otherwise there will be patterning gaps or blind spots, unintegrated elements, etc., which reduce mastery), technological applicability and reliability (otherwise the system cannot practice external mastery, thereby defeating the practice of internal mastery also), and so on.

Note especially that Piaget's conception quite naturally couples together practical success with cognitive understanding. Indeed, the one grows out of the other. Success in external mastery leads, via assimilation and accommodation, to that rich internal organization that can properly be labeled understanding, and vice versa, understanding provides the behavioral basis for extending mastery. (One of Piaget's books is titled *Success and Understanding*, Piaget 1974/78; in like spirit, chapter 7 of Hooker 1987 is titled "Understanding and Control.") Piaget's position here thus matches, or rather complements, the account given by Rescher and elaborated in chapter 4.

In contradistinction, the passage omitted from within the Kitchener text quote above says in part that "Piaget has a two-tiered theory of rationality. The earlier and logically more primitive level is rooted in the concept of *praxis*, whose sole epistemic criterion is success. The second and higher level is that of propositional thought in which one's concern is, by contrast, *understanding* the reasons for things happening as they do." The second level, Kitchener says Piaget says, develops out of the first and is superior to it. I have no complaint about the idea of a development of reason—to the contrary (see below). But Kitchener creates the impression that, once one has attained understanding, one *leaves behind* success as a relevant criterion of rational conduct. The continued quote above goes on, ". . . equilibration, a notion incorporating both means-end rationality (in the earliest stages) and formal propositional rationality (in the later stages)." I think this is a wrong conception of the relation between the two aspects of reason; neither is eliminable at any stage. Rescher has the interconnections roughly right, I suggest (see chapter 4). But then, it turns out, so does Piaget. The correct reading of Piaget, I suggest, is that the internal criteria become increasingly complex and conscious as development proceeds, and that is all; the relevance of environmental feedback never vanishes, except perhaps in some ideal never-actually-reached limit, and even this is dubious. (Kitchener at least agrees, later in his text, that logic is present even at the praxis stage. And see also his insistence that Piaget is fundamentally a pragmatist! - Kitchener 1986, 94–8.)

No actual equilibrium can be expected to be perfectly complete, even within its own terms. Some accommodations may not fully succeed, and so there will be mismatch signals accompanying some assimilations (some scientific concepts may fail to apply under some conditions or apply only approximately), and there may be some lack of either match or connection between locally successful assimilations (science as a whole is not unified or even fully coherent, and neither are some particular theories). These imperfections will not disturb if their occurrence is suppressed (cf. Piaget 1975/77, 149–50), or judged not relevant (cf. Piaget 1975/77, 18–19), or if, while judged relevant, accommodation is too costly and they are set aside for the time being (cf. Piaget 1975/77, 66). Though all these responses may be temporarily reasonable in the circumstances, unconditional acceptance of the imperfections would represent defects in healthy function, in reason.

So equilibria typically don't last. For creatures born into ignorance but with open learning systems, a complex world is sure to upset each tentative equilibrium. Then the macroaccommodatory passage to the next equilibrium ensues. For Piaget, recall, this is not a passive process, but rather one actively prosecuted through regulated exploration (section 5.I.1). The transitoriness of equilibria is intrinsic to them; autoregulation includes destabilization as well as stabilization. Reason is not an equilibrium housekeeper only; rather, it is primarily the driver of development as search for the next wider equilibrium. This aspect of the process is of special interest here.

There is a natural feedback from equilibrium to disequilibrium. Two significant sub-processes can be distinguished, one runs through environmental exploration and the other through environmental reaction. Either may be initiated by, and will certainly be reinforced by, any imperfections in the existing equilibrium but may also proceed independently of that factor. In the former process (exploration), external mastery associated with a given equilibrium allows the organism (system) to extend the range of accessible environments and/or to manipulate those accessible to it in new ways. Infants who acquire a new behavior, like crawling, are like scientists who acquire a new instrument; in each case, the immediate context stimulating the acquisition—perhaps, respectively, the desire to follow a parent or the desire to answer a specific laboratory question—is quickly enlarged. The information or satisfaction originally sought is often quickly submerged beneath the welter of new information and needs generated through use of the new mastery. But this raises the probability that a feature will be encountered that cannot be assimilated. So present mastery is the cause of passage beyond it.

In the latter process (reaction), it is the reaction by other systems to system behavior (including its explorations) that initiates the disruption of equilibrium. Other regulatory systems in the environment will respond to system behavior in ways based in their own autoregulatory capacities and designed to continue to maintain their own transequilibration process. But these new behaviors often will in turn prove to be ones that the exploring organism (system) cannot assimilate within the current equilibrium. An improvement in flight strategy for gazelles, for example, may be met by an altered hunting strategy by lionesses to which existing gazelle flight planning has no adequate response, triggering any or all of new gazelle learning, ecological succession, evolutionary shift. A new theory designed to solve one problem often leads to experiments or new demands on existing theory that in turn generate new information or questions that cannot be assimilated into the existing science, leading to any or all of new learning for some scientists, changes in roles and emphases across the institutionalized pursuit of science, changes in accepted scientific theory and/or practice.

> (PQ20) Any knowledge raises new problems as it solves preceding ones. This is evident in the experimental sciences where the discovery of the causality of a phenomenon raises the question of the cause of the causality and so forth. . . . any finished structure can always give rise to new requirements in fresh substructures or to integrations in greater structures. (Piaget 1975/77, 30)

These intrinsic feedback processes producing disequilibrium from the resources provided by an equilibrium are only strengthened for cognitive systems. First, curiosity ensures that the environment will be explored as fully as possible. If the environment has surprises to offer, then, ceteris paribus, curious systems are more likely to encounter them. Second, cognitive systems have the capacity to direct an examination of their own extant constructions in the search for incoherence or incompleteness, thereby increasing the likelihood that the limitations of the current equilibrium will be revealed. I shall often use *exploration* to cover both external and internal searches. Third, when the environment of a cognitive system contains other cognitive systems beside itself, then its communicated ideas and methods will be met with curious exploration and critical scrutiny, typically going beyond its present competence. This raises the probability that new problems will emerge if there is any scope for them to do so.

We may note that just as equilibria cannot in general be expected to be defect-free, the creation of disequilibrium is not guaranteed. The environment (both cognitive and non-cognitive) may be too circumscribed, for example, to produce any disturbances, or the organism (system) may not have either the capacity or the will to explore it appropriately, or it may ignore, suppress, or otherwise set aside the disturbing feedback it receives. But we have just concluded that this is less likely to happen with cognitive systems, ceteris paribus, and increasingly so as their acquired information and methods increase. For rational systems, disequilibrium is the overwhelmingly likely outcome of any equilibrium. It is quintessentially rational that knowledge should reveal its own limits.[26] This provides a second component to Piagetian reason.

A normal or healthy regulatory system responds to disequilibrium by constructively searching for a new, wider equilibrium that will assimilate the disturbing input along with all that had already been assimilated (but perhaps in ways accommodated to the new circumstances). The key cognitive instruments employed for this purpose are reflective abstraction and constructive generalization (section 5.I.1). Reason organizes or directs these instruments. If the changes across equilibria are substantial enough and represent expanding operational adequacy, we can speak of development. Roughly, and in the terms of chapter 2, microaccommodation = regulatory refinement, while macroaccommodation = development = regulatory ascent.[27] The organization of the search for, and construction of, new wider equilibria is the third component of Piagetian reason (cf. PQ8). Its continued activity as the highest autoregulatory level provides for a specific directionality to developmental change, namely toward increasing endogenous (i.e., self-) regulatory adequacy, which includes (cf. PQ9, PQ22) increasing knowledge. Taken together, these three organizational capacities of reason can all be seen as components of a complex but ultimately unified and powerful homeorhetic drive for increasing equilibration (transequilibration). This unifies otherwise disparate remarks of Piaget's and opens the way to a unified regulatory theory of complex living systems.

Once again, this conception of development corresponds to a fairly standard conception of scientific change. Science is institutionally organized to use its existing knowledge and resources to explore its environment, to develop new instruments for doing so and to analyze its current edifice for incoherence and incompleteness, thereby probing critically its theories for inadequacies in the light of all this (chapter 2). This makes it likely, ceteris paribus, that existing scientific knowledge will reveal its limitations through scientific mastery being disrupted by disturbing inputs, whether it be the disruption of external mastery by unassimilable (unpredictable, unexplainable) exogenous observational inputs or the disruption of internal mastery by endogenous inputs signaling lack of internal coherency or completeness. When disruption—or rather, sufficient disruption, so that it cannot be set aside as anomalous—occurs scientists initiate a search for a more adequate structure (theory, paradigm), one that will successfully assimilate (predict, explain) the disturbing input along with all that had already been assimilated (but perhaps now reconceptualized and corrected according to the new theory or paradigm). If the changes across these scientific equilibria are substantial enough and represent expanding empirical (operational) adequacy, we can speak of scientific development. Since successive equilibria represent mastery over successively wider ranges of environments, this provides a complementary account to that given for epistemic progress in chapter 2.

Thus far, we have three components to Piagetian reason, intra-equilibrium organization, systematic search for the limits to equilibria, and the search for and construction of new equilibria. Development, we noted in section 5.I.1, is the manifestation of increasing autonomy, and reason has a special role to play in this. The shift to greater autonomy is a shift from reliance on exogenous inputs to reliance on endogenous constructions, and this is for Piaget, we have just seen, central to reason.

But, for Piaget, reason itself has a particular autonomy:

> (PQ21) . . . reason was built up by stages and continues to evolve, not without reason or reasons but in such a manner that not only is the evidence transformed but even that which appears logically demonstrated or rigorous at any stage may subsequently appear doubtful and may give rise to considerably greater degrees of rigour. (Piaget 1970/73, 34)

Reason is a fallible construction, logic included; even so, reason only changes for a reason. The regulation is entirely endogenous.

It is not immediately obvious what Piaget might intend here. The key to understanding Piaget's intention, I suggest, is contained in the special role given to the "seeking of reasons" in Piaget and Garcia 1983/89 and discussed briefly in section 5.I.1 above. The process is one whereby reason itself includes a search for those constructions that would provide a genuine explanation of, and hence a principled assimilation of, a phenomenon. This is just the search for equilibrium again, but more stringently specified. The force of "genuine explanation" and "principled assimilation" here is that the conceptualized (assimilated) phenomenon must be seen to fit an intrinsic pattern for objects of its kind, that is, to be an instance of a relevant structure that is demonstrably complete or eqivalently, operationally closed (and so possesses "necessity"; see below). Only then do we possess genuine laws that could provide the foundation of a genuine explanation (as opposed, for example, to the mere deduction of the phenomenon from an inductively constructed pattern). Thus reason is not only characterized by the seeking of reasons, it also specifies what kinds of constructions could count as satisfying the search—and does so with increasing efficacy as knowledge and experience increase. In this sense, reason specifies (constructively) the reasons for each development, as Piaget says, including (self-reflexively) the theory of reason itself. This reinforces not only the ubiquity of reason, but also its nonformal character.[28]

Piaget's conception of reason is that of an active self-organizing capacity providing an intrinsic drive for development and autonomy. Piaget often speaks of intrinsic drives for wider equilibria:

> (PQ22) . . . the expansionist factors which in the realm of behaviour push assimilation and accommodation combined toward what appears to constitute behaviour's dual goal on all levels: to widen the environment and to increase the living organism's capacities. (Piaget 1976/78, 142)

And see especially PQ19 above. The search for limits of current equilibria must be actively prosecuted; nature does not automatically deliver the appropriate disturbances, and our built-in inertias must be overcome. And the instruments of reflective abstraction and constructive generalization are actively wielded by reason to achieve reequilibration.

The rational developmental process comprehends, as Kitchener says, the whole spectrum of regulatory change: changes in accepted evidence and in theoretical concepts and principles, fundamental and derived, certainly, but also changes in principles of natural order, in relevance structures (including logics), in metaphysics and in methodology, even in the virtues constituting valuable theories. (Cf. the schematic for the Aristotelian/Galilean shift in diagram 2.3, chapter 2.) Under a broad Piagetian scheme, all of these kinds of changes can count as rational in the appropriate circumstances.

Moreover, they may all interact with one another in the process of equilibration.

> (PQ23) . . . the transcendental, according to philosophers, refers to . . . the instruments that are necessary to render experience possible. . . . All sciences of nature, therefore, involve transcendental aspects . . . that are in constant movement and construction. . . . In fact, there exists a reflexive progress in the sciences (which is indissociable from their progress in extension). It consists of the constant delineation of new conditions of intelligibility, which are transcendental with respect to the content of later experience. (Piaget 1978, 649)

By contrast, the search for a set of rules for rational theory change, the dominant model for rationalists and empiricists alike, is extremely narrow and constrained; under its terms, the history of science must remain a mystery (Feyerabend 1978a, but cf. Hooker 1991a).

Thus Piaget's approach long ago captured two central insights into scientific development, the importance of normative as well as descriptive change (specifically methodological change) and the importance of the interaction between normative and descriptive elements in the process. (See also section 5.II.4 below.) For a naturalist account of norms as regulatory controls and of normative/descriptive interaction, see chapter 1.

Reason then names that fundamental autoregulatory capacity of organisms (systems) that organizes equilibria, organizes the testing of limits to equilibria, organizes the search for and construction of more adequate successor equilibria, and organizes these activities autonomously. How in detail is all this possible? What kind of organizational capacity is this? How powerful need it be? Unfortunately, Piaget does not have much to say that would help us to understand the answers to these questions. For this, he can perhaps be forgiven, since we have only very recently acquired any nonsuperficial insight into the (self-)organization of complex systems and we are as yet a very long way off from understanding even its rudimentary implications for cognitive science (cf. chapter 2 and Hooker 1994e).

For most of his career, Piaget instead followed another, simpler approach that made good strategic sense in the circumstances, namely, to study the interrelations among equilibria as products to see what they might reveal. But this took Piaget toward a formalist structuralist theory, perhaps not least because it held out the promise of bypassing the dynamical or process details entirely on the way to rendering reason self-transparent (see note 17 and text). This was unfortunate, because it was too simple, because it was formalist, and because it pushed Piaget toward rationalism, which ill-fitted his naturalist constructivist framework. Because of its influence on the image of Piaget's conception of reason, I shall review it briefly.

Piaget wanted to characterize development by a unique, purely logically determined sequence of formal structures that have no formal intermediaries, so that each defines a unique, clearly demarcated equilibrium or stage (see Piaget 1970/72, chapter 1). To this end, he co-opted formal axiomatics (Bourbaki school), aiming to provide a unique sequence of formal structures where the sequence transitions would be purely formally specified so that sequence order is given internally by the logical relations among the contents of specific stages. This would then characterize development independently of specifying the developmental dynamics. Of course Piaget agrees that the future segments of a developmental sequence cannot be predicted by the organism (system) whose sequence it is, since the organism (system) is discovering what the sequence is as it is creating it. But the fact that development does form a formal sequence, which can be formulated with hindsight, shows the sequence to be nonarbitrary, indeed to have some kind of necessity.

Further, Piaget held that the developmental sequence consists of conservative accommodations in the sense that each new equilibrium "generalises" and "completes" the old operational structure. In general, causal explanations are modified, not rejected wholesale; categories are added, extended, modified, and made conditional, not rejected wholesale, and so on (cf. Piaget 1975/77, 67; Piaget and Garcia 1983/89). Old knowledge reappears in the new suitably generalised and suitably completed. The child first learns to spatially displace objects in single moves, but then learns to perform displacements serially and to reverse the operation, so completing construction of an abstract displacement group and generalizing the concept of spatial displacement. Relativity theory completes the operational analysis of space and time measurements found in classical mechanics and introduces the Poincaré Group of space-time operations of which the Galilei Group of classical mechanics represents a special, limiting case. Piaget introduces a conservative principle of "maximum integration" to describe these interrelations (see Kitchener 1986, 203–10).

Unfortunately, this is the part of his theory that has received extensive criticism (see sections 5.I.1 and 5.I.2.2). There must be something right about the idea of developmental sequence; not every collection of modifications to a complex regulatory system will cohere stably and likely most will not, various collections of operations form groups (e.g., spatial displacements), and some group structures are special cases of others, and so on. The truth, however, is evidently rather more complex than Piaget's co-option of Bourbaki mathematical axiomatics permits. We can gain some idea of what would be involved from consideration of sequences found in the history of science—the schematic for the transition from Aristotelian to Galilean mechanics in chapter 2, for example, which shows more regulatory complexity than Piaget considers.

And while quantum mechanics shows a much more radical departure from classical physics than any "integration" would suggest, yet there is a substantive sense of generalization through which they are connected, which Bohr called "rational generalization." (However, it raises subtle issues about the nature of objectivity and even of reason itself; see Hooker 1992, 1994a, c and references.) Even the intertheory relations among post-Galilean space- time theories are quite subtle, far more complex than a vague law of maximum integration would suggest, yet again there is a substantial notion of rational generalization applicable (cf. Hooker 1992 and references). Piaget's position in this respect is substantially modified in Piaget and Garcia 1983/89, where this principle is dropped and process recognized, but it is not replaced by any detailed analysis of intertheory relations (cf. note 9). In sum, there is both regulatory continuity and change across development of an intricate regulatory kind; it is not random, yet not simple inductive either, indeed not a formally specifiable sequence (notes 24, 28).

With this, I propose to leave Piaget's stages behind (cf. section 5.I.1 close), and with this a chief manifestation of his rationalist tendency. That returns us to the issue of rational dynamics. What can be said about the dynamics of Piagetian development has been said in section 5.I; it is not negligible, but not sufficient by far. Reason is the organizational capacity to develop; a more detailed regulatory account of this capacity beyond what is given elsewhere in this book will have to await the development of the sciences of self-organization (a satisfyingly naturalist Piagetian conclusion).

There is one further facet of Piagetian reason to consider: its relation to an account of cognitive ideals. While rational developmental changes are justified externally by their pragmatic success in delivering wider equilibria, they are also to be justified internally in terms of a theory of ideal regulation.

> (PQ24) Reason is an ideal immanent in all acts of thought, as well as in all practical operations. . . . As an ideal, reason is a form of equilibrium towards which all cognitive systems tend. (Piaget 1933/77, 257; quoted in Kitchener 1987, 363.)

Cf. PQ8. This ideal is not the rationalist or empiricist one of a collection of certain truths, but a regulatory ideal, that is, one specifying an ideal regulatory equilibrium. An ideal equilibrium is one that is ideally complete or closed; it cannot be disturbed any more, because it is already able to produce all the relevant responses to the circumstances that any accessible environment in this cosmos can produce. It would allow the organism (system) complete autonomy (while at the same time achieving and expressing it would require total system interactive openness). We come to recognize such an ideal by coming to recognize that successive equilibria represent successively more complete operational structures, that is, have a compensatory operational response available for increasingly many inputs. Regulatory completion across increasingly many environments provides direction to the sequence and so meaning to the notion of progress toward the ideal (cf. chapter 2). It is perhaps similar to the manner in which we come to recognize the existence of a mathematical series given only some of its terms and then come to investigate whether it has a limit (see below).

The ideal is represented as the end product of an ideal trans-equilibration, an ideally completed real trans-equilibration process. This latter grades back from cognitive trans-equilibration to find its roots in biological trans-equilibration and ultimately in the closure characterizing self-reproduction, which is the ultimate precondition for, and expression of, autonomy. The ideal, we shall see, also exhibits a generalized kind of self-reproductive closure modeled on biological self-reproduction. In this way, Piaget is able to construct a thoroughly biologically grounded account of norms.

One strives toward an ideal but never attains it. One may, however, make progress toward it, approximating the ideal more and more adequately; cf. PQ9 and the following:

> (PQ25) The existence of this [ideally known] object constitutes the only possible explanation of the directed approximations, even if one is never certain of having attained the end and even if the knowledge acquired in the course of this history prevents us from believing in all ultimate characteristics. (Piaget 1967, 116, quoted at Kitchener 1987, 349.)

This last quote shows also that Piaget is fundamentally a realist, as his naturalism requires.[29] The ideal is approached analogously to the convergence of a mathematical series on a limit (Piaget 1950, 1: 38; 1950, 2: p. 69; 1970/72, 17, 82). But we also disturb the world when we probe it for its properties; depending upon the character of the disturbances (e.g., classical versus quantum), this may place bounds of either an epistemic or principled kind on both the convergence and on the accuracy of the resulting best available representations.[30]

There are two coordinated improvements manifest in cognitive development: improved knowledge of real objects and improved knowledge of reason (e.g., a better representation of the ideal of reason). In the above mathematical analogy, the more terms one has in a series, the more constraints one can put on its form (knowledge), and one regulatory order up, the more experience one has with mathematical series, the more powerful is the abstract machinery for characterizing limits (reason). So it is with reason, the more experience we have with epistemic development, especially historical experience within a scientific field, the better we are able to characterize both reality itself and the relevant ideal of reason.

What ideals are actually involved? PQ24 speaks about "a form of equilibrium towards which all cognitive systems tend." Let us ignore its inaccessibility for a moment and consider this ideal cognitive end state. In this idealized equilibrated state, there will be a completed or closed collection of internal operations that suffice to assimilate all possible exogenous inputs and deliver all required outputs and to pass all required internal coherency and completeness checks, without generating any endogenous mismatch signals, no matter how far and rigorously exploration is extended. The corresponding conceptual contents will thereby constitute ideally complete concepts of the objects of this world, they are the ideal objects of PQ25 (see also PQ8, PQ9). The conceptual relations between them will constitute ideal knowledge of those objects, knowledge that is neither erroneous nor partial. The ideal of truth is realized in this knowledge, for truth holds open the permanent possibility of error or gap in all cognitive states short of ideal completeness. Thus we may associate the truth ideal with the regulatory property of operational closure. Finally, there is the organizational property itself of complete equilibrium under all possible explorations, the "form" of the equilibrium. This property (rather more complex than it may appear) Piaget identifies with the ideal of reason itself. There are then at least five ideals worth distinguishing in Piaget's conception (ideal concepts, objects, knowledge, ideals of truth and reason). Here I shall comment briefly on the last two.

Piaget's concept of operational closure provides, I suggest, a plausible naturalist account of the construction of the correspondence truth ideal. A naturalist philosophy needs a strong correspondence notion of truth. Its primary purpose is to ground systematic epistemic fallibilism and open-endedness by holding open the permanent possibility of error at all regulatory orders (chapter 1). The ideal of truth then is a theoretical term in a theory of cognition, including epistemology, and its several roles therein derive from its primary purpose (see Hooker 1987, chapter 8, and chapter 6 below). The question then arises as to the accessibility of such a notion to natural creatures. Humans certainly experience the failure of operational closure and/or operational adequacy at each disruption of an equilibrium, and they succeed in repairing these for the disrupting phenomenon at each new equilibrium. According to Piaget, they also have the capacity for reflective abstraction. So reflecting on the development process, it is plausible that they will abstract the notion of operational closure itself. Moreover, they will then be able to abstract the idea of successively wider closures across successive equilibria.

The closure notion then becomes the primitive truth surrogate; ceteris paribus, a theory succeeds (i.e., cannot be disturbed) because, and to the extent that, it expresses closure (cf. is true), and a theory expressing a wider operational closure (cf. a truer theory) will succeed more widely. Our notion of truth then emerges from the reflective separation of our selves, including our self-representation that contains the closure property, from that of our environment; in this process, the fundamental relation becomes that of theory success because of truth, and closure is understood to be a truth surrogate. (Plausibly, the abstraction/separation occurs in something like the manner described by Piagetian developmental psychology, for example, in intimate relation to the achievement of object constancy and so a three-dimensional world in which we may each represent our own body as one among many, cf. Hooker 1992.) The truth ideal is then the direct extrapolation of this construct under the relation of increasing closure width.[31] It is an ongoing improvable construction.

> (PQ26) . . . progress indicating an ideal orientation, which would not antedate this evolution but which is emergent and which will partly control the development to come. (Piaget 1924, 602; quoted at Kitchener 1987, 345.)

See also PQ18. In this way, we may make naturalist constructive sense of a realist truth ideal.

The ideal of reason is the ideal of a stable self-reflexive global organization. The collection of closed operations needs to be stable under the organization of limitation-seeking exploration (= criticism + experimental processes in science). This connects directly to the characterization of the rational ideal given in chapter 6 (see section 6.II.2). Note that the complete collection of operations also includes the representation of method, metaphysics, and other higher regulatory components of reason itself, such as conditions of intelligibility (see PQ23 and chapter 4). And limitation seeking exploration includes seeking alternative rational organization of these. So the ideal organization has itself a global closure property analogous to the global closure property of self-reproduction possessed by all living organisms (cf. the related discussion in chapter 6). The Piagetian ideal of reason is an ideal self-regulatory organization.

With respect to characterizing the components in the ideal of reason, logic and meta-logic are the most advanced, corresponding to the formal sciences. But reason cannot be reduced to formal logic; there are many reasons why this is so (notes 24, 28) that may be summarised under the rubric that logic is static and cannot determine process. None of the processes of exploration and construction can be captured as formal logical sequences. Rather, logic contributes internal consistency and the metalogical notions of formal soundness and completeness as component virtues of the regulatory ideal (see chapter 6). Other natural sciences all show significant rational development (i.e., in method, metaphysics, conditions of intelligibility, etc.). So although the ideal result cannot be reduced to logical necessity, it includes it when logic is understood naturalistically (cf. Bickhard 1988). The operational structures subserving ideal knowledge are as necessary for us in our cosmos as anything can be, since they can neither be avoided or found defective. But since we never actually attain an ideal, not even for logic (see PQ23), all of our present characterizations stand as approximations to the ideal itself, and the virtues those approximations exhibit are our best current proxies for characterizing the ideal. This too corresponds to the theory of ideals given in chapter 6.

Caution: There is a certain unrealistic monolithicity to Piaget's conception of rationality, every component fits neatly together to form a tightly knit coherent whole. This is no doubt because a unique logic was seen as the crown structure, the immanent ideal, and formal mathematical structures defined unique global closures. Yet reason represents a much more complex regulatory process than this. There is internal competition among pursuit of the proxies for truth (e.g., between pursuit of empirical adequacy and explanatory power), or equivalently among the conditions for regulatory refinement and ascent (chapter 2). And as noted above in the discussion of quantum theory, the development of knowledge itself can reinforce internal tension between competing specific models of reason itself. Furthermore, these complexities point to new aspects of reason as trans-equilibration, since they require reason to sort out. Natural rationality is a regulatory process far more subtle and powerful than anything we would have invented a priori. So Piaget's rationalist tendency to formal monolithicity must be countered by recalling his own doctrine that all knowledge, including our representation of reason itself, is an active, unfinished construction.

Thus far then, this is the Piagetian conception of reason as ultimate self-regulatory organization. It is the conception of a total developing regulatory structure, increasingly autonomous, creating and revising its representation of its environment and of itself as it goes, including its own rationality, articulating its own conception of ideals and its own exploratory and critical activities in pursuit of those ideals in the process. Reason becomes the central organization of cognition, with knowledge its crowning product.

5.II.4. The Normative Nature of Genetic Epistemology

Reason and epistemology are normative notions. Piaget needs a characterization of normativeness that fits with this regulatory conception of reason and knowledge.

In this regard, Piaget himself is of little help. He often seems to avoid the issue. Kitchener remarks that Piaget has a general and a restricted notion of genetic epistemology, with the latter simply assuming that our current normative criteria for knowledge are correct, while the former opens them also to scrutiny, and that Piaget avoids philosophical discussion by seeking refuge in restricted genetic epistemology (see Kitchener 1986, 148–9). But Piaget also speaks confusingly of various principles as if they were now descriptive, now normative, or both normative and descriptive:

> (PQ27) . . . the two conceptions of causality as a law of nature or a requirement of reason: this disjunction appears to us to be both non-exclusive and reducible to a logical conjunction. (Piaget 1970/72, 91)

Piaget also speaks of the phenotype as a "norm of reaction" (1970/72, 85), which "expresses the limited number of possible variations open to a genotype when it is confronted by a wide range of values in an environmental variable." There is a corresponding "norm of accommodation," but in "later developments . . . [this] will no longer be simply a measure of its own flexibility, but rather of the number of interactions and coordinations it has been able to enter into with other schemes" (Piaget 1974/80, 103). Finally, Piaget also describes his genetic epistemology as a whole as both normative and descriptive:

> (PQ28) In short, all epistemologies, even those which are anti-empiricist, raise questions of fact and thus implicitly adopt psychological positions. . . .
>
> Epistemology is the theory of valid knowledge, and, even if this knowledge is always a process rather than a state, this process is in essence the change from a lessor to a greater validity. It follows from this that epistemology is by its very nature an interdisciplinary subject, since a process of this kind raises questions of both fact and validity. If it were a question of validity alone, epistemology would merge with logic. . . . If epistemology were only a question of facts, it would be reduced to a psychology of cognitive functions. . . . It is therefore only through [the interdisciplinary collaboration of psychologists, logicians, mathematicians, cyberneticians and others] that the requirements of fact and validity can be equally respected. (Piaget 1970/71, 4, 6)

See also PQ17 in this regard and Piaget's talk of a parallelism between the psychological and the rational (see PQ12).

Thus Piaget on norms. Unsurprisingly, his writings confuse even his philosophically trained commentators. Kitchener speaks of Piaget's "progressive developmental transformation of merely causal facts into rational norms" (Kitchener 1986, 23), of "hybrid laws" of genetic psychology that are both empirical and normative (Kitchener 1986, 171-2) and of "nomic isomorphism" between rational and causal sequences (Kitchener 1987, 359).

But there is available a natural and straightforward conception of norms that arises in a regulatory systems context, namely the conception of the regulation of a domain with a regulatory process equipped with an ideal construction. On the one hand, this conception correctly models the directive and critical function of norms. On the other hand, these regulatory functions evidently exhaust the nature of normativeness. (See chapters 1 and 6.) This conception has in effect been employed already throughout Part 5.II. According to this conception, it is a distinctive feature of all norms that they are also descriptively explanatory; indeed, it is precisely their explanatory success that endows them with normative (i.e., regulatory) force. It is in exactly this naturalistic regulatory sense, I think, that Piaget says of regulatory development that the end result is "a normative form" and that

> (PQ29) . . . the regulations entail from the start the distinction between normal and abnormal . . . and at the behavioural level their effect is normative necessity itself inasmuch as operations form the limiting case of regulations. . . . (Piaget 1970/72, 58.)

Thus Piaget's insistence on a dual normative and descriptive role (e.g., PQ28 above) can be read as exactly correct, as can Kitchener when he says:

> If the concept of equilibration explains epistemic change, it does so largely in virtue of its normative character. . . . Piaget's concept of equilibration is both a normative standard of evaluation and an explanation of scientific change. (Kitchener 1987, 360.)[32]

It is precisely this duality that allows us to learn about normative (i.e., regulatory) systems and do so as an integral part of improving our descriptive knowledge, just as Piaget requires (see PQ23, PQ26 above). And it points the way to understanding Piaget's term *norm* of *reaction*, since regulatory structures are operationally determined in terms of system responses to input variation (this is, in fact, the most general form of system estimation in traditional control theory, see Hooker et al. 1992b for analysis). Finally, it is a conception that makes good sense of what Piaget does say about normative force; in particular, it explains his seeming to be ambivalent about the status of genetic epistemology.

Certainly any theory of epistemology is to be normative; a fortiori genetic epistemology is to be normative. It proposes a theory of how organisms (systems) ought to develop, of what healthy cognitive development is. But genetic epistemology is also a very wide or encompassing theory. Traditional epistemologies are essentially applied logic (i.e., rationality) conjoined with a theory of perception (i.e., input) to provide premises or data, with the normative theory of logic (rationality) and the descriptive theory of perception (input) coming from elsewhere. But genetic epistemology includes within its own structure a general theory of developmental psychology and a theory of reason (cf. PQ28, PQ17). Among the positive consequences of this is a wider normative role. (And not just in epistemology but in ethics and other areas as well, but I set this aside here.) So then, how is genetic epistemology normative?

First, genetic epistemology is offered as our best theory of how knowledge progresses and is therefore normative for judgments about knowledge claims (chapter 1). This account is reflexive; genetic epistemology is itself a part of our knowledge and so should itself be open (an open system) in the manner it specifies. Second, genetic epistemology provides for normative assessment for knowledge claims in terms of epistemic value, as measured by our current proxy virtues, including pragmatic success. This corresponds roughly to the traditional philosophical account of assessing scientific theories, except that it comprehends a far more interactive process of normative judgment than does traditional analysis—in particular interaction between normative and descriptive components as both proxies (truth and reason representatives) and theories codevelop. This enriched trans-equilibration process proceeds via the pragmatic assessment of the practical success of our epistemic state. Practical success reinforces particular cognitive frameworks and practices and practical failure initiates the search for improvements in them, including the highest level representations of reason. It grounds the process in reality.

Though Piaget gives little attention to methodological analysis, the methodological conception that thus emerges is essentially that of evolutionary naturalist realism (chapter 1). This itself needs to be elaborated by (a naturalized version of) Rescher (see chapter 4) and an account of method as a regulatory system (see note 9). The larger philosophical framework of naturalist realism is needed to make systematic sense of Piaget's position, especially of its normative dimension. But thus augmented, Piaget's naturalist regulatory constructivist framework provides a rich and powerful framework.

5.II.5. Piaget: Rationalist or Naturalist?

Nonetheless, to many the Piagetian conception of reason and cognition as set out above will still appear primarily rationalist—a not-unreasonable assessment. Anyone reading many of Piaget's passages (e.g., PQ3, 4, 11) and familiar with European thought will not find it hard to see Piaget as a French structuralist seeking to specify an abstract, universal rational form or structure, whether of process or product (cf. Bickhard 1988, Smith 1992). Moreover, the structures themselves evidently have a strong necessity of some kind, which Piaget attributes both to the ideal and, by derivation, to the sequence of stages approaching it (see PQ29; cf. PQ6, PQ10). The superficially nonmodal characterization of the ideal of reason in PQ24 above apparently leaves its features a contingent matter, but its modal force lies in its "all" and the following modal formulation is a clearer expression of Piaget's position: An ideal equilibrium is one that is necessarily complete in the sense that it could not be disturbed by any possible input. Reason, we might conclude, produces strong necessity as a central result of its process, and this necessity grounds genuine laws and explanations.

Further, there is the tightness of closure in the ideal of reason, suggesting a fully self-transparent reason whose completeness is necessary. The model here is again logic and mathematics generally (especially group theory). Somehow, one is supposed to be able to know that a specific logical structure will be part of the ideal of reason because it is intrinsically provable that this structure is complete for the relevant operations (cf. the theory of finite groups). At this level, the autonomy and self-consciousness of the organism is complete; it no longer needs to refer to the environment to characterize operational completeness and it is fully self-aware of its characterization (though all this came about only as a result of a long sequence of interactions with the environment; cf. Piaget 1974/76). While all this has a transcendental rationalist flavour,[33] at other places, Piaget seems to say that the reason the necessity is guaranteed is only that we made it so; we have, for example, so constructed our operations that logic is necessarily an ideal of reason (Piaget 1970/72, 72-4; 1976). In any event, the necessity seems self-transparently intrinsic, guaranteed.

At this point, it is indeed easy to see how someone lacking the regulatory systems perspective could arrive at a rationalist reading of Piaget. Here is reason directing development "from ahead," directing it toward an unrealized abstract, formal, and intrinsically necessary ideal, through a necessary sequence of stages. Moreover, reason is a ubiquitous drive that is manifested in all living systems, evidently related to Bergson's elan vital or "creative impulse," (Kitchener 1986 describes well Piaget's early involvement with Bergson's vitalist biology and with the neo-Kantian rationalists, but we should note that Piaget also explicitly rejects these positions, see below.) Ally this reading of Piaget with a fully or quasi- Lamarckian reading of Piaget's biological dynamics and of rational scientific change, and the result is a thoroughly rationalist position.

This is, for example, an important component in how Kitchener 1986, 1987 reads Piaget. First, there is a rationalist biology:

> In his biological writings [1950, 1976/78, 1974/80] Piaget criticised orthodox neo-Darwinism ("Selection Theory") for its commitment to the notions of randomness and chance. Piaget's underlying (and rather uncritical) assumption, here, inherited from 19th Century thinkers such as Bergson, is that either evolution is *contingent* and *irrational* (on the one hand) or it has a *direction* and is *rational* (on the other). (Kitchener 1987, 345; Kitchener's emphases.)

Then Kitchener goes on to explicitly equate Piaget's regulatory development with a Popperian World 3 abstract rational process, a sequence determined by objective formal reason:

> Piaget is not offering an explanation of why a particular individual person abandons one theory and adopts another one. His theory of epistemic development concerns what he calls *the epistemic subject* and not the *psychological subject* or *individual subject*. The epistemic subject is an idealised abstraction, namely, the set of underlying epistemic structures common to everyone at the same level of development. . . . [Piaget's] account would be about why the epistemic subject—what one could call the Scientific Mind or Scientific Reason—made the epistemic transition . . . about why an idealised subject—Historical Reason—made a particular transition. Thus, an

> equilibrium explanation really concerns the epistemic relations between two cognitive or scientific theories and the epistemic transitions from one of them to the other. . . . [It] is closely related to Popper's World 3 and not merely to World 2. (Kitchener, 1987, 355–6; Kitchener's emphases.)

The basis for this interpretation is Piaget's characterization of development as "an autonomous and endogenous affair—one of autoregulation—having an internal logic of development all its own" (Kitchener 1987, 358.) Kitchener quotes Piaget on development as follows:

> (PQ30) . . . that the development can be explained without necessary reference to various factors which undoubtedly do play a part in its concrete realisation, e.g. maturation, learning and social education, including language. For the key to its development lies in the concept of equilibration. . . . (Piaget 1933/77, 24; quoted at Kitchener 1987, 358.)

Kitchener goes on to read this distinction in terms of the standard dichotomy between reasons and causes, with equilibration belonging to "internal" rational explanation and the other factors belonging to "external" causal explanation "of something irrational" (cf. Kitchener 1987, 358).[34] The result is to turn Piaget into a Popperian critical rationalist:

> This principle of equilibration, as an abstract, quasi-Platonistic World 3 phenomenon, characterises what it is to be rational and it can be exemplified in the individual solitary Cartesian ego, in a network of social relations . . . or in the biological realm. Wherever it is exemplified, however, equilibration and adaptation characterise what it is to be rational. In short, rationality is neither individual, nor social, but a set of logical relationships. (Kitchener 1987, 362–3)

This is, I believe, a basic misreading of Piaget. But Kitchener has studied Piaget's writings extensively and carefully, and he can be relied on to provide an accurate "phenomenological" portrayal, as it were, of Piaget's doctrines. If he can arrive at a rationalist reading of Piaget, it is understandable how others have also.

Yet the evidence that something is wrong with it as a basic reading of Piaget lies all around; witness Piaget's rejection of Lamarckism in biology (section 5.I.1), his complete silence about it in science where it would still more plausibly fit (contrast Rescher, chapter 4), his insistence that philosophy and even logic and mathematics are revisable constructions (see PQ21, PQ23), and his own explicit rejection of rationalism. Piaget constantly denied he was a vitalist (see Piaget 1967/71, 154; 1970/72, 92), and he offered a naturalistic teleonomic approach to choice:

> (PQ31) . . . selection has ceased to be envisaged as an automatic sorting process: it is seen more and more as bound up with processes of regulation and even of choice, in that the organism is capable of choosing its environment before suffering or confronting its constraints. Organic selection is thus bound up with devices or regulators, each showing a certain flexibility and, above all, a cybernetic teleonomy. As such, it is certainly much closer to a system of choice than is selection imposed by the external environment. (Piaget 1974/80, 52)

Indeed Piaget rejects both reductionist materialism and vitalism on just the same grounds as he rejects Darwinism and Lamarckism, namely, that they ignore the autoregulatory constructive activity of the organism.

Kitchener himself supplies evidence antithetical to his own reading. He is forced to recognise, for example, that with respect to the ideal of reason, Piaget explicitly rejects an Aristotelian rationalism of final causes:

> (PQ32) The "vections" or orientations [of development] . . . are not due to a direction impressed prior to evolution, otherwise we would fall into a priorism. Even less can one translate this into the language of final causes. (Piaget 1955, 45; quoted at Kitchener 1987, 346)

And see especially PQ26. There is a corresponding concession concerning teleology:

> . . . [Piaget] even employs . . . *teleonomy*—to distinguish teleology (which he rejects) from teleonomy (which he accepts). . . . The directional form it [development] takes is due only to the internal requirements of equilibration itself. . . . (Kitchener 1987, 346–7)

These remarks and those cited earlier ill fit Kitchener's rationalist reading.

What has gone wrong? The problem is the absence of an adequate regulatory systems perspective when reading Piaget. What Piaget says has a straightforward regulatory systems interpretation that does not involve rationalism. For these remarks are directly in keeping with Piaget's earlier insistence that knowledge is neither a priori and innate nor a posteriori and impressed into the organism by the external world, but is instead a construction by the active organism. And recall that, for Piaget, a characterization of the ideal of reason itself is not given a priori, but unfolds as development occurs. We learn about reason as we learn about the world. This is a view at odds with rationalism (and empiricism) but quite amenable to a naturalist realist reading—and quite central to the evolutionary epistemology herein (cf. chapters 1 and 6).

The case for Piaget's rationalism centers on the abstract formal characterization of processes and stages. Several points are in order. (1) Piaget is committed to the view that all cognitive systems share a development process in common, so a specification of that process will quite properly abstract from the detail specific to each system. Any systems dynamic description of regulatory processes will typically abstract the regulatory features of interest from the welter of detail; for example, electrical circuit theory of feedback in amplifiers ignores heat dissipation, noise, etc. (2) One should not mistake the distinction between functional and causal specifications of systems for a rationalist/scientific distinction. That an abstract electrical circuit theory exists (a functional theory of circuits) does not commit engineers to some dichotomy between rational and causal effects in electrical circuits or to any similar ontological reification. Reasons belong to a (normative) functional specification of cognitive systems (to the information representation—chapter 2). Rationalism does not follow simply because causal specifications cannot be directly inferred from functional ones, they cannot be so inferred even for simple engineering circuits. (3) Piaget's insistence on the autonomy of the process is part of his emphasis on the centrality of the constructive activity of the organism, and the same conception as (we noted) explains his remarks about the status of the ideal in the process; all of that is fully compatible with naturalism. The only incompatible component, already noted, is the attempt to specify cognition in terms of formal, universal logico-mathematical structures, whether for processes or products; this must be relinquished.

The other support for a rationalist reading of Piaget comes from claims about his critique of neo-Darwinian biological theory, especially that he is committed to Lamarckism. But it is possible to interpret the major part of Piaget's biology satisfactorily so that it has no Lamarckian components (see above and especially Hooker 1994d, section 5). Moreover, anyone who wishes to include the theory of the phenotype in biological theory on regulatory systems grounds will be led to make essentially the same criticisms of strict neo-Darwinism as does Piaget; Waddington did so, as would even a systems engineer who has no stake in biology per se (see also this chapter, paragraph 5, and chapter 2). Piaget's central criticisms of orthodox neo-Darwinism follow just from his regulatory conception.

While investigating these issues, let us inquire about the naturalist status of Piagetian reason. Do any of its organizational capacities require the postulation of some non-natural "vital impulse" or rationalist metaphysics? Certainly the kind of creativity involved in searching for and constructing new equilibria is the kind of capacity perennially seized on by philosophers as non-naturalist. In his examination of creativity (Piaget 1970/71, chapter 3, IV), Piaget argues that neither Platonism, Kantianism, nor empiricism suffice to explain creativity, but he says little else about the process. I make only the following few remarks here (see also chapter 6). It seems clear that in itself the drive to restore equilibria does not require any non-naturalist account, elementary engineering control systems show this feature. Thought of as individual activities, neither reflective abstraction nor constructive generalization seem to require a non-natural principle to be at work. We might suppose that they are some kind of random variation and selective retention processes. But there are other possibilities; for example, the regulatory ascent may result from the "retuning" of some neural net (cf. Churchland 1989, Hooker 1994e; Hooker et al. 1992b). At present, we are very largely ignorant of these processes. The point is that they may turn out to be new applications of common naturalistic regulatory processes, and at this time, that seems the most reasonable hypothesis to adopt.

So Kitchener's Popperian conclusions do not follow.[35] Kitchener's rationalist mis- reading, while common, is importantly misleading for it suppresses the whole naturalist regulatory framework of Piaget's thought, his most distinctive and valuable intellectual contribution to theorizing. With the regulatory systems framework omitted, Piaget's work falls apart into some odd biological speculation, some empirically interesting but ordinary psychology, and fragments of a rationalist philosophy. Instead, Piaget's lifelong work is variegated but of a piece, an attempt to think through a unified regulatory systems perspective on the whole of life, with all its radical intellectual consequences.

To be fair to Kitchener, he does elsewhere acknowledge the importance of autoregulation to Piaget, and he does agree that there is a basic naturalism running through Piaget's writing (see section 5.II.3). To some extent, he may be misled about the extent and importance of rationalist elements in Piaget by still common anti-naturalist but erroneous assumptions about the specification of complex systems (see above and chapters 2 and 6) and to some extent misled by omission of a biological perspective. But, I concede, to some extent the tensions in his writing reflect those in Piaget himself. I would make two concessions to a rationalist reading of Piaget, a general and a specific one.

The general one is that there seems no question but that Piaget was heavily influenced by various non-naturalist positions found in his own developmental milieu (e.g., Bergson and the Kantians and later the Structuralists). The development of Piaget's own position has evidently been something of a "dialectical" interplay between these influences and the radically naturalist tendencies inherent in his biological and regulatory systems ideas. (Cf. again Piaget's rejection of Bergson's vitalism but his insistence on a universal regulatory process driving development.) The attempts at eclectic synthesis crop up everywhere; see, for example, the discussion of his philosophy in section 5.II.2. (Self-reflexively, Piaget seems to have tried to accommodate and then assimilate all the intellectual currents in his milieu.) Nonetheless, a consistently naturalist reading of Piaget can be given and is the most satisfying reading of him. I propose then to excise recalcitrantly rationalist elements in the reading of Piaget I shall use herein.[36]

The particular concession is to recognize that Piaget did attempt a formal characterization of cognition, and of stages in particular, around which the most recalcitrantly rationalist elements seem to cluster. As noted above, this is also the most empirically and formally controversial of the Piagetian doctrines. Moreover, there are good reasons for a naturalist not to want to confine reason to formal structures (notes 24, 28, and chapters 1, 6). I agree that there is a rationalist flavor to Piaget here, but I have argued that formal characterization nonetheless represented an attractive scientific research strategy at the time and that there must be something right about it since development is coherent (section 5.II.3). Well then, set aside the rationalist element, and retain what can empirically be retained of an account of structural constraints on development. Doing so will not adversely affect Piaget's basic regulatory systems framework—to the contrary.

The notion that we have already attained at least some core of the formal, logical ideal of reason, to which Piaget gives special place, provides the central case in point. There has this century been a veritable explosion in new systems of logic. These range from "generalisations" of classical logic (meaning Boolean logic, not Aristotelian syllogism— already a damaging shift for a rationalist), for example, various modal and many valued logics, and higher-order logics, to attempts at "completion," e.g. various inductive logics, tense logics and free logics). But they also now include increasingly "deviant" structures (e.g., relevance logics, quantum logics, and paraconsistent logics). The point is not that the formal study of logic(s) should be abandoned—to the contrary—but that there is less and less reason to think of logic as a finished subject that could provide an intrinsic self-guaranteed ideal of reason. Piaget himself is ambivalent on the point, on the one side often taking logico-mathematical "necessity" as immutable, and on the other side, recognizing developments in logic (and the appearance of the great constraining theorems of Church, Gödel, Lowenheim-Skolem, etc.) as evidence for a continued construction. (Contrast Piaget 1970/72, 67, with 52, 72.)

Accepting Piaget's naturalist constructivist assessment, we conclude that there is in history no attainable human-made necessity, even in logic and mathematics, since everywhere we have also been forced to unmake it in the pursuit of development. (See here forcefully PQ21.) There is a pragmatic necessity that represents the status of the currently most entrenched epistemic commitments (i.e., regulatory structure), and there is the necessity of the ideal, but even this inaccessibly perfect condition does not have any transcendent force, only the force of operational completeness (section 5.II.3).

Developments this century provide increasing reason to represent our situation exactly in naturalist regulatory terms; the more we are coming to learn, in all subjects, the more we are able to explore the nature of (even formalized) reason, and we do this through the creation of fallible theories, including conjectures about the proxies for ideal reason. So while some part or aspect of Piaget's doctrine has to go, or ascendancy be given to some passages over others, the basic framework, the (auto)regulatory naturalist constructivist core of the doctrine, remains.[37]

5.II.6. Conclusion

Understood aright, and despite the difficulties and defects, Piaget's work provides a consistent, wide-ranging and interesting naturalist regulatory systems approach to all aspects of life. It is in good keeping with the naturalist realist approach I am seeking to elaborate. Despite all the charges that have been made against Piaget over the years, of being Lamarckian, rationalist, or worse, the greater part of Piaget's work can be defended from these charges—and ought to be defended, because of the intrinsic naturalist interest it has within the new regulatory systems theory revolution.

Furthermore, Piaget's regulatory approach provides fruitful conceptions of knowledge and reason. It provides, for example, a conception of reason of the right kind, one much broader and more adequate than the traditional formalist conception, and one that can fruitfully be integrated with a naturalist approach, as I try to do in chapter 6. And it provides, a biologically grounded conception of knowledge that shows how to construct an adequate conception of the complex relations between knowledge and reason. (And I have not explored all the aspects of these here. There is, for example, the relation of the social to the rational in knowledge formation; see Kitchener 1989, chapter 2.)

Nonetheless, I have had to reject or reshape certain portions of the Piagetian corpus in order to extract the naturalist regulatory framework. In particular, his quasi-Lamarckianism in biology and his quasi-rationalism in cognition (if I may call his focus on logic and necessary relations by this name) have had to go. It should be emphasized, however, that in biology the major part of what Piaget says can be retained, and similarly for the treatment of logic and abstract contents in cognition (but always within the constraints of the regulatory naturalist framework).

Finally, for the twin reasons given above, the overall naturalist regulatory framework Piaget provides seems to me the proper framework within which to situate evolutionary epistemologies, especially those taking a regulatory systems approach. Piaget's framework provides for a natural enrichment of evolutionary epistemologies, especially in respect of the roles of reason and norms, and also by other compatible extra-Piagetian components, in particular an integration of Rescher's methodological architectonic and a regulatory account of methods.

Chapter 6

Naturalized Reason

Introduction

A systematic naturalism is committed to naturalizing both epistemology and reason. An evolutionary epistemology is one specific form for the naturalized epistemology, an alternative with a natural attraction to it. But in general, connections between evolutionary and naturalized epistemologies are much weaker than this. Naturalized and evolutionary epistemologies are logically independent of one another; one may be committed to the one without any commitment to the other. It is possible, for example, to adopt the view that our brains have evolved through natural selection, while being committed to a non-naturalist conception of epistemology. It is even possible to adopt the view that knowledge itself progresses in a way analogous to Darwinian processes whilst nonetheless maintaining that in epistemology, this structure is expressed through a non-natural collection of rational rules. In short, any of the first three evolutionary epistemology positions identified in chapter 1 can be developed in a manner compatible with rejecting naturalization of epistemology. Of course they can also be developed from within a naturalized framework. But only the last position, Embedded Evolutionary Epistemology, requires a naturalized framework.

There is a specific reason why many philosophers are loath to naturalize their epistemology, even when it is an evolutionary epistemology: Epistemology is essentially applied rationality and it is considered impossible to naturalize reason.[1] The non-natural character of reason emerges because of reason's non-natural necessity on the one hand, and because of its intentional focus on truth on the other. In its essence, the necessity of reason is the necessity of logic. The laws of logic constrain those that would aim at truth, and these constraints are apparently quite independent of any material conditions in this world. Moreover, science intentionally aims at the truth and hence deploys logic to that end. But causal systems simply do what they causally do in virtue of their moment-by-moment natural properties, among which truth does not figure. Hence the very capacity to grasp truth as a goal seems to lift the mind beyond this world. All told, therefore, the view that scientific selection must be non-natural, even as part of an evolutionary epistemology, seems the right one to take, precisely because it is rational and reason is non-natural. This is the popular view held by Popper, Rescher, and many others.

These considerations raise in an acute form the problem of understanding the place of reason in nature. If reason is conceded to be non-natural then not only is the embedded conception of evolutionary epistemology flawed, but the evolutionary epistemology program as a whole reduces at best to pointing out an amusing cross-"world" reason-cause analogy (cf. Popper's worlds, chapter 3). This analogy might offer the odd practical suggestion for the furtherance of knowledge (such as the creation of some competition) but otherwise makes no contribution to any fundamental understanding of processes on our planet. If this circumstance is to be avoided, as evolutionary naturalist realism supposes, then reason too must be naturalized. Given the non-naturalist view expressed above, the prospects for naturalizing reason seem daunting. And there are recent explicit arguments that it is doomed from the start (see Putnam 1982).

The systematic naturalism that is espoused here has already been outlined in chapter 1 above. In this context then, to naturalize knowledge and reason will be to understand them as natural phenomena, as arising naturally during the evolution of complex systems of our kind (at least, and possibly of many other kinds as well). The issue before us is what precisely naturalization involves.

6.I. Naturalizing Reason

6.I.1. How Not to Naturalize

The ideas of Jean Piaget, a major pioneer of a biologically grounded epistemology, serve as inspiration for this chapter. Shortly after a research stay with Piaget, Quine published his "Epistemology Naturalised" (1969), a paper that has set the terms of much of the debate in the English-speaking world. But Quine's argument is different from, and more simplistic than, Piaget's complex normative-factual constructions (see chapter 5). It is important to begin a discussion of naturalization by freeing ourselves from Quine's constraints.

Reduced to essentials, Quine's argument runs as follows. (1) Either epistemology is normative, or it is purely factual, and it cannot be both. (2) The only normative epistemology that could be acceptable is one that carried out an essentially empiricist program, that is, one that provided an adequate theory of science as a superstructure generated purely by logic operating on an observational base. But (3) no such empiricist epistemology is available. Therefore,

(4) epistemology is not normative, or what comes to the same thing, the ideal of creating a normative epistemology should be abandoned. Therefore, (what remains of) epistemology is purely factual, a branch of applied psychology and perhaps applied sociology. Note that we could as easily substitute reason for epistemology in this argument and reach a similar conclusion.

Following this conclusion as advice is tantamount to committing intellectual suicide, and twice over. First, because it wipes out any distinctive content to philosophy, pretending that the problems that have driven it over the millennia can all be simply reduced to empirical issues. Second, it leaves us blind with respect to the critical appraisal of science and any critical understanding of the history of science. Only an empiricist straitjacket could push one into such a position. The proper response is to reject empiricism, and there are good reasons for this, some supplied by Quine himself (see Quine 1963, and further see chapter 3 above).

But is not the rejection of empiricism already represented by premise 3? And then we are left precisely with Quine's conclusion after all. The point is, however, to reject empiricism but, unlike Quine, to reject it thoroughly, including its metaphilosophy (cf. Hooker 1987, chapter 3). And then both premises 1 and 2 will be rejected, while premise 3 is affirmed. Premise 1 expresses a sharp normative/descriptive dichotomy that appears as a central component of empiricist metaphilosophy. It turns out that its rejection is a necessary condition for developing an adequate naturalism. This is precisely not a rejection of the normative, but an attempt to theorize it in a naturalist realist manner (see chapter 1 and below). Premise 2 goes, because empiricism will no longer be the only acceptable form for normative epistemology. This premise derives from an empiricist metaphilosophical assumption that epistemological structure is exhausted by logic and logic delivers a unique epistemology, namely, empiricism (see chapter 3, section 3.I). The naturalism espoused herein retains a theory of norms as both genuinely normative and yet informed by empirical experience, so it can avoid Quine's conclusion even while rejecting empiricism. The way is then open to reconceptualize naturalization.[2]

An effective program for naturalizing reason and epistemology must both treat reason and knowledge as natural phenomena and do justice to their central normative characters. To this end, it will be helpful to pause and consider briefly the Western philosophical tradition of reason. This will provide us with a clearer focus on what it is desirable to accept from that tradition and what it is essential to reject.

6.1.2. A Perspective on Reason

6.1.2.1. The Western Rational Project.[3] The essential idea of Western intellectual culture is that man is a rational animal and that reason is what distinguishes men from other animals. The essential project of reason is transcendence, for example, transcendence of cognitive limitation and imperfection: ignorance, prejudice, bias, egocentrism, speciescentrism (anthropocentrism), and projection (anthropomorphism). In a larger context, transcendence of moral and aesthetic limitation and imperfection are included. The application of reason brings about progress in objective understanding, cognitive, moral, and aesthetic, and is expressed in objective knowledge, ethics, and aesthetics.

Thus the Western tradition. It makes philosophy the alternative to religion, reason the alternative to faith and grace. The aims are comparable in category; what differs are the instruments (reason versus faith) and the nature of the promised transcendence (transcendence into reason versus transcendence into salvation, and perhaps beyond reason). It is unsurprising that reason was construed non-naturally. But my aim here is a naturalist construal. This is possible, I believe, though not easy. While the goals at which reason aims typically were given a non-natural twist, they are certainly compatible with naturalism. But reason will have itself to be included in the improvement process, for our grasp of it is as fallible and subject to limitation and imperfection as is our grasp of knowledge, ethics, and aesthetics. Nothing was said about this in characterizing the Western tradition; there is a bias in Western thought toward an already perfect, hence non-natural reason. In what follows, at least an initial case for this naturalist fallibilist view will be argued, restricted to reason and knowledge.

Naturalism requires a certain nicety of formulation, since the sharp contrast between natural animals and rational humans in my initial expression of the Western tradition is no longer acceptable. We are distinguished by various degrees, not metaphysical kind, from other life forms. We are distinguished then by degree of reason, not possession of reason. (And we do not always possess the higher degree in every intelligent attribute; in selected respects selected other species outperform us.) So the transcendence reason provides is to be construed as a transcendence within natural life, not of natural life. It follows that the process of improvement will be distinguished within development as a whole by its particular pattern, not by entry of a special non-natural force. That pattern has to do with the internal regulatory structure induced by the pursuit of naturalistically constructed ideal goals (see chapter 1 and section 6.II below). Nonetheless, the essential drift is the same. Reason is a natural instrument for alleviating our natural deficiencies, for improving our native capacities. Certainly reason itself is a native capacity, so this is a self-improvement exercise (a matter of self-organization), one in which reason is also improved.

This may sound impossible for us finite imperfect natural creatures. But as examples of progress in objective knowledge, reflect on the amazing construction of mathematics and science that has occurred over the past two millennia, involving constructions of great abstractness, coherence, and empirical power, and of the concomitant transformation of our scientific self-image as an example of improving on our anthropocentrism and anthropomorphism, and of our rapidly increasing understanding of methodology, rational action, institutional design, and the like as examples of the improvement of (our grasp of) reason itself. That all this has happened demonstrates the power of reason. (That morally imperfect humans have often put it to evil use is a, not quite, separate issue.) The problem now is to understand how it came about naturally.

For evolved animals are focused down on their own biological life agendas. How then can an animal possibly transcend the native limits and imperfections of its own capacities? By contact with an external reality that thereby conveys truth to that animal, and by being equipped to make proper use of that contact. What is the mark of such a contact? The involuntariness of the contact. It is precisely the involuntariness of the conveying of content to us that convinces us that we are not here fooling ourselves with anthropocentric and egocentric projections, wish fulfillment, distorting the truth with bias and prejudice, etc. The fact that the content is pressed upon us at all ensures that it is alleviating our ignorance.[4] For the rationalist, contact is with some transcendent world of abstract truths or forms; for the empiricist, contact is with the external material or phenomenal world.[5]

The Platonic world of forms transcends this world and thereby confers on rational agents the opportunity to transcend their native condition. (Of course in Plato's case, it is more a question of the transcendent soul regaining its erstwhile status, liberated from its fleshly prison of finitude and imperfection, than it is a case of redeeming this bodily existence.) For the empiricist, it is the autonomy of the natural or phenomenal world from our psychological machinations that offers us release from grubbing around with our noses to the dirt.[6] Platonic rationalism and empiricism between them mark the poles of the distinctively philosophical approach to the human condition within Western culture.

For those within the Western tradition of reason, a further issue needs to be resolved: How is the content or information derived from external contact to be organized by us humans as objective understanding? If the understanding and use of the information is itself infected with our native limits and especially our imperfections, then the project will still fail. What is required is a system of understanding organized through a faculty of reason having these three features: (1) It is normative, that is, capable of separating out legitimate understanding from all the others (truth from falsehood, right from wrong, beauty from ugliness). (2) It should be guaranteed not to introduce limitations and imperfections into its operation. (3) That conditions (1) and (2) are met should itself be transparently recognizable by us.

The dominant philosophical answer has been that this system is logic. At least, logic lies at the central focus of such systems, which may be held more or less extended, for example, to mathematics or to other synthetic a priori truths supported by demonstration. Rationalists hold that these broadly logical truths are informative but necessary and hence contain no errors. Empiricists hold that logical truths are necessary but vacuous and hence contain no errors.

To complete the space of possibilities for Western culture, another axis should be added orthogonal to the first and again containing two poles, a religious pole and an anarchist-existentialist pole. These positions join in emphasizing the centrality of the human spirit, unfettered by the constraints of reason (at least as any position on the rationalism- empiricism axis would understand the term reason). But where religion transcends reason with authority of another kind, anarchism-existentialism denies all authorities that would constrain human freedom. The religious pole also recognizes the external and involuntary as the mark of truth and self-transcendence as the essential impulse of the human spirit. The transcendent reality is anchored in the deity, and the involuntary contact is mediated through the soul, which has the capacity to grasp truth outside of the limits of the exercise of reason. The anarchist-existentialist pole emphasizes an areligious, naturalist metaphysics of free creation, both the anarchic and existentialist positions denying that there is any privileged communal or species transcendence project, religious or philosophical; though at any one time various cultural practices will be flourishing and various of them fading, there are no winners, there are only practices that are engaged in or not. In this way, both poles defining this axis break away from the distinctively philosophical tradition of Western culture. (For some further discussion, see Hooker 1991a.) Rather than directly pursue its challenges, which are important, I return to the discussion of reason in the Western tradition; once we have completed the task of developing a richer conception of reason, we may find ourselves in a better position to relate it constructively to a richer conception of persons, but that is work for another occasion, the point here being that the conditions laid down above for reason have been the more readily accepted because their religious parallels were uncritically accepted. The religious person must also be inwardly prepared to understand and receive the revelation, and this state is called grace or enlightenment, but because grace is from God, it is readily assumed to meet the three conditions parallel to those laid down above for reason. (The anarchist-existentialist owes us an account of the nature of human creativity and freedom and of the inner conditions that support its development.)

From this perspective, empiricist epistemology and rationality is simply one working out of the Western philosophical tradition, and classical rationalism another. But this will not do for a systematic naturalism. We have already seen that a naturalization program will abandon, not just one or two tenets of empiricist philosophy, as Quine does, but its basic metaphilosophical dichotomy between norms and facts as well (section 6.I.1 above). The same kind of response is made to rationalism; naturalism denies both rationalist claims to necessity, for example, and the normative/descriptive dichotomy, which reappears in rationalist metaphilosophy. And naturalism, being systematically fallibilist, will reject conditions (2) and (3) on an acceptable theory of reason. From the naturalist perspective, we humans have barely begun on a species quest for rational understanding, including understanding of rationality itself, that has already taken millennia and during which any particular current aspects may be subject to radical change. Is this sufficient freedom for naturalism to flourish? Do these rejections of basic assumptions provide sufficient room to develop a defensible naturalist theory of reason? I believe not. Naturalism is led to abandon also the Western emphasis on formal logic as the essential nature of reason. The argument for doing so is both negative and positive.

6.I.2.2. Beyond Formal Reason. In outline, the negative argument runs as follows: Wherever the Western formalist project for reason has been tried, it has bogged down in paradox and impotence; it should therefore be regarded as a degenerating research program and another substituted in its place. A chief example of this negative claim is the fate of formalism in philosophical theory of science, summarized in chapter 3, section 3.I. Briefly, the critique of empiricism comes to this: Formal reason (logic) lacks the resources to construct the edifice of scientific theory from any plausible observational foundation, and it also lacks resources to critically determine membership in the foundation, so science cannot be understood as a product of empiricist reason. Empiricism has a radically incomplete characterization of scientific rationality which illustrates the difficulties generated from pursuing a formal conception of reason. Popper was a severe critic of empiricism yet did not take this lesson of its failure to heart. His own methodology was equally formalist, with the result that it has analogues of many of the empiricist difficulties (e.g., for observation), plus it is riddled with additional ineliminable appeals to strategic human judgments and decisions. After Popper, through Kuhn, Feyerabend, and all the others, the appeal to community widens rapidly—so rapidly, that all of these latter have been accused of abandoning reason. As noted, this response, along with those who leap to the conclusion that science is merely socially caused and the like, has the formalist program of reason as a suppressed premise, one that the negative argument supports abandoning. The negative argument against taking logic to be the foundation of rational procedure is set out in chapter 1, section 1.I, and chapter 3, section 3.I.

The positive argument has the following outline: Naturalism understands reason as a natural feature of living systems, evolving as part of intelligence; intelligent systems are best characterized in terms of regulation or control processes, not logico-symbolic ones; these latter are late arrivals, specializations of more fundamental processes; therefore, reason too will have a basic characterization in regulatory terms. Moreover, initial studies of regulatory cognitive models make it plausible that such a characterization can be given and is empirically defensible. Here the studies of chapters 2, 4, and 5 will serve as examples.

The two parts of the argument are brought together to conclude that it is only in regulatory terms that an adequate account can be given of the full range of rational activities. Note that it is the formalist program that has been criticized, not the Western tradition of reason as such. My present view is that the core western tradition should be preserved (i.e., the centrality of reason to intelligence and the conception of reason as focusing on transcendence of limitations and imperfections).

Now under the formalist program, the fundamental terms for a theory of intelligence were foundational statement and formal inference rule, but with the rejection of formalism, it is necessary to find new fundamental terms for theories of knowledge and reason.

We will see that the concept of judgment must play a fundamental role, and that notions of relevance, risk, and efficiency will all play key roles.[7] These are all concepts that "grade off" suitably back along the complexity ordering that corresponds to overall evolutionary sequence. The relinquishing of the detailed internal control supplied by formalist reason—each inference was justified by foundational rules—also demands a replacement. This leads to a renewed importance for a theory of objectivity, which was almost an incidental by-product under formalism, since reason is the capacity to transcend limitations and its product is objectivity. There is yet more involved (see 6.II below), but this suffices to indicate here the overall theoretical issues. We are now in a position to return to the problem of naturalization.

6.I.3. Naturalization

The aim then is to treat reason and knowledge as natural phenomena while at the same time retaining their essential normative characters. How is this to be achieved? Essentially, by treating both of them as theories. That is, one treats the concepts of reason and knowledge as theoretical terms entering a theory of rational cognition.

Why take this approach? Essentially because, as outlined in chapter 1, naturalism allied with our current evolutionary understanding leads to the conclusion that all of our conceptions are in the nature of fallible theoretical conjectures. This is to be applied to philosophical conceptions equally with scientific conceptions. Moreover, it turns out that this conception provides a natural account of the nature of the normative character of philosophical theories (see below). At the same time, since they are fallible theories, one has the basis for understanding how they grow out of and interact with our general empirical understanding of the world. This approach then will provide the right kind of framework for naturalizing reason and knowledge. Of course, establishing this general framework is far from sufficient for carrying out the task adequately. Before entering further into the detail of the task itself, the case of naturalizing epistemology will be briefly considered. The lessons learned will prove helpful when naturalizing reason is tackled. The examination begins by considering the treatment of truth as a theoretical concept and then moves on to epistemology proper.

6.I.3.1. Theorizing Truth. In Hooker 1987 (section 8.3.1), I wrote along these lines: Truth, properly conceived, is a theoretical posit of cognitive theory. Much current philosophy, however, invites us to assume instead that truth is at best a term appearing solely within some theory of semantics, and it usually invites the additional assumption that truth is reduced to, or explained wholly by, reference or satisfaction.[8] But from an evolutionary naturalist perspective, it is poor procedure to start with semantics. Language is only a part of our cognitive capacities; we do not understand it well.[9] To give semantics primacy, or exclusive rights to basic cognitive concepts, is to reverse reasonable procedure. Within evolutionary naturalist realism, both method and theory must be understood as related to reality, indeed in some sense map that reality, yet neither is reduced to the other. The representative of reality in cognitive theory is truth; thus within cognitive theory, truth remains related to, but not reduced to, theory or method.

A basic framework for a realist defense of truth is the realist distinction between truth and rational acceptance. The question at issue is whether this distinction can be shown to play the kind of fruitful theoretical role that would justify its retention in philosophic theory. Even if scientists aim at the truth, they aim at it through proxies. Scientists aim directly at rationally acceptable claims instantiating values of various sorts, claims that are: secure, reliably confirmed, empirically adequate, explanatorily unified, widely applicable, of metaphysical or natural ordering depth, precise, intertheoretically fecund, and so on.[10] But if proxies do all the work, why not scrap truth, which is inaccessible as an autonomous criterion, in favor of values that can be made cognitively accessible? The question is given added bite by the Popperian thesis that, if one is beginning from ignorance, one cannot even aim simultaneously at security (likelihood of truth) and informativeness (content). A defensible answer has to show that without those basic distinctions that the concept of truth underwrites, indeed without the metaphysics and semantics of realism, one cannot develop an adequate theory of cognition and in particular theories of cognitive rationality and epistemology. The position then is that truth is cognitively accessible only via proxies, yet indispensable. Truth then should be understood as an essential theoretical term in cognitive theory.

What are the basic contrasts that truth marks? T_1: Truth marks the contrast between sensory appearance and reality; illusory and hallucinatory judgments are characterized by their lack of correspondence with reality. T_2: Truth marks the distinction between meaningfulness and vacuity; what is meaningful has truth conditions and a truth value. In particular, having a truth value marks the distinction between theoretical sentences being meaningful and their being instrumentally construed, while the kind of truth conditions they have marks the distinction between realist and empiricist semantics for theories. T_3: Truth marks the contrast between meaningfulness and valuableness; what could possibly be true is distinct from which conjectures prove most valuable to us cognitively and otherwise in our current circumstances. T_4: Truth marks the contrast between acceptance (on whatever grounds) and correct acceptance (acceptance of what is in fact true). A judgment may be accepted because of a wide variety of factors: fear, worship, ignorance, training, usefulness, and so on, even conformity to some methodological rules; among those that are accepted, only some are in fact true. In particular, T_5: Truth marks the distinction between rational acceptance and successful rational acceptance. The justification of a judgement will in general be a function of the causal relation between judger and judged as well as the cognitive context of judgment, while the truth of the judgment concerns that judgement's correspondence to reality. T_6: Truth grounds the distinction between error and error-freeness; error arises because judgments do not correspond to reality. In particular, truth grounds the distinctions, within error, between imprecision, partialness, approximateness, and referential failure.[11] T_7: Truth is a ground of the distinction between rational and nonrational cognitive structures. Deductive logic, for example, is basically the theory of truth-preserving inference; inductive logic must yield maximum probability of truth.[12]

From the naturalist realist perspective, the major motivation for these doctrines lies not in transcendental theses and the like, but in the general evolutionary picture of ourselves science offers; it lies, in short, in awareness of our own cognitive limitations and idiosyncrasies vis-a-vis our speculative and linguistic abilities. The argument for retaining a correspondence theory of truth is simply that without that theory, the open-ended texture of cognition based on the foregoing distinctions cannot be captured (see chapter 1). We learn, learn how to learn, learn about learning and about learning how to learn, etc. Every component must be open-ended for us, since we began in ignorance of them all. Our actual history confirms these dynamics. And since we can and do construct very rich cognitive models of ourselves of this interactional sort, whatever the alleged cognitive inaccessibility of the truth relation be, its intelligibility seems clear. Metaphysics theorizes a framework for epistemology without which epistemology is blind and arbitrary. The only metaphysical framework that adequately provides for thoroughgoing naturalism (i.e., that allows every philosophical doctrine conjectural status) is one within which the notion of truth is not tied of conceptual necessity to any fixed cognitive construct.

What empiricism and idealism share in common is an insistence on a conceptual (as opposed to a causal) connection between cognitive conditions and truth, between knower and known. It is just this doctrine that naturalist realism denies. Recently, it has become fashionable to try once again to retain the attractions of both worlds. The formula is still that truth is (by definition) rational acceptance, "idealised" rational acceptance (see Ellis 1985; Putnam 1981). These are rediscovered ideas with a long history; Putnam's formulation, for example, can be found, very closely, in Churchman (1948, 169-170), who traces his inspiration explicitly back to Peirce's pragmatism. (However, Churchman independently provides an insightful analysis of the complexity of notions of risk, quality control, and experimental design in scientific methodology quite at variance with this analytic tradition.)

It seems to be assumed by this fashion that appeal to idealization removes objections of the sort just leveled against pragmatist definitions. But either it is claimed we know (really assuredly know) these idealized rules of rational acceptance, or it is agreed that we don't. If we are held to assuredly know them (or can come to assuredly know them through some specific process P), then this is both a non-naturalist (presumably rationalist) position and a covert version of pragmatism after all. In the event it is agreed humans can't assuredly know these rules (or P)—and this is certainly so on a naturalist account, since they remain conjectural theories—then of course the definition can't be pinned down to some specific pragmatist criterion. But then we may ask how the development of methodology itself is to be understood. In the realist case, the answer is provided via an external reality that acts as anchor while all cognitive structure is fallibly explored, but what could ground the corresponding exploration when truth itself is a human artifact? It is hard then to see what sense can be given to the notion of methodology as a conjectural theory. Indeed, it is hard to see what sense can be given to an idealized science, methodology, or anything. Rather, the whole process of development is in danger of being ultimately arbitrary vis-a-vis truth, a mere wandering around, now reinforcing one values-methods-theory mix and now another. Thus reason is emasculated along with truth.

There are several themes emerging from this discussion of truth that will be picked up in the subsequent discussion of reason and knowledge. Most obviously, there is the point just emphasized: the treatment of truth as a theoretical term. This theorizing does not deny that truth is a normative notion, and it does insist that it is not reducibly definable in other terms, otherwise the exploration of truth could not be intrinsically open-ended. Truth then is a theoretical term with a rich theoretical role; it is a fundamental concept in our conjectural theories of cognition, reason, and language. The discussion has also illustrated how that treatment can release one from currently dominant or even unexpressed philosophical assumptions that limit the available conceptual strategies. And it has thrown up two critically important features of our cognitive structure: (1) The necessity for theorizing it as an open-ended inquiry system, both thoroughly fallible in status and always understood to be capable of further elaboration in both structure and content. (2) The ineliminable tension that arises out of the necessary distance between truth as a regulatory ideal and our attainment of its cognitively accessible proxies, the tension which expresses open-endedness.

I shall aim to develop a theory of reason in which its open-ended character is central, as befitting its fundamental characterization as that by which we strive to transcend our limitations. And reason will be theorized as a regulative ideal, hence as having a normative capacity and one that exhibits an ineliminable tension arising out of the distance between that ideal and its cognitively accessible proxies. It will be that tension that expresses open- endedness. Objective knowledge about rational process is the result of rational development.

6.1.3.2. Theorizing Epistemology Naturalistically. The general naturalistic epistemology I advocate has already been sketched in chapter 1 and outlined in some more detail in Hooker 1987. The discussion here will, therefore, be quite brief. The essential point is that knowledge is treated as a theoretical term in a theory of cognition. As such, one looks to develop a theory of cognition that has its roots in the biology, and psychosociology, and cognitive history of the community of knowers, ourselves. This requires that we develop an evolutionary epistemology.

The fundamental motivation for this is the development of a unified conception of life; one wants cognition to grade off properly along the evolutionary sequence, so that its current forms can be seen as the current manifestation of more fundamental processes. Thus, for example, cognitive judgment grades off to conditioned response, then to reaction, and ultimately to physical action. And scientific institutions grade off through simpler forms of cooperative strategy based on information sharing and are ultimately rooted in biochemical (e.g., cellular) cooperative systems. It is also required that this epistemology provide an account of the psychogenesis of knowledge in individual humans and the cognogenesis of knowledge in the human population historically, and that its normative demands be rooted in our best scientific (psychological, sociological, etc.) understanding of our actual capacities.

One important upshot of these requirements, combined with the failure of the formalist program for a logic of science, is the switch to a decision theoretic model of epistemology, as sketched in chapters 1 and 2. Theorizing knowledge claims as claims for rational acceptance relative to a range of cognitive utilities that serve as proxies for truth has a number of advantages: (1) It brings the model into productive interrelation with the account of truth offered above. (2) It provides a theoretical framework within which it is intelligible how epistemology and scientific theory may interact and mutually inform one another. After all, it is a quite general fact about intelligent agents that they both use their present values to intelligently explore the world and use the experience deriving from that exploration to intelligently change their current values. In this way, we attach normative learning in epistemology firmly to a quite general feature of intelligent agents that will have to be recognized by any empirically adequate theory of them. So (3) we have thereby provided the theoretical framework within which we can give an account of learning about the nature of knowledge and the nature of rational cognitive norms and hence provide a historical theory of cognitive dynamics, progress in the theory of knowledge. The formalist theory of reason and epistemology precisely cuts us off from this crucial feature of our own capacities and historical experience (chapter 1). (4) The decision theoretic structure also provides an appropriate framework for theorizing the function and dynamics of cognitive institutions, the essential dimension to scientific knowledge that is rooted in our social capacities. As noted in chapter 2, institutions are both a direct extension of our individual capacities and essential to the nature of objective knowledge.

There are two systematic complementary reasons for developing a theory of cognitive institutional design (chapter 2). The first is to complete the characterization of science as a regulatory system, a regulatory subsystem of the total regulatory system characterizing this planet. This provides a unifying perspective from within which to understand the roles of individuals vis-á-vis their institutional groupings, the evolution of observational methods vis- a-vis technology and that of technology, vis-a-vis theory and practice, the relationships between science and society, and so on. The second reason for developing institutional theory is to provide a theorizing of the implementation of reason within the scientific regulatory sub-system through theorizing objective processes. Here the result to be emphasized is that both individual and institutional rationalities play their roles and that no consistent theorizing of science would be possible without both, and further, that no understanding of the dynamics of science is possible without an appreciation of the patterns of mutual conflict and reinforcement that these various rationalities display, engendering complex patterns of consensus and dissensus across scientific institutions. This entire perspective is subverted and suppressed by a logic-centered formalist theory of reason. A decision-theoretic theorizing of reason as strategic action, by contrast, provides a rich and promising framework for capturing these complex features of science. More positively still, the decision theoretic framework actively encourages an appreciation of the complexity to individual and institutional reason.

To complete a naturalized epistemology requires providing a theory of objectivity showing how the resulting structure of objective knowledge processes and accepted objective knowledge content are not only rooted in our actual biological history and present capacities, but also are explanatorily adequate for understanding our actual cognitive history. This is much too large a task to be attempted here, and in any case it would distract from the focus on naturalizing, and naturalizing reason in particular; part of it is sketched in chapter 2, section 2.II.1.

It is time to turn explicitly to the theorizing of reason. Just as Quine earlier proved a useful foil for developing a notion of naturalization, here it will be quite useful to begin by considering Putnam's recent argument that reason can never be naturalized.

6.I.4. Putnam against Naturalized Reason

Putnam 1982 attacks the proposal to naturalize reason, and he attacks evolutionary epistemology in particular. I shall argue that his attack is instructive but unsuccessful. An examination of Putnam's discussion will instead lay the groundwork for a critical discussion of the requirements of a naturalistic program for reason.

Putnam begins by considering two naturalized definitions of reason. The first, which he labels "evolutionary epistemology," defines reason as a "capacity we have for discovering truths. Such a capacity has survival value; it evolved in just the way that any of our physical organs or capacities evolved. A belief is rational if it is arrived at by the exercise of this capacity." To this conception of reason Putnam levels two objections. The first is that it requires "at bottom, a metaphysically 'realist' notion of truth—truth as correspondence to the facts. . . . And this notion . . . is incoherent." To this charge, Putnam adds another: "We have no way of identifying truths except to posit that the statements that are currently rationally acceptable (by our lights) are true. . . . This characterisation of reason has thus no real empirical content." (8). The second objection is to the effect that the notion of a capacity is not well defined because it does not pick out any neural function that can be identified as such; rather, whatever neural functions support a given capacity of this kind "can only be separated by looking outside the brain, at the environment and at the output behaviour as structured by our interests and saliencies."[13]

The second position criticized is the reliability theory of rationality (Goldman 1986). On this view, a rational belief is defined to be one that is arrived at by using a reliable method. The latter is one that leads to some suitably high frequency of true beliefs in the long run. Again, Putnam makes two objections, of which the first is that this definition also requires some realist correspondence notion of truth, which is incoherent. The other objection is simply that reliability by itself is not enough, since one may come to beliefs by methods that are reliable but that are not intentionally employed because they are reliable. In short, even if reliability suffices, the assessment of reliability itself must be rationally based, thereby rendering the whole notion circular.

The conclusion Putnam draws from these two criticisms provides a theme for his entire article.

> What I am saying is that the "standards" accepted by a culture or a subculture, either explicitly or implicitly, cannot *define* what reason is, even in context, because they *presuppose* reason (reasonableness) for their interpretation. On the one hand there is no notion of reasonableness at all without cultures, practices, procedures; on the other hand, the cultures, practices, procedures we inherit are not an algorithm to be slavishly followed. . . . Reason is, in this sense, both immanent (not to be found outside of concrete language games and institutions) and transcendent (a regulative idea that we use to criticise the conduct of *all* activities and institutions). . . . Our task is not to mechanically apply cultural norms, as if they were a computer programme and we were the computer, but to interpret them, to criticise them, to bring them and the ideals which inform them into reflective equilibrium. (Putnam 1982, 14; Putnam's emphasis)

This is the main basis on which Putnam rejects other views. Following the criticisms of the two positions above, Putnam goes on to discuss cultural relativism and the cultural imperialism to which it leads, and both Quine's positivism and his epistemology naturalized. Putnam argues that cultural relativism is self-refuting and cultural imperialism is dangerous and, in our culture, self-refuting as well. And about positivism, he says that, quite aside from all the valuable activities positivism excludes as nonrational, it is not self-reflexive. It "produced a conception of rationality so narrow as to exclude the very activity of producing that conception" (ibid., 18). Quine's naturalizing is included here, as one example of an eliminationist position—eliminate reason as a normative category entirely—which Putnam concludes is tantamount to "attempted mental suicide". Putnam concludes his paper as follows:

> If there is no eliminating the normative, and no possibility of reducing the normative to our favourite science, be it biology, anthropology, neurology, physics, or whatever, then where are we? We might try for a grand theory of the normative in its own terms, a formal epistemology, but that project seems decidedly over-ambitious. In the meantime, there is a great deal of philosophical work to be done, and it will be done with fewer errors if we free ourselves of the reductionist and historicist hang-ups that have marred so much recent philosophy. If reason is both transcendent and immanent, then philosophy, as culture-bound reflection and argument about eternal questions, is both in time and eternity. We DON'T HAVE an Archimedean point: we always speak the language of a time and place; but the rightness and wrongness of what we say is not *just* for a time and a place. (Ibid., 21; Putnam's emphasis)

Now it is time to assess Putnam's position. There are three kinds of remarks made by Putnam: (1) his specific rejection of evolutionary and reliability epistemologies, (2) his generalized argument for non-naturalism from nondefinability (see the opening sentences of each of the last two quotes), and (3) his other general remarks summarized in the two quotes above. I shall criticize and reject (1) and (2). But I shall accept (3),[14] indeed argue that it points the way to a satisfying naturalism! In the long run, this is all that matters.

Consider (1). First, Putnam's rejection of candidate naturalized theories of reason is less than convincing once it is understood that his arguments that the correspondence theory of truth is incoherent presuppose confining truth to some narrow semantic function; contrast section 6.I.3. (Though evidently there is some problem in formal semantics to resolve.) Second, why should it be objectionable if reason turns out to be a complex relative, or a complex relational, capacity of humans-in-an-environment?[15] What would be intrinsic to the organism is whatever regulatory structure is requisite to being able to develop these capacities in various environments. Like all epigenetic processes (e.g., the capacity of a plant to grow toward the light, wherever it comes from, not just to grow up in a fixed pattern), these will be higher-order regulatory capacities. Small wonder Putnam did not find them among the simple formulations he considers. Though I do accept his second objection to reliability epistemology, Putnam's other criticisms of his candidates for naturalized reason are unpersuasive.

His choice of candidates themselves is, however, even less persuasive. Both of the alleged attempts at naturalized reason he considers conceive of reason in terms of a *product*, and a product *guaranteeing truth* at that. Neither of these features is plausible from an evolutionary naturalist viewpoint. What we have learned of ourselves, our origins, and our universe gives us good reason to accept that we have accumulated some insight into the nature of things, truths within suitable error bounds, and good reason to reject any notion of a guarantee of this, to accept instead systematic fallibilism (cf. the discussion of openness in chapter 1 and above). And the core characterization of reason, that it be our capacity for transcending limitations, characterizes a process, not a product. Looking to what is produced instead of to the open-ended process through which it is produced is looking in the wrong place to understand reason.

Reason regulates how we go; where we end up is also a function of where we began, the equipment we bring to the journey and the character of the environment through which the journey is taken. A product focus also prevents recognition of learning about, and developing the structures of, rationality as we go, but this is a crucial feature. Like the baby learning to walk, as our journey proceeds successfully our control over our going also improves. So even had Putnam's specific objections to his two candidates been convincing, this would have been irrelevant to the kind of theory of naturalized reason that I would regard as plausible.

Caution: These remarks do *not* make success irrelevant to an account of reason. How else could we ultimately naturalistically judge the adequacy of a theory of rational procedure except through its capacity to organize successful action? How else do we do so? Consider progress in scientific methodology (chapter 2). Reason remains our best chance of succeeding (in any except perhaps the most pathologically contrived universes). But this claim is quite different from claiming that reason guarantees delivery of a certain proportion of truths, or even that it in fact delivers them. These are not only different kinds of claims, they also point in the wrong direction. Recall Popper's point that aiming at security (read reliability) and informativeness conflict. Theorizing reason in terms of reliability leads to an emphasis on securing truths rather than on rational risk taking. The grandmaster who loses a chess game has not therefore played irrationally; the same applies to the lesser skilled who struggle to win but most often lose.[16] Even if (pace Popper) both security and risky content play a role in rational cognition, in any evolutionary setting risk taking remains fundamental to the exercise of reason (see section 6.I.2.2 above and Hooker 1987). But if success is not necessary for rationality, neither is it sufficient, since a procedure or strategy may succeed for other causes. Nonetheless, rationality defeasibly entails success: rational practices should normally confer (improvements in the rate of) success; if a set of them does not do so, then that is prima facie grounds for doubting their rationality, because otherwise a specific explanation for their failure is required.

In traditional terms, the alternative to a direct reliabilist conception of reason is not an attempt at an indirect reliabilism via evolution, but a reflective equilibrium account. (Putnam makes his task easier by simply not considering this alternative.) But reflective equilibrium at least fixes on a regulatory process as the central feature characterizing reason. Of course a naturalist will not want to give commonsense judgments a privileged role in this process (pace Goodman, thereby relieving Stich 1989 of some of his problem; see section 6.II.6 below). And it will be natural to see in philosopher's accounts of reflective equilibrium a quasi-formal conception that needs replacing with a thoroughly regulatory one. Within this process, internal criteria, even formal criteria like consistency, can play their important regulatory roles, for example, in the general manner outlined by Piaget (chapter 5). And again, success is not eliminated, for ultimately the improvement of internal criteria must itself be driven by overall success (cf. Rescher's analysis, chapter 4).

Next consider (2), Putnam's generalized argument for non-naturalism from non-definability. Here is the opening sentence of each of the last two quotes: "What I am saying is that the 'standards' accepted by a culture or a subculture, either explicitly or implicitly, cannot *define* what reason is, even in context, because they *presuppose* reason (reasonableness) for their interpretation." "If there is no eliminating the normative, and no possibility of reducing the normative to our favourite science, be it biology, anthropology, neurology, physics, or whatever, then where are we?" Evidently we are left with constructing normative principles of rationality, ultimately universal formal ones or fragments thereof. What are we to make of these assertions in the context of a critique of naturalizing reason? They evidently appeal to the following underlying argument: Either normative reason can be eliminated outright or it can be reduced to science or it is normative and non-natural; any attempt to eliminate or reduce normative reason leads to vicious circularity. Therefore, reason is normative and non-natural.

Notice first that in this form the argument has the same structure as Quine's earlier argument for the opposite conclusion about epistemology (section 6.I.1). There we saw that, while an intermediate premise was acceptable, a thorough naturalism would reject the opening disjunction as well as the conclusion. So too here. Hidden in the first premise is a leap from being irreducibly normative to being non-natural. We may recast the argument to make it explicit: Either normative reason can be eliminated outright, or it can be reduced to science, or it is irreducibly normative. Any attempt to eliminate or reduce normative reason leads to vicious circularity. Therefore, reason is irreducibly normative and therefore, reason is non-natural.

Now both premises are accepted. But what licenses the last move? So second, we see that ultimately Putnam's argument rests on the undefended supposition that there is no naturalized account of an irreducible normative possible. But this is another hidden piece of empiricist metaphilosophy that should be rejected here just as it was for Quine's argument. (Putnam's argument parallels Moore's argument for a non-natural account of goodness and fails for similar reasons.) And along with that, goes rejection of the implicit assumption that normative reason will be formal (section 6.I.2.2). More precisely, the proper response is to insist that reason be *theorized*, that rationality be treated as a theoretical term in a theory of cognition that represents reason as regulating action (including inner action) under an ideal. Then, as set out in chapter 1, in virtue of the understanding conferred by cognitive theory, rationality will have a normative role.

We are now left with (3), Putnam's general comments about reason summarized in the two quotes above. These can now be accepted. Perhaps surprisingly, they are essentially compatible with, indeed supportive of, evolutionary naturalism as conceived herein. It is essentially only Putnam's supposition that what he has to say is antinaturalist, which is to be challenged. It will be helpful, then, to extract from Putnam's comments some constraints on an acceptable theory of reason. Putnam insists that:

C_1: Reason is normative and the normative cannot be eliminated. Hence a theory of reason is not to be wholly eliminated or to be eliminated through reduction to some science.

C_2: Reason is an autonomous capacity; it is not to be defined solely in terms of other normative capacities, such as truth.

C_3: Reason is to be conceptualized as a regulative ideal, among other ideals (such as truth), an ideal that we are permanently exploring but that we never fully attain. The exploration of reason is open-ended.

C_4: At any given time, however, the specification of reason is context-bound, the relevant contexts being those of our cultures, traditions, practices, and procedures.

But it now follows from a naturalistic theory of theories and norms (chapter 1) that theorizing reason already satisfies Putnam's four constraints. Since the normative is retained, C_1 is met. Since the concept of reason plays an irreducible role as a fundamental theoretical term in cognitive theory and is not eliminated either in favor of other fundamental theoretical terms or by reduction to other theories, C_2 is met. Moreover, this is a functionally appropriate requirement, since it is functionally important to creatures that their decision processes achieve operational closure and so issue in specific responses. Next, the cognitive theory within which the concept of reason is embedded is a theory of an intrinsically open-ended dynamics governed by regulatory ideals, and reason is theorized as one of these regulatory ideals, so C_3 is met. Again, from a naturalistic perspective, this is an appropriate requirement. From an evolutionary point of view, as illustrated in the discussion of the normative concept of truth above, it is central to our conception of norms not only that they provide determinate decisions (e.g., by declaring what is rational and what not on some occasion), but also that they make our decision-making framework permanently open-ended. This is the kind of regulatory structure that creatures evolving from evolutionary ignorance but are able to learn require if they are to go on exploring their world and to go on exploring it with ever greater facility. The ability to bring closure, to force a determinate decision, is a feature of normative systems most emphasized, but from an evolutionary point of view, it is their capacity to transcend their present judgments, to add new regulatory orders to the evolving system that is their most important feature. It is precisely this open-endedness that the third of the requirements asserts. And finally, since at any given moment in history, particular theories are always expressions of the particular intellectual reflection and concrete experience of which they are the current historical culmination, C_4 is met.

Moreover, these are characteristics required for any basic (irreducible) feature of life about which we are able to learn; they are not special to reason. (They will apply, for example, to Moore's goodness, and to truth.) Reason is a basic cognitive capacity about which we are able to learn. We learn about it by theorizing it in as large a unified scientific context as we are able to create. Thus, for example, we relate it to the organizational capacities of biological systems on the one side and to economic theories of choice behavior on the other. (The resulting normative-descriptive interactions, so important to current scientific development, are unintelligible to empiricists; see chapter 3.) So then, the very characteristics that Putnam sees as ensuring that reason cannot be naturalized, these are the very characteristics that link it most intimately to its roots in broader natural phenomena. We are now in a position to turn to a positive account of naturalized reason.

6.1.5. What Is It to Naturalize Reason?

The following four requirements are surely necessary steps in the development of a naturalistic theory of reason:

1. Develop a naturalistic theory of the normative in general.
2. Show where reason fits into a naturalist theory of human agency, and into naturalist philosophical theory generally, and with what consequences.
3. Show how reason can be embedded in a broader biological setting as a natural phenomenon, and with what consequences.
4. Show how reason is in fact concretely, historically implemented, and what are the consequences.

Now add to these four steps a fifth:

5. Steps 1 through 4 are to be taken in such a way that the theorizing of reason can meet the four constraints C_1 through C_4 which followed from Putnam's discussion.

I suggest that these five steps together constitute a sufficient condition for a naturalized theory of reason. It is not really necessary to discuss steps 1 and 5 in any great detail. Step 1 is already accomplished (see section 6.I.4 above and chapter 1). And a good general idea of how the strategy operates has already been given in previous sections, including how it leads immediately to the general satisfaction of step 5. The remainder of the chapter will therefore be occupied with steps 2, 3, and 4. Each of these represents an enormously complex set of issues. There will be space here only for partial analyses. Moreover, our current ignorance (mine anyway) requires that many issues be left open. Nevertheless, the next section provides major components for step 2, and the subsequent sections provide briefer treatments of steps 3 and 4.

6.II. The Nature of Reason

6.II.1. Reason, the Regulation of Judgment

In order to make progress in completing the remaining three steps, it is necessary to have a more specific characterization of the nature of reason. Let me emphasize at the outset that what follows is an outline or a sketch of a theory of reason, and as such, it ultimately needs detailed empirical investigation and must find support from that investigation. At the same time, it will have to show its mettle in insightfully explaining what is found. That is the nature of naturalized, normative theory. But there is no hope of undertaking that defence of the theory here, and that thrice over: (1) Much, perhaps most, of the relevant empirical evidence and theory is simply not available yet; (2) what is available shows our cognitive activities to be excruciatingly complex and ambiguous in many of the relevant respects (e.g., driving Cherniak 1986 to a proper rejection of traditional idealized models of rationality; cf. Hooker 1994c) and (3) the task is well beyond the reasonable scope of this book. So here let the theory rest on its prima facie plausibility, even reasonableness.

Intelligence is the capacity to form appropriate judgments. Forming a judgment is simply coming to a conclusion. And coming to a conclusion is, at its crudest, no more than a change of internal state relevant to some response behavior (including internal responses). Understood in this way, judgments begin with simple reflexes (e.g., the amoeba's avoidance reflex to particular chemical concentrations over a range of pH values), where a simple external stimulus produces a change of internal state relevant to behavior.

Thereafter, judgments become successively more complex as biological creatures themselves become more complex. The feature bundles of the stimuli that evoke judgments become successively more complex and differentiated, while the internal state changes that the stimuli evoke and their relationship to subsequent behavior do likewise. As a crude measure, we judge the intelligence of an organism by the number of nested conditionals on which behavioral response depends. Reflex responses are unconditional; creatures with simple nervous systems show at least one order of Pavlovian conditioning, and we ourselves typically deploy many orders and can arrange to deploy indefinitely many orders should the occasion demand. Of course capacity to form (and exercise) judgments depends not only on conditionalization, but on input and output channel capacities as well, that is, on perceptual and behavioral capacities (including their modifiability as well as their fixed engineering features).

So far, nothing specific has been said about the regulation of judgments. It is compatible with the basic conception set out above that trees exercise judgment, since they make internal chemical and functional responses to external environmental stimuli and these in turn are relevant to their subsequent leafing, rooting, and like behaviors. This initial breadth is an advantage of this conception, since it shows how the concept grades off down to elementary biological processes and hence how it can be integrated as part of the natural world. But within this range, it is necessary to distinguish the kinds of regulation of judgment that we recognize as characteristic of mental intelligence. One important differentiation is that of self-correcting judgments, judgment formation processes that employ tests for error, generating a sequence of judgments that converge upon a stable final judgment. Such devices range from elementary engineering control systems and their elementary biological counterparts to the sophisticated operation of human visual scanning and information-processing procedures through which invariants are extracted from the incident electromagnetic array. The cognitive face of such processes is critical judgment. And there is a complex structure of judgements about other judgments and about judgment formation subprocesses (e.g., about our color judgmnts when the light is dim). Again, the organization of this structure is a crucial factor in the kind of intelligence possessed. It is here that reason enters.

To give judgment this fundamental role is certainly appropriate to theorizing reason. However it is theorized, the product of every exercise of reason is judgment (i.e., a decision or decisions). Which pieces of evidence we accept and which evaluation of a theory we accept, for example, are themselves matters of judgment, as we say, and the acts of acceptance or rejection follow from such judgments. Brown (1988) has emphasized the similarity of rational cognitive judgment formation and skilled practical judgment formation. The latter are not arbitrary; they are learnable, improvable, criticizable and are ultimately constrained by empirical success. The capacity to exercise them is not equally distributed across people, and even the experts are fallible. So a community of practitioners shows the same general characteristics as a community devoted to the systematic improvement of more abstract knowledge. Yet improvement in the practical skill across the community does not depend upon following rules; to the contrary, practical skills are typically marked by their context-dependent complexity and by innovation, thereby forcing rule descriptions to trail practice and to remain as crude approximations. It is marked instead by complex patterns of judgment formation under evolving community practices. So too it can be for abstract judgment. Indeed, from a naturalist, nonformalist perspective, there is no longer reason to dichotomize the two.

Moreover, judgment has the right character to underlie theorizing all the activities of reason, not just the pursuit of truth. For judgment is the basic act that transcends the distinction between fact and value:

> In spite of the impression created by philosophical reflection upon adult life, there is probably no deep affective/cognitive division in the young child. To the contrary, one of the major cognitive achievements of the child is that of establishing affective order within him or herself. Conversely, the extending of a coherent affective response framework to increasingly large and complex environments is a major driving force behind the development of intellectual framework. The judgement is the basic entity; a factual judgement has a normative dimension, for every factual judgement involves a deliberate selection, a deliberate design of the observing instrumentation; conversely, a normative judgement has a factual dimension, for every normative judgement presumes an actual situation which structures the factors which are relevant and determines the terms in which the judgement is made. We explore what it is to be a person in essentially the same manner in which we explore the world into which we were born.
>
> The primary form learning takes is the evolution of an increasingly rich and differentiated framework—cognitive and/or affective. Actual relationships among the elements of

> the framework take a secondary place, being for the most part either subconscious or only vaguely expressible. Thus in visual perception, for example, the crucial kind of progress in the young child is the evolution of perception of an external world of relatively stable objects. In this evolution it is the form of the organisation of the sensory information itself which is the important thing. . . . the same process applies in all other fundamental learning as well . . . (Hooker 1987, 241–2.)

It is this basic model that will be developed here. Roughly speaking, reason is a capacity to modify, organize and extend judgment formation so as to be able to make progress in transcending limitations and imperfections through moving toward realizing the principal regulatory ideals: truth, goodness, beauty, and reason.

Reason includes the regulation of intelligence in the pursuit of improved reason. This apparent circularity clearly needs comment. Reason as itself a regulatory ideal is included in order to capture this basic feature: We are able to develop and modify, not only the cognitive, ethical, and aesthetic contents, but also the reasonableness of our cognitive processes. We use reason to learn about and improve reason. (Or once again, we prevent ourselves from doing so only at the cost of cutting ourselves off from our full capacities and self-realization.) Reason then pursues its own objective understanding and also the balance between particular forms of reasoned life. This is the self-reflexive character of reason.[17] An improved structure for reason is what emerges from the improvement of languages, relevance structures, scientific methods, and risk strategies and so on. Reason is a kind of organization; the ideal of reason is an ideal organisation, and the improvement of reason through reason a particular self-organization. This marks off reason from other ideals only in one respect; what is aimed at in their cases is an ideal product (truth, goodness, beauty), but rationality aims at an ideal process or organization.

Reason then is to be theorized as a particular kind of regulatory structure to intelligence, grading off down the evolutionary sequence in rough proportion as intelligence itself grades off. A mouse shows less capacity to reason about the location and obtaining of food than does a primate, and a lizard still less so. But intimate though their interrelation may prove to be, we should not simply identify reason with intelligence. Judgments may be detailed, numerous, and even complex without being organized in a reasoned manner. A frog responding to moving objects and shadows, both food and foe, may well have judgment formation structures as complex in some respects as those exhibited by a mouse, but they are organized far more crudely in respect of their reasonedness. Computers and scientific instruments provide other examples of systems with complex judgment formation and low reasoning capacity. Our everyday distinction between cunning and reason as two forms of intelligence reflects the same awareness. Indeed, it is well known to be advantageous in a range of circumstances not to reason about one's responses but simply to develop very complex but highly specific unreasoned judgment formation processes, for instance, in all of the motor skills of the sports and arts (e.g., making a badminton stroke or playing the violin).[18] Reason then designates a particular regulatory character, that which makes us capable not merely of complex behavior per se, and not merely of behavior with survival value per se, but behavior characteristic of a creature whose intelligence can ultimately only be theorized in terms of regulatory ideals.[19]

Reason works with conceptualizations, for that is how judgments are internally structured and interrelated. Concepts may be tentatively theorized as relatively stable particular ways of forming discriminations, perhaps as particular patterns of node connection weights in some neural network (cf. Hooker 1975, 1993d). The formation of a judgment is always a discriminatory act; certain features of the incoming information invoke the relevant internal change, and it in turn is relevant to particular features of subsequent behavior. The concept that characterizes that judgment may be associated with the particular processes that instantiate those discriminations. Once again, concepts grade off in complexity with intelligence, and it is necessary to characterize the particular kinds of processing that characterize concepts in the neurally organized form of intelligence to which we are heir. But the point here is to show how we may go about the job of providing a naturalistic foundation for the theory.

Nonetheless, this generalized characterization of concepts already shows how we may give an account of the information-sensitive organization of intelligence without being committed to placing language at the centre.[20] Indeed, it offers the opportunity to develop a theory of intelligence in which language is seen as a particular specialization of a normally rational, sufficiently complex intelligence. And this frees us from the compulsion to look only, or even primarily, to the structure of language for the organization that represents reason. If language stands at the center of intelligence, then reason inevitably becomes focused on formal symbolic structures, principally logic. The activities adumbrated in the next paragraph strongly suggest that this is far too confining to embrace the manifold regulatory capacities that make up our rational capacity.[21]

But with a more fundamental characterization of intelligence the way is open to theorize reason in terms of complex regulatory systems capacities, including the capacity to develop and modify languages. Thus this is a framework of much wider scope than the formal-symbolic one provides.

The importance of this wider perspective is underlined by everyday skills, whose significance is indicated in their connection to contemporary developments in cognitive psychology. On many occasions, even a propositional representation of a problem together with its logically constrained computational solution algorithms may not be the most effective problem-solving approach. There is an important contrast between computed and fitted solutions to problems. A problem is solved by computed solution when it is represented symbolically (i.e., digitally) so as to be amenable to logico-mathematical modeling. Logico- mathematical algorithms are then applied to compute a symbolic representation of the solution. A system exhibits a fitting solution to a problem, by contrast, if it achieves a solution by causally fitting itself to a set of constraints. Carpenters often prefer to fit and mark a board for cutting rather than take measurements and calculate the locations of cuts. The fitting method is often simpler, faster, and more accurate, especially if the shape is complex. Control system engineering is filled with examples where it is more efficient to choose the control signal by fitting (e.g., by mechanically following a surface) than it is to compute its description. (Here a literal model is worth a thousand words.) All this shows that it will often be more efficient to employ fitting strategies over computational strategies. And the cases extend; the sun, moon, and earth simply are a fitting solution of the gravitational three-body problem, and whether or not an analytically representable, or even a computable, solution exists for this problem. Significantly, many trainable connectionist or neural networks may represent fitting solutions to cognitive tasks, at least they will do so if they resist cognitively relevant computational decomposition. They may do so if the causal implementation of the relevant tuning algorithm (e.g., the back propagation of error algorithm) for the learned adjustments leads to such resistance, as it seems it might. And indeed, analog processes generally may achieve fitting solutions by literally becoming causal models of the relevant system.[22]

It is not in fact necessary to my present purposes that any highly specific regulatory hypotheses concerning the nature of our intelligence should be set forth and defended here. And in view of our present ignorance, it would be unwise to attempt to do so.[23] (I think it is unwise to even assume that the difference is purely regulatory, and not also qualitative; cf. Hooker 1978, 1993d.) Nevertheless, intelligence certainly must have a distinctive regulatory structure, and it will be helpful to at least indicate here in a rough way the range of regulatory activities comprehended under the general conception of reason proposed above (i.e., to appreciate the complexity and diversity of reason). The discussion will be confined to the pursuit of truth. The pursuit of truth takes the form of constructing objective knowledge, which in turn requires capacities to: create, modify, and extend both theories and observations and to organize their suitable storage in memory; to create, modify, and extend relevance structures (including saliency, deductive inference, and statistical and ampliative inference structures) and risk-taking strategies; to create, modify, and extend relevant value structures and goals; to create, modify, and extend the diverse context-dependent web of efficient strategic methods for the pursuit of cognitive goals; to create, modify, and extend formal and informal language structures suitable to express what needs to be communicated; and to create, modify, and extend cognitive institutions whose roles support and reinforce the foregoing activities. (Something of the detailed complexity of these activities is provided in the more extensive examination in Brown/Hooker 1994.) Finally, reason includes organizing, modifying, and extending the mutual interaction of all these components so that they form a self-correcting system capable of improving both objective knowledge and this system of activities itself. This last is a very complex activity; for example, problems may be solved, or dissolved by transcending their initial specifications or resolved through ineliminable but temporary compromise, and reason plays an essential role in our deciding which kind of problem we are faced with.[24] Similar, though not identical, lists of facets would result from an examination of the pursuit of goodness and beauty.

One of these facets of reason deserves a further comment: risk. Products of evolution have evolved from a condition of ignorance. Their need to learn is profound, for they need to learn even what learning is, let alone how to use their own bodies to learn, let alone learn about the rest of the world. We are always only dimly aware of what is really involved; it has taken millennia to become conscious that we need to learn about learning. How does empiricism look as a cognitive strategy in these circumstances? Very poor. Why trust the senses? They may be partial, biased, and erroneous in all sorts of ways of which we are ignorant. Use them we must, but not uncritically. Why restrict science to the products of observation and logic? Caution? Fear of error? But caution is a poor policy for creatures born in profound ignorance and trying to find the truth. If they risked no theoretical ideas at all about the unobserved world, they would in effect be placing a blind faith in the accuracy and adequacy of their own bodies as observing instruments. Such a faith would have quickly mired us in unnoticed biases, errors, and limitations—we see only the color spectrum, not the rest of the electromagnetic spectrum; our perception is full of illusions. In addition, we would be quickly mired in dogma, for it would be difficult to develop penetrating alternative views of our situation with which to criticize those we accepted. As Popper taught us long ago, we have no choice but to risk bold theoretical conjecture about the real structure of the unobserved world (chapter 3). We must then attempt to contain our risks by boldly seeking to criticize conjectures, whether about theory or method, through critical comparison with alternative conjectures and through comparison with (conjecturally guided) experimentation. On the other hand, we must also prune our initial conjectures and their testing to within the limitations of our resources; this too is a risky undertaking, since better conjectures may be eliminated before they have been evaluated. Risk avoidance through rationalism provides an equally poor platform for cognitive policy, and on grounds paralleling those set out here for empiricism. We have no choice then but to accept risk as fundamental to rationality.

Reason has both a constructive, creative function and a destructive, critical function. The critical function of reason guarantees the perpetual open-endedness of the search for understanding through the revision and transcendence of current judgments. Balancing this critical function is a constructive function comprising all the constructive activities listed two paragraphs back. The outer envelope of this constructivity comprises judgments as to the appropriate constraints within which reformation of the existing system should take place.

Reason in the pursuit of reason should express itself as an objective theory of reason. What does this theory look like? We have as yet no really coherent idea, only fragments. Formal logic is one fragment of this theory. But even here, there is controversy about the nature of formal logic and massive proliferation of alternative formal logics at the current time. And there is the embedding of formal logic itself in the much richer structures of mathematics (e.g., as a characteristic structure of each topos). So even in this ancient and central fragment of objective reason, we are still exploring the nature of reason and indeed, scarcely seem to have begun on that exploration. We also have potential fragments of objective reason in various theories of optimal control, information processing, and economic decision. As we shall see shortly, optimization, expressed as generalized economy or efficiency, finds a central, if derivative, place in the theory of objective reason. (In this sense, it parallels the position of formal logic.) In each of these three areas too, we are learning what it is to efficiently and powerfully organize, modify, and extend judgment so as to increase objectivity, but once again, each of them is characterized by neotenous diversity, ambiguity, vagueness, and controversy. Consider the controversies in the psychology of reasoning (see Kornblith 1993; Rips 1990) and in the foundations of decision theory (see Hooker et al. 1978; McClennen 1990; Elster 1983, 1986). The stark contrast between the formal structures of logic and decision theory, our clearest fragments of reason, indicates the poverty of our present position. (I argue for the priority of the decision theoretic approach, but one substantially wider than that found in current economic models; see chapter 1 and section 6.II.4 below.) We have really only recently begun to seriously explore these features of intelligence.

6.II.2. Reason and Regulatory Ideals

Reason then drives dual processes of improvement in both knowledge and in reason. Or more accurately, since from a regulatory systems perspective both knowledge and reason belong to a single regulatory system with many internal, interrelated components (see chapters 2 and 4), reason drives improved content and organization across this system. Chapter 2 models improvement naturalistically as a process of regulatory superfoliation, a process occurring across the spectrum of regulatory systems. How does this process take place? Alas, we again have no detailed scientific conception of it. We have some beginning ideas about increasing order in dissipative systems and self-organization, about tunable neural nets and control processes and the like, but little else. To repeat a phrase, these areas are characterized by neotenous diversity, ambiguity, vagueness, and controversy. On the other hand, there is some progress that can be offered in other respects. I turn now to setting out a naturalist philosophical theory of the ideal of reason and its relation to the truth ideal (for which, see chapter 1), and this will be followed by a naturalist account of how the construction of ideals is possible for us. (Cf. here chapter 1 and Piaget's concept of developmental closure, chapter 5.) In the longer term, one hopes that these accounts can be unified with our developing scientific theory of self-organizing regulatory systems.

Reason is to be theorized as that capacity in virtue of which, within our finite resources, we transcend our imperfections. Reason then is an instrumental capacity to organize, modify, and extend judgment so as to systematically replace ignorance with information, prejudice, and partiality with critical appraisal, and so on. The ideal of reason has to be an ideal of *use* of this instrumental capacity not, as in the case of truth, an ideal of *destination*. The destinations are correlative to the imperfections (e.g., complete truth is the ideal corresponding to the imperfections of ignorance and error) and provide criteria for what are the proper ways of replacement (they should, for example, be relevant to striving for the truth), but do not determine our tools or capacity to use those tools in pursuit of them. That is the business of a theory of reason. This is also the lesson that emerges from the critique of Putnam earlier: Rationality is to be characterized in terms of process (use), not product (destination).

It is easy to miss these crucial distinctions. Traditional or "received" rationality sets down formal global requirements for agent rationality; for example, all beliefs must be consistent, and all deductive consequences of rational beliefs must also be included among rational beliefs. These requirements make no essential mention of any features of actual agents, so all reference to agent capacity and process, hence to use, is lost. The ideals are simply given and ought to be achieved; *ought* implies *can*, so it is assumed that agents can achieve them (somehow), at least "in principle"; all the rest can then be set aside as "psychology." In doing so, the traditional account implicitly assumes that rational agents have certain formal capacities, such as being able to check for global consistency of their beliefs, which finite agents cannot have, because either these capacities are not formally accessible at all or would require infinitely fast processing times, infinite memory capacity, and so on. In consequence, conversely, the traditional account assumes that certain highly relevant capacities that finite agents do have are in fact irrelevant to being rational (e.g., the capacity for nonformal judgments about context-dependent standards of evaluation, algorithm use and so on). As a result, the distinction is lost between rationality as characterizing a process, namely the use of certain tools to (partially) transcend our imperfections, operating within our finitude, and the ideals toward which that process is directed (e.g., complete truth). Instead, certain formal conditions, specifying properties of a logic or logics, are taken as both giving the ideals of reason and as specifying necessary conditions for achieving the truth ideal.

Since for finite, imperfect agents, contrarily, these formal logical conditions cannot be met, the process of striving toward the ideals under the constraints of limited resources, fallibility, and the like becomes crucial for characterizing rationality. Rational capacities center around the context-dependent organization, modification, and extension of resource-limited, more-or-less risky heuristic strategies for fallible but correctable nonformal judgment formation in the pursuit of such ideals as truth via explanation. But taking formal properties of logic as the ideals of reason suppresses this, collapsing out these fundamental features. Traditional rationality theory is degenerate; its psychology-free, context-free, risk-free, computational-cost-free degeneracy hides/distorts the fundamental character of reason for simply finite creatures, let alone for imperfect finite creatures.[25] In a less degenerate account, logic becomes one tool among others, with characteristic advantages and disadvantages, and while global consistency and the like remain necessary conditions for achieving the truth ideal, it can no longer be expected that they will provide the primary characterization of rationality. Here formal tools have an instrumentally useful role to play but only within this larger framework. This strategic structure is neither sui generis nor immutably necessary, but evolves in dynamic interaction with our actual knowledge, social organization, technological capacities, and so on.

Why do philosophers want to impose the "all" conditions in the first place? Consider the requirement that all beliefs be consistent. Is not this ultimately because consistency is a necessary condition of truth? If T claims to be true, then T must at least be consistent. Likewise for deductive closure, which is necessary for complete truth and so knowledge completeness. But while consistency and completeness may be ideals we pursue (however we formulate that complex idea), they are primarily ideals of knowledge (i.e., of some form of rational belief) rather than of rationality per se. What philosophical theory offers is really a theory of the knowing agent (ethical agent, etc.) that is a composite account involving complete truth and rationality. Moreover, it is an account of only the end-state, idealized knowledge, not of the process of arriving. Of course beliefs that were not consistent, etc., could not be true and complete, so could not realize idealised knowledge. So it must be rational to require these conditions. Hence the usual rationality conditions. But the considerations of the preceding two paragraphs provide reason to rethink how to theorize the matter.

The general conception of normative ideals appropriate to finite imperfect creatures is essentially the Kantian notion of a regulatory ideal (i.e., of a goal or end to be striven for), whether or not it can be actually achieved, because it is judged valuable and attempts to move in its direction are feasible and have at least some beneficial consequences. The principle of a fully explanatory science is a regulatory ideal in this sense, since while it evidently cannot in practice be achieved, it is valuable and improvements in the explanatory scope, precision, and power of our theories are feasible and the search for these improvements has beneficial consequences (Hooker 1994c). And truth is a regulatory ideal, which grounds the explanatory ideal.

From within this framework, we need a rethought conception of the regulatory ideal of reason. A formulation along the following lines seems most appropriate. The ideal of reason is the perfect use of the instrument by which we seek to transcend present limitations; it is to have canvassed all relevant reasons for each action or belief, to have assessed them using all relevant assessment, arguments or procedures, and to have subjected these reasons, their assessment and the grounds for these in turn to all possible criticisms. The fallible, improvable, learnable tools available to us in pursuit of this ideal include observation, processes of criticism (e.g., carrying out tests, proposing alternative explanations) and such formal tools as various logics and decision theories.

It should be clear that this is a regulatory ideal. First, to be ideally rational is desirable, because we would not only be as sure as possible that our conclusions were correct but also have a maximal reflective understanding of why they were correct. Second, it is possible to move toward the state just described; that is, we can extend the rational basis for our beliefs by extending the range of evidence considered, improving our assessment procedures (e.g., by improving observational power or statistical inference), enriching the criticism (e.g., by examining alternative hypotheses), and so forth. Note that all this includes reflexively improving our ability to be rational, and in a quite concrete way; the instruments, institutions and the like we use to elaborate the regulation of our judgments are real features we have created. Rationality is an instrument, and like other instruments, it may be improved (within natural limits). Cognitive progress is centrally concerned with the improvement of instruments, especially with the improvement of the instrument of progress itself (see also Hahlweg/Hooker 1989a). Third, the attempt to improve our rational instrument has desirable consequences, since these improvements amount to improvements in the bases for our actions and beliefs together with improvements in our understanding of these bases.

This is the right kind of formulation of a rationality ideal for finite imperfect creatures, since it emphasizes the agent process as the heart of being rational. It captures a minimal normative rationality requirement as a specification of the maximal movement toward this ideal feasibly available to humans, context by context: to have canvassed all feasibly available relevant reasons for each action or belief, to have assessed them using all feasibly available relevant assessment arguments or procedures, and to have subjected these reasons, their assessment, and the grounds for these in turn to all feasibly available criticisms, maximizing cognitive value return against assessment process cost.

The old ideals of reason—global consistency, deductive closure, and the like—are, as noted, necessary conditions for achieving a more encompassing ideal, the knowledge ideal, which includes the complete-truth ideal as well as the rationality ideal. They are properly derivative ideals of the complete-truth ideal. For finite imperfect creatures, all these derivative ideals are of course also ideals in the regulatory sense; that they are unachievable does not preclude them from occupying this role so long as they are valuable and attempts to move in the direction of satisfying them are feasible and have at least some beneficial consequences, which is surely so. In our pursuit of complete truth, we find it rational to try to meet these derivative ideals, within our constraints as expressed in the minimal normative rationality standard. Insofar as they are partial necessary conditions for complete truth, the derivative ideals could stand as proxies for it, but proxies ought also to be actually accessible, and the derivative ideals are still inaccessible ideals. So the proxies remain security, explanatory and predictive power, scope and precision, and the like.

There is a further complication: because of their connection to the complete-truth ideal, the derivative ideals also express the formally desirable properties of a major formal tool we use in the pursuit of the rationality ideal, deductive logic, the theory of truth-preserving inference. If one ignored the cognitive imperfections of our actual constitution and our finite capacities, then one might come to believe that logic alone sufficed to meet the explanatory ideal (as empiricists tried to do; see chapter 3). In which case, one might come to conflate the two roles for the idealized rationality conditions and take them as expressing the ideals of reason instead of (derivative) ideals of complete truth and so of their product: complete knowledge.

Let us take stock. Traditional analytic theory presents an account of the knowledge ideal, composed of ideals of rationality and complete truth, and it assumes an idealized theory of agents in which they achieve that ideal. A realistic theory of finite imperfect agents has to recognize that the degenerate idealization involved collapses out essential structure. In consequence, it is required to (1) distinguish the two ideals, (2) reformulate the rationality ideal, and (3) embed the logical tools for specifying the achievement of the degenerate ideals in the richer tools for striving for the rationality and complete-truth ideals.

The methodological line pursued here can be schematically presented as follows. According to the traditional analytic view, being rational consists in meeting a set of conditions (completeness, consistency, etc.) I whose achievement specifies an ideal rational state R. Then we add the hypothetical imperative: If I is to be achieved, then do action A (check for completeness, consistency, etc.). Since R is an obligatory ideal, this provides the norm: Agents ought to do A. Indeed, doing A is taken as both necessary and sufficient for achieving I. But finite, imperfect agents cannot possibly do A, so cannot possibly achieve I or be R and, since ought implies can, that they are not obligated to do A or achieve I or be R.[27]

Instead, the theory of rationality should follow a methodological norm M drawn from science. Instead of the traditional theory, construct I′, R′, A′ for rationality such that (i) R′ is the achievement of I′ and specifies the ideal rational state and I′ has I as a degenerate special case in the limit where agent capacities are idealized; (2) the hypothetical imperative—if I′ is to be pursued (not achieved) then do A′—and consequent norm—do A′—are defensible, in particular that pursuit of A′ is possible and beneficial for finite imperfect agents and A′ has A as a degenerate special case in the same limit; (3) retain I as a necessary derivative ideal of an ideal of complete truth. The argument for following M is that, from naturalism, philosophical theories are like scientific theories in being fallible, learnable, improvable constructs and that science properly accepts an explanatory ideal E in which degenerate theories are recaptured as limits of the their less degenerate successors, and that M follows from, or is in context the best available way of realizing, E (see note 25).

There remains the question of proxies for ideal rationality. Construing these similarly to proxies for the truth ideal, we look for accessible, rationally warranted (not guaranteed) indicators of progress toward I′ (i.e., measures for the achievement of the minimal normative rationality requirement). One might consider some measure of adaptability, degree of closure (in Piaget's sense, chapter 5), widths of various kinds of cross-situational invariances, capacity for error identification (criticalness), or efficiency. However, I know of no neat collection of properties of cognitive contexts and/or contents to offer as such measures that stand out in the way that, say, explanatory completeness does as proxy for the complete-truth ideal. Perhaps this is because these measures themselves are so very theory dependent, which means they are cognitive context dependent.

What counts as total relevant evidence will be a factor, not only of the methods used, but also of what our theories say about the connectedness of the world. (Compare in this regard the very different connectednesses provided by classical and quantum mechanics or by classical Darwinian and nonlinear dynamic systems biology.) What counts as the most important evidential assessment argument depends on the current status (which is constantly changing) of the foundations of the theory of statistical inference and like theories, and practical wisdom dictates choosing different statistical methods to use from among the competing varieties according to the circumstances. And so on. In any event, I shall not attempt even a partial theory of rationality proxies here, contenting myself to note that in any specific context there clearly are proxies for proceeding rationally.

A theory of reason has been proposed in terms of progress toward some regulatory ideals. It would have been less onerous to have theorized reason simply as economy in the pursuit of some utilities. (These latter would in effect be fixed current proxies for the regulatory ideals at each given time slice.) The present choice of this more complex theoretical form is preferred because of the involuntary character of the regulatory ideals and the way we change our proxies for them as we learn more. We do not choose which ideals we find ourselves forced to recognize. And once having recognized their existence, we can only shut ourselves off from their pursuit by doing some violence to our intelligence, indeed essentially by freezing our development at some stage of relative egocentricity and/or homocentricity. We can, however, dispute them as more sceptical naturalists like Churchland 1989 and Stich 1989 do; this much is part of fallibilism and rational learning. We have, as Plato might say, rational appetites for the regulatory ideals (though not entirely natural appetites; the appetites grow as the life of reason is practiced). This is the position I would defend as "most fair to human experience," even though it evidently places considerable strain on the naturalization program. Indeed, there are circumstances in which I should be prepared to drop naturalism.[28] Nonetheless, it seems to me that we can have this much of Platonic insight within a naturalized theory. Let us see how.

One way to proceed would be to look for reductions of the regulatory ideals to more clearly naturalist features. We might, for example, consider the ideal of truth as emerging out of the requirements for efficient communication, with our desire for this latter driven by its immediate material benefits. Similarly, we could understand the operation of the ideal of goodness as deriving from the requirements for a negotiated stable society, perhaps along Hobbesian lines. And these approaches do offer some fundamental insights into the genesis and function of the regulatory ideals. But I doubt that any convincing reduction can be achieved. For these theories in turn both tacitly embed and also presuppose norms and ideals for their critical development (cf. C_1, Section 6.I.4). Also, no one context or function is rich enough to account for their regulatory roles, it seems to me, in particular for the open-ended, self-correcting character of intelligence; cf. those for truth, section 6.I.2 above.[29]

Suppose then that the regulatory ideals remain irreducible. Even so, so long as they are part of a unified, fallible theory and not given any transcendent status, why can they not be accepted? Most plausibly, one might theorize them along Piagetian lines as derived from concepts of operational completeness or closure that arise as extrapolations of the actual sequence of (imperfect but) achieved regulatory equilibria; the parallels between the account offered here and that of chapter 5 suggest this as an obvious strategy. According to Piaget development takes the following rough general form: It is initiated by failure of assimilation and (micro-)accommodation, which sufficiently disturbs homeostasis (equilibrium), this leads to a search for deficiency among current cognitive operations with a resulting development of higher order operations over these operations, and these in turn provide through reflective abstraction and completive generalization a new, improved set of assimilations and accommodations supporting function over a wider range of inputs (i.e., across a wider range of environments). Such processes might help us to understand, for example, how homeostases and homeorheses might come to be represented as goals, rather than simply be causally operative, and so support a system of cognitive interrelations among cognitive activities.

This idea is explored in chapter 5, section 5.II.7, where it is argued that it provides a natural basis for the construction of an internal representation of the truth ideal as an extrapolation of this constructive closure process. Moreover, the organization of the search for, and creation of, new operations provides a specific way in which the very organization of reasoned intelligence at some given stage is itself active in searching for and promoting the superfoliatory passage to a yet more organized stage. Further, the ideal of reason given above also clearly invites a similar account of how it comes to be internally represented, namely, as an extrapolation over sequences of imperfect closures on the operations of marshaling relevant evidence, constructing relevant assessment arguments, etc. In this way the internal representation of rational and cognitive ideals would be understood as a natural part of the internal regulatory order that self-organizing regulatory systems of the cognitive kind spin out as they develop.

An account of this sort follows the Piaget-Popper line of emphasizing endogenous construction and autonomy. But another explanatory option was also noted. Whilst it is always tempting to assume that the whole explanation for the creative capacities of reasoned intelligence must lie solely in some hidden inner organizational structure that we already possess, another theoretical avenue that needs exploration is the view that actually our internal mechanisms are relatively much simpler and it is the organizational coherence of the environment itself that accounts for our capacity to develop as we interact with it. (In that case, much of the metaphysics of development will lie in the world rather than inside us; cf. Brooks 1991.) This must be true to some extent, no reinforcement learning would be possible, for example, unless the world generated signals with some minimal regularity, hence predictability, but (pace Brooks) creatures with effective anticipatory capacities must store internal representations of their world and not simply react to it (Rosen 1985).

The truth then lies in some combination of these factors. Since how any of this may happen precisely is as yet the subject of only dim theoretical intuitions, the issue will be explored no further here (cf. Hooker 1993d). The development of self-organizing systems theory is really at too primitive a stage to make it profitable to pursue this or other lines of inquiry further here. Since it is a fallible naturalist account, it is open to criticism and revision. Were it to turn out, for example, that there was no theoretical basis for this kind of self-interaction within self-reproducing, self-organizing systems, or even simply not within those of reasoned intelligence, then this would make a strong argument for restricting reason to economy or generalized efficiency and removing reason itself as a regulatory ideal in the theorizing of intelligence. Meanwhile, Piaget's model provides an interesting, empirically connected, naturalistic process from which to work. While the issue remains open, it is clear how a naturalist would go about seeking its resolution through substantive theoretical insight. The point is that, however construed, the regulatory ideals are still theoretical terms in our conception of cognition, and the necessity that we may feel once having recognized them need be no more than the necessity we may feel once having recognized that there are electromagnetic as well as gravitational forces. And their construction and use need be no more mysterious or antinatural than is the construction and use of ideals in science, where they play a powerful role, one that also provides insight into rationality theory (Hooker 1994c).

6.II.3. Contexts of Rational Action

A given person carries out many different roles in many different institutional settings (e.g., music lover, parent and spouse in a family, manager and employer in a business, or member of a political party). In each of these contexts, alternative roles may conflict, certainly, but also the requirements of different institutions may conflict with one another across contexts. Institutions have their own functions and responsibilities and hence interests and rationales, and through representing agents, they too may carry out roles in yet other institutions. We may define rational behavior for them as well as for individuals, though the spectrum of functions they exhibit will typically be much narrower than that for an individual. (A business, for example, may have as its primary goals profit maximization and security of market share, and it will develop rational methods and theories in pursuit of those goals. But it may also be a member of a national business council where other goals and constraints apply.) In this setting, for an individual to be rational is not simply a matter of carrying out each role efficiently, or of maximizing some autonomous personal utility function. (All too often, personal integrity is undermined by the role conflicts produced.) Rather, in parallel with the corresponding biological problem, it has something to do with constructively changing this totality, sometimes struggling to modify institutions and sometimes struggling equally hard to modify oneself, so that overall both oneself and one's society progress toward truth, beauty, goodness, and reasonableness, all the while compromising in a way that holds uncertainty and conflict within endurable bounds. (Vickers 1968, 1970, 1980, 1983 provide insightful accounts here.) And the same applies to institutions, though with appropriately narrowed conceptions of the regulatory ideals. (A market business, for example, is dominated by very narrow proxies for goodness and reason, viz., calculation over priceable inputs and outputs to maximize profits, it has substantially no proxies for beauty; within these, it may strive to alter its marketplace, or itself, in order to reduce uncertainty and conflict and improve profits and/or stable market share.) If something like this is correct, then it is essential to recognize the institutional context dependence of rationality and so a variety of context-delimited conceptions of rationality.

Conflict among context-delimited rationalities is but one species of incompatibility between regulatory systems and/or subsystems and as such grades back to a biological basis. Each regulatory level of a complex regulatory system has, within characteristic limits, its own conditions for stability and quasi-autonomy and has its own regulatory functions. The collection of these conditions and functions only needs to be sufficiently mutually compatible to ensure system viability. It does not follow that regulatory requirements must all be identical or even that there are no trade-offs to be made among them. Even in the simplest case of a population without significant social structure, not all phenotypes are identical, say, copying the locally most efficient (though this might be in the interests of each individual involved), and the resulting distribution of genotypes typically increases the adaptability of the population. The optimal survival conditions for individuals, groups, populations, and ecologies may, and often do, conflict (without one being "right" and the others "wrong"). To theorize reason adequately, we shall have to recognize that discrepancies of these kinds across regulatory contexts play essential roles in the drive toward transcendence. An account of this kind has already been set out in chapter 2, section 2.III.1.

6.II.4. Reason and Efficiency

The theory of reason presented thus far has scarcely mentioned the notion of efficiency. This would be understandable if reason were thought reducible to logic or to some similar formalism, since that formalism would specify correctness independently of any consideration of scarce resources and hence of the notion of efficiency. Nor do we need to appeal to efficiency to understand the emergence of logic and propositional processing. Any complex of operations that process geometrical features only has to include projective structure to enable abstraction of a logical structure.[30] But chapter 1 argued for the rejection of this conception of reason and for its replacement by a decision theoretic conception of rational action, and efficiency is central to decision theory, in the form of utility maximization and related principles. And to the extent that fitting solutions more generally predominate over computed solutions in our most effective cognitive procedures (see section 6.II.1), the focus will be on the wider efficiency considerations in choosing them. Moreover, efficiency can certainly be related to a regulatory systems conception of evolutionary development, and natural selection will penalize inefficiency, at least up to functional equivalence in ecological context. And the notion grades back through optimization to extremal path dynamics. So it is important to give some account of its place in the nature of reason as proposed here.

Despite its superficial attraction and ubiquity, considerable caution needs to accompany the introduction of the notion of efficiency to an account of rationality. The essential difficulty with the standard notion of efficiency is that insofar as it is well defined, it too is formal and noncreative; it works on whatever utilities (values) and constraints (facts) are fed to it. Thus it is not immediately suited to encompass the constructive process itself, which includes construction and modification of utilities and constraints. It is not immediately suited even to encompass lower-order processes structured by given regulatory ideals, because it lacks still resources to understand modification of constraints (choice of context) and context-specific goals. This is so even in those bastions of reason-as-efficiency, neoclassical microeconomics, and decision and game theory. For we not only purchase commodities, make investment decisions and play games efficiently, we also design the economic purchasing environments, and through diverse individual and institutional processes, we choose which games to play. Further than that, we ourselves construct a theory of efficient behavior in fixed settings drawn from these areas. All of these latter activities involve quintessentially the exercise of reason. And by comparison, following the efficiency algorithms once they have been worked out for a given context is no exercise of reason at all, simply one of memory and attention; but constructing an algorithm and demonstrating its effectiveness is a significant exercise of reason. If we try to rely on some simple formal notion of efficiency, we will lose our grip on how and why we create the contexts for our decisions (including ultimately the creation of rationality theory itself). Of course we can always create an efficiency model for any given rational activity, but with a cost. Efficiency is essentially able to explain reasoned intelligence only through the device of postulating a hidden framework transcending the judgments involved, the framework in which the formal efficiency calculation is made. This merely postpones the problem one step, since we can now ask again about the development of that crucial framework.

These are only preliminary considerations, but they make it evident that the relationship between reason and efficiency is complex. Still, efficiency is clearly important. What follows offers a preliminary positive characterization of the place of efficiency in the structure of reason.[31]

Given the regulatory ideal of truth and given a collection P of proxies for truth in the form of cognitive utilities, reason prescribes efficiency in the pursuit of truth. Let us call this narrow cognitive efficiency (NCE). NCE requires at least this: Increasing where possible objective content, through increasing valuable content (by maximizing/satisficing a structured expected cognitive utility function, if available), increasing the power of objective methodology and decreasing process costs (e.g., in the use of experimental laboratory resources)—all subject to some constraints, principally natural laws and features, including constraints on human capacities for reason and resource constraints (including relevant economic allocation decisions). In the actual situation of science, where it is embedded in a wider institutional network, rational decision making will also have to take into account constraints imposed by society, for example, some minimal acceptable level of technological reliability for scientific applications, perhaps some minimal level of technological progress (as payoff for economic support of research), and so on. These additional constraints cannot even be dismissed as wholly external to the intrinsic process of science, for applied technology is an important part of that intrinsic process and so, one may hazard, is the longer-term transformation of society around knowledge-creating institutionalized processes (see chapter 2).

NCE is already a complex requirement, subject to a number of competing internal demands. Thus the pursuit of various cognitive utilities (e.g., predictive precision and explanatory scope) will often compete with one another. The pursuit of direct increases in expected cognitive utilities may often compete with the pursuit of objective method (e.g., when further development and testing of a theory outruns the establishment of stable cross-situational invariants in existing experimental programs). Even pursuit of the various components within objective method may mutually compete. And of course, minimizing or even constraining process costs may compete with any of the previous goals. Finally, there are various conflicting prescriptions for rational decision principles that may equally compete for rational allegiance: for example, maximum expected cognitive utility, minimizing maximum cognitive regret (in the face of risks of error), satisficing (sufficient improvement within resource constraints), and so on.

So NCE is not sufficient, there has to be a wider rational function focused on the choices that the foregoing conflicts force on us. And as noted in previous sections, there are more choices still. We explore alternative cognitive utilities, alternative proxies for truth, learning about these as our scientific experience progresses. Choices of these belong here. More broadly, the notion of objectivity itself is complex, and what may be involved is still under debate; our understanding is at a preliminary stage. Choices among theories of objectivity belong here. Recall too that we are also able to learn about the nature and structure of reason itself and of how to improve it. The development of formal logic, statistical inference, and rational decision making, of their now multifarious applications and of the striking feedback that is now going on from those applications to models of reasoning itself (e.g., in computer science, economics, and control engineering), all bears vivid witness to the capacity of reasoned intelligence to explore the nature of reason. The choices of what to develop and how to deploy it also belong here.

Finally, these several sets of choices must themselves be brought into coherent inter-relationship; and not an arbitrary coherence, but one designed to pursue the regulatory ideals. Let us call efficiency in this largest superfoliatory development process "wide cognitive economy" (WCE). (There is a further complication: Specific WCE decisions may conflict with specific NCE decisions, requiring still further choices. But let us understand these to be included in WCE.)

What does postulating WCE gain for us? An ultimate efficiency model of reason, to be sure, with the convenient representational formalisms that brings with it. It also opens the possibility of explaining how we are able to recognize improvements to reason in a simple and naturalist manner, namely, through any one of the many machine-implementable ways of testing for optimization. (It does not require this explanation, the real process might, for example, simply be a homeostatic one that happens to be equivalent to optimization because of feedback relations between us and our environment.) Whatever these gains are worth, introducing WCE does not do away with the necessity to confront all of the component decisions, so it does not empty reason of its wide scope or creativity or involvement with regulatory ideals. Ultimately, it is the limit-transcending focus to reason that determines its overall structure. Coming to grips with all its component processes and decisions is still the real substance of reason. Only after we have characterized that overall regulatory process is an efficiency problem well defined, if at all. A fundamental feature of reasoned procedure is, for example, the exploration of differing cognitive utilities and of what form the principle of rational decision should be from context to context, and without these no optimization problem is well defined. Finally, there is the consequence that attempting to impose an efficiency model on this creative process threatens to necessitate postulating a regress of ever more powerful hidden rational structures with their attendant utilities over which optimization can be calculated. This problem has already been noticed, and it is the general penalty paid for trying to understand our self-organizing capacities purely in terms of an ultimate efficiency model. (Cf. further related issues raised by Slote 1989.) But more than this I cannot venture here, since I have no specific model of WCE that I am prepared to propose at this stage.

Reason is then not mere efficiency. Nor is reason mere formal rule following. Neither of the two formalist models of reason, logic and utility maximization, suffice in themselves to capture the complexity of our reasoning capacities, though both play their roles in the resulting structure of reason. However it is that we are able to engage in superfoliation of this kind, the naturalist needs to remain open-minded and look toward improvements in our theoretical understanding of the process in the future. One part of this naturalist strategy is to try to embed these regulatory development processes within a much wider domain, as has been attempted throughout this book.

This essentially completes what I have to say about the nature of reason. Taken together with chapter 2, it forms what I am able to contribute to step 2 of the naturalization program (section 6.I.5). Although the discussion does not guarantee that a naturalized account of reason can succeed, at the very least it should remove the assumption that it obviously must fail. In view of the ability to integrate reason with a larger set of system capacities that apply across a much broader spectrum of contexts, the naturalist regulatory systems approach to reason seems a fruitful program to follow.

6.II.5. Naturalist Reason and Creativity

But it does hinge on defending its naturalism. So here a potential difficulty, creativity, is confronted. Reason is the central factor in the process through which we construct more adequate conceptions of our world, including conceptions of our normative theories. Reason is central, for example, to the process through which we construct the self-transcending world of three-dimensional space with its objects, including our own body as but one among others, and later extend that construction to the construction of those abstract representational spaces within which all physical systems are but ones among others (cf. chapter 2, section 2.II.1). Giving to reason this creative-constructive role departs from the Western formalist tradition; it is widely held, though not uncontroversially, that both logic and efficiency and related formal systems are uncreative. It also poses the problem of a naturalist understanding of creativity. In my judgment, there is no satisfactory account of creativity to be had at the present time, naturalist or otherwise. But it is at least possible to investigate the relevant activities (Nickles 1980a, b provide promising relevant examples), and it is at least possible to see how at least some components of the required account might go.

Consider, by way of illustration, the process of concept formation. One way in which we may come to acquire new concepts is through the exercise of formal similarity judgments. Suppose that we have experience of several relations that share some higher-order properties in common. ("Is taller than" and "Is to the left of" are both irreflexive, antisymmetric, and transitive relations.) Then if we are able to form what is common to them, we shall have ascended the abstraction ordering and simultaneously put ourselves in a position to recognize new instances elsewhere. But such a capacity seems to involve, besides the fundamental capacity to form higher-order regulatory structure, only the capacities for sameness and difference noticing, which we can conceptualize as filters, and hence to be within the bounds of a naturalistic understanding. In more traditional terms, this model makes formal analogy (similarity degree) or metaphor, its generalized version, the basic mechanism for the abstraction bootstrap. The formal analogy bootstrap in individuals is intimately connected to the operation of Whewellian consilience of inductions in public science, which is in turn intimately connected to the rational structure of science.

An alternative process for acquiring new concepts might be through the tuning of tunable connectionist networks, in the manner described by Churchland 1989. Roughly, a collection of nodes changes its internal connectivity until it comes to support an appropriate input-output function (e.g., one representing some relevant discrimination). Here again, conceptualisation is intrinsically linked to generalisation in just the way required, for thereafter inputs sufficiently similar to those on which it was trained (its paradigm cases) will elicit a relevantly close (to paradigm) output discrimination. This procedure is not symbolic and does not presuppose any formal logical or linguistic machinery. And in principle, it can operate with any signal to create new discriminations, it does not rely on prior conceptualization. It can therefore operate in much wider regulatory contexts than can the formal analogy bootstrap. Churchland 1989 argues that it too may provide a framework for rational method in science, a wider framework than a formalist method can provide. This latter procedure has only emerged as an alternative in the last decade and is still being brought into theoretical focus (see note 24 and references). For other, related approaches to concept formation, see Holland et al. 1986, Mitchell 1993 and Hooker et al. 1992b. And the sheer fact of the existence of these new models may encourage us in the view that there may also be other naturalistic processes of concept formation that we have yet to discover.

And notice that the nature of creativity is potentially quite different as between formal analogy and fitting. For the formal analogy bootstrap, the key locus of creativity is not the structural similarity judgment—that could be implemented with any suitable comparator, including a tunable network—it is the provision of the conceptual categories into which the resulting higher-order concepts can fit, the creating of a sufficiently large conceptual space. If this is theorized as "hard-wired," then we recover a no-creativity role for reason here, but if not, then some account must be given for the higher-order construction and its relation to reason (cf. Piaget on reflection, chapter 5). For the tunable networks, on the other hand, the conceptual space emerges from their systemic interconnections and the problem of creativity is the problem of understanding the principles by which those interconnections are modified. (Note that understanding for cognitive theory will be given in the information representation, while models of connection dynamics will be given in the causal representation.) Here there is no longer an obvious boundary available to demarcate a noncreative reason from whatever creative factors are operative. So how we think of reason will depend critically on how the overall design of minds turns out. (Contrast here, Cunningham 1972, Powers 1973, and Lloyd 1989.)

In any event, it is clear from even this brief excursus into just this one fragment of creativity that it is premature to attempt to theorize reason in relation to creativity any more definitively at this time. We can say this: From a naturalistic perspective, creativity is but the most intricate expression of a still wider problem, the problem of how regulatory systems superfoliate, creating new regulatory orders that subordinate those previously in existence. This process is certainly a plausible candidate for a naturalist treatment, since it occurs across the entire spectrum of life and is not confined to some domain of mentalistic or abstract operations. But since we do not yet have any satisfactory scientific characterization of this process, it is too much to hope for a solution to the specifically human creativity process. On the other hand, once the problem is stated in these terms, it becomes not unreasonable to suppose that the whole can be theorized within a naturalistic framework.

6.II.6. Biology and Reason

Step 3 of section 6.I.5 asks for a biological embedding of reason so that it can be seen how it is a natural aspect of life in our world. In point of fact, the preceding discussion of the nature of reason, allied to the chapter 2 discussion, has already furnished the essential substance of the naturalist response to this demand that I am able to provide, for both biology and reason are there embedded in a unified regulatory systems framework. Cognition is theorized as the information face of biological self-organization, at sufficient regulatory sophistication, with judgments the basic units of organization. Judgment itself is the cognitive face of selective, self-correcting reactivity, and reason is the regulation of judgement. The transcending process that is central to reasoned intelligence is the cognitive aspect of the general process of superfoliation in self-organizing systems. Objectivity, the product of reason, is theorized as the cognitive face of invariance in homeostatic/rhetic regulation. Efficiency structures in reason are the cognitive dimension of optimization processes, themselves grading back to extremal path physics. Logical organization is a special case of relevance or association structures that in turn reflect information processing optimization. In particular, logic is accessible through selection for processing spatiotemporal features, and its use will be (imperfectly) selected for as a function of the utility of the various inference structures in different contexts. (Context dependence, a pervasive feature of reason, grades back to response conditionality.) Conflict among organizational contexts of rational action is the cognitive face of conflicts in conditions for homeostasis/rhesis among organizational contexts of regulatory systems.

There is still more detail that ought to be provided: for example, for risk taking (grades back to response precedence ordering?). But there is neither the space nor the evidential justification for proceeding further. At this time, we do not have a sufficiently developed account even of adult intelligence capacities in regulatory systems terms to make the attempt reasonable. We understand even less about regulatory models for biological systems and still less about a general theory of abstract regulatory systems. (But these literatures are currently undergoing an explosive development; cf. chapter 1.) I have tried to provide just enough content to indicate how the naturalist can proceed. It has been my aim to see reasoned intelligence, as it appears along the mammalian evolutionary line, as a specialized case of much more general and ubiquitous processes. And although our understanding of many of these processes is far from complete, indeed rudimentary, there seems at this point every reason to believe that step 3 in the naturalization program points to a fruitful theoretical interaction and a deepening of theoretical understanding. The task is not merely defensively possible; the prospect is promising at this time.

The preceding discussion has made no explicit mention of consciousness. This is not because it should be avoided; to the contrary, there will need to be a naturalist theorizing of consciousness so that it takes its place in an integrated account. One popular, and naturalist, approach is to theorize consciousness too as an organizational feature of mind (e.g., as a selective attention gate or as a public reporting capacity). (The reader will find four or five major models of this kind in the philosophical and psychological literature; for example, Piaget's model of an operation's becoming conscious when it becomes the object for higher- order operations is a conception of this kind. See chapter 5.) But these models are unsatisfying, both because they ignore the qualitative character of consciousness (cf. the "missing qualia" argument against functionalist/programming accounts of mental states) and because of the way they confine consciousness to very simple functions. Consciousness poses deep challenges to theorizing, naturalist or otherwise. On the other hand, there is no argument from features of consciousness to non-naturalism, that I know of, that is persuasive (cf. Hooker 1978), while there is evidence that consciousness needs a naturalist theorizing (e.g., the evidence for its synthetic character; see Churchland 1983). On balance, therefore, I conclude that it is rational to push on with a naturalist investigation.

There remains one important issue, that of an evolutionary rationale for rationality itself. From the fact that reason is extension of evolutionary regulatory processes, it does not follow that the capacity for reason has been brought about by evolution. Is rationality itself a product of natural selection? If it is, does evolution then provide a guarantee that humans are (predominantly) rational? Unsurprisingly, discussion focuses on the role of reliability here as well. The search is on for that feature of reason, if any, in virtue of which rational actors survive best, and the obvious feature on which to focus is reliability. If natural selection can be shown to favor reliable methods for belief formation and reason is the possession of such methods, then the evolution of rationality can be understood. But reliable methods yield generally true beliefs, and true beliefs conduce to survival, so natural selection plausibly should favor rationality. Both Sober 1981 and Stich 1989 present evidence that some such argument is at least implicit in many philosophers. And Sober goes on to (cautiously) argue that, among equally practically reliable methods, the theoretically reliable ones are selected because of their greater economy or efficiency.

Yet there are those who object. Putnam 1982 (p.6), for example, finds it obvious that "there is no contradiction in imagining a world in which people have utterly irrational [unreliable] beliefs which for some reason enable them to survive, or a world in which the most rational beliefs quickly lead to extinction." Munz 1989 argues that false beliefs may not only contribute to social solidarity, but their very falsity may be their important feature in this role. Sober 1981 notes several ways in which selection may fail to yield the optimal trait, arguing for agnosticism about reason at present. And Stich 1989 explicitly attacks the whole enterprise, arguing that natural selection can provide no defensible basis for a coherent theory of reason.

To proceed properly, several issues need to be distinguished. In order from specific to general, the issues are these: First, is there a case for an evolutionary account of a reliabilist conception of reason? Second, is the reliabilist conception of reason the only or primary one for a naturalistic evolutionary theory of reason? Third, what other conceptions of reason might plausibly be advanced, and how do they fare in relation to selectionist explanation? I shall first discuss why it is indeed difficult to establish a selectionist case guaranteeing a reliabilist rationality along the lines of the argument sketched earlier. But I shall argue that this is of no importance in itself since a reliabilist account of reason is not the only, or even the primary, one for naturalism. I shall then discuss the remaining issue for just the case of the theory of reason advanced in this essay.

When Sober 1981 proposed the idea that rationality might be favored by natural selection because it might prove to correspond to the most economical internal organization of organisms, he also issued several cautions against taking this proposal as obvious. Among these, there is the fact that evolutionary processes may fail to select the most economical internal organization for manifold diverse reasons: Selection for other features might have been more urgent and genetic pleiotropy resulted in suboptimal organization; selection can only be among alternatives actually available, and perhaps the really efficient organizations were not genetically represented; not all features are selected for but may result from genetic drift, etc.; or perhaps the selecting environment was such that less reliable but organizationally simpler strategies would result in equal fitness. What such arguments show is that evolutionary processes don't guarantee optimality. Insofar as reason is tied to optimality, either through reliability or via internal economy, then one cannot employ an evolutionary argument to guarantee rationality. This conclusion is unavoidable. But it does not follow that optimality never results from evolutionary processes, nor even that it is never selected for. So it doesn't follow that reason, even reason as optimality of some kind, is not the result of evolutionary processes, or even that it probably is not. Rather, what account is given will depend upon the precise nature of reason and the actual details of the evolutionary process involved. As Sober says, short of these, agnosticism is the proper attitude.

An evolutionary process cannot be expected to do more than select from the alternatives available what is "good enough" in the circumstances. In that case, reliabilist accounts of reason have particular difficulties to face, since good enough may not demand a high degree of reliability overall, only in whatever particulars immediately affect fitness. In a food-rich environment, for example, a strategy of treating all fruits similar to a noxious fruit as also noxious produces many false beliefs but may be good enough for survival, and no doubt offers a simple internal organization to boot. Stich 1989 makes much of arguments such as these. (Stich 1985 represents an earlier version of the same attack, one carefully criticized by Feldman 1988, who nonetheless agrees that evolution cannot be used to guarantee rationality.) Stich is in pursuit of larger conclusions, namely, to undermine the view that reason is a unique, universal trait and that our rationality is distinctively valuable. Discussion of this larger agenda will be postponed for a little. But note here, contra Stich, that only very modest conclusions follow from the evolutionary considerations above. At most, we might conclude that simple reliabilist accounts of reason were in difficulty. In fact, we can only reasonably reach a still weaker conclusion: To the extent that fitness does not require reliability in belief-forming strategies, to that extent those strategies may not be reliable. It is still open to argue that in fact fitness requires that our central belief-forming strategies be generally reliable. If central belief forming strategies are those that are widely employed (e.g., perceptual search, cross-modal sensory correlation as a test of veridicality, context-bound straight rule of induction, and so on) then this is surely not an implausible position. (See the optimism expressed by Kornblith 1993.)

But from the point of view of the characterization of reason in sections 6.II.1 and 6.II.2, the whole of the foregoing argument appears importantly misposed, thrice over. First, it restricts explanation to natural selection, but this may not be the only significant alternative. Second, it represents reason as a single separable trait, like hair color, but reason is the cognitive face of a much more profound and pervasive regulatory feature of life than this, more so even than mere internal economy or efficiency. Third, it directs attention to simple reliabilist accounts of reason, while reason is more adequately theorized in regulatory terms. These three defects are connected, the last being the easiest entry point.

The concept of reason is that of a complex process, not that of a product. By contrast, reliability characterizes a product or outcome of a process (cf. section 6.I.4 discussion). For a process to be reliable is for the products of that process to meet some criteria of goal achievement (e.g., show more than some percentage or probability of truth). But this requires guaranteeing the outcome in the relevant way. However, there can be no such guarantees consistent with a thoroughly naturalist evolutionary approach. The whole of the process of reason is fallible and risky. The systematic fallibilism stemming from this orientation is but bolstered a little by the evolutionary considerations Sober and Stich adduce.

But to suppose that this latter is the heart of the matter is to miss the point. The important feature of reason here is that rational processes are learnable and improvable (i.e., self-correcting). It does not matter that different parts of the process are unreliable to varying degrees. What matters is this: (1) The unreliabilities of reason are as a matter of fact tolerable in our circumstances, so we survive while we are learning. (This need not have been so, is not always so—cf. deaths resulting from various investigations of the atomic nucleus—and need not be so in future, for example as our planetary experiments and impacts scale up.) (2) The unreliabilities of reason are discoverable, and the process can be improved. And it is just this latter feature that science exhibits in abundance; from the investigation of the illusions of perception and the replacement of perception by machine recording to the investigation of reasoning processes across many disciplines (e.g., mathematics, economics, and psychology), we are now busy successfully understanding and (where appropriate) correcting the deficiencies in our reasoning processes (cf. the role of perception in the science system discussed in chapter 1 and the discussions in Hooker 1987, chapters 7, 8). There are clearly limits of various kinds to our reasoning capacities, some formal and some pragmatic (note 28); these are compatible with, and on the whole to be expected within, a naturalist account. But it does not yet follow even that there are intrinsic limits to our self-correcting capacities. Such limits may exist, but this is at least a subtle and unobvious matter; I shall remain agnostic about the matter here.

Self-correcting regulation goes back to the very foundations of life, for example, to Eigen's hypercycle in prebiotic chemical organization (Eigen/Winkler 1981). If reason is conceived along the general lines of Piaget's self-regulating account (chapter 5), then it is central to all intelligence. Insofar as intelligence (i.e., behavioral adaptability) is selected for, it is then plausible to suppose that reason has been selected for. (But of course no particular optimization, for any given regulatory function, is thereby guaranteed.) Note too that it is not necessary to confine discussion to selectionist explanations of reason. If, for example, it came to be plausible to believe that self-organizing regulatory development was basically a causally realized thermodynamic process that did not depend primarily on Darwinian chance variation, then it would also be plausible to explain reason as the direct outcome of such processes. Various recent ideas about self-organization might be understood as moving in this direction (see chapter 1). Such frameworks might also provide a plausible reading of Piaget's intentions (cf. chapter 5). At the present time, I remain agnostic about these matters; here I only note the potential relevance of widening the scope for reason-evolution relations. Beyond this, it is still necessary to consider the particular character of reason exhibited by humans. Again, judgment should await the provision of the relevant details. But if human reason shows something like the universal characteristics, albeit fallibly, that led Piaget to call it necessary, this would be a reason to believe that it is an unavoidable, or somehow preferred, end product of developmental-evolutionary processes.

The last paragraph briefly summarized the general relationship between evolution and the kind of naturalist account of reason proposed and so addressed the last of the three issues with which the discussion began. This discussion does not exhaust the issues. There is, for example, a larger agenda in Stich 1989, which attacks various grounds for theorizing reason as a unique, universal capacity and for holding truth as an ideal of reason. In this, he is re-tracing ground that Feyerabend has also crossed (see chapter 3, note 26). Stich concludes by defending a version of pragmatic relativism. Clearly debate is widening rapidly here, far beyond the scope of this chapter, so the discussion will be confined to brief remarks on Stich.

There is much in Stich's particular criticisms with which to agree. At one point, for example, he argues, against Quine, Dennett, and others, that there are no convincing substantive a priori or conceptual constraints on what rational organization is or the degree to which intelligent creatures must exhibit any particular version of it. At another point, he argues that it is implausible to measure rational strategies against commonsense intuitions. These criticisms seem to me to be well taken. But his overall argument fails to convince. In particular, within the internal structure to reason given above, allied with the regulatory accounts of Piaget and Rescher, an account has been given of how practical success relates to truth and why both are essential. The ground for holding that this account is the most adequate characterization available has two sides, its adequacy to our experience, as critically reflected in theory, and its integration into a larger regulatory account embracing all disciplines. Stich argues only piecemeal; he has not attempted a critique of an approach of this kind. The larger issues he raises are beyond the scope of this chapter, but the manner of response to Stich is well illustrated in the discussion of the limits of the selectionist critique of reliabilism above.

6.II.7. The Historical Manifestation of Reason

We come then to step 4, section 6.I.5, the last step to be completed in the naturalization program. To properly complete this step would require delivering a complete account of the historical development of knowledge together with a persuasive case that it is explained by the naturalist theory developed in this book. This task too is well beyond the scope and resources appropriate to the present work. But various fragments of the task are in hand, including the situating of historical progress in scientific method and objectivity in the framework (see chapter 2) and the pertinent discussions in Hooker 1987, chapters 5 and 8, and 1992.

The simplest picture would be an accumulative one. Over the course of the last few millennia, humans have been slowly developing scientific knowledge, including empirical information, theories, methods, metaphysics, and philosophical theories of knowledge, reason, etc. The development has occurred unevenly across these regulatory orders as well as across time, but roughly the lower-order regulations develop ahead of higher-order ones. Thus pre-Greek science is characterized by accumulation of empirical information but attenuates rapidly as one ascends toward general theory; despite the Greek accomplishments in mathematics and astronomy, it is only two millennia later that modern mathematical science emerges, and then it does not spread significantly beyond physics for another two centuries. During this time, explicit theories of scientific method emerge, along with formal logic itself, and begin to develop. And so on. In short, regulatory superfoliation pushes regulatory ascent in just the complex ways discussed in chapter 2.

Of course this simple accumulative picture is too simple in lots of ways. Superfoliation can include hindering as well as pushing ascent. Changes in higher-order regulation ("revolutions") can lead to modification of those beneath (deletions of "facts," re- shaping of empirical generalizations, etc.) High-order concepts can enter "horizontally" from mathematics or other scientific domains rather than emerging from steady regulatory refinement. And so on. Finally, the concomitant development of scientific institutions and their increasingly intimate engagement with the larger society introduces a host of further complications. None of this threatens the basic regulatory conception (to the contrary, it is made understandable by it), but it does make history of science a demanding study.

There are, however, at least two issues concerning the development of science that are of particular interest here, and these are potentially more disturbing. The first concerns whether science as a whole is developing a common shared higher order regulatory structure (e.g., a common general theory of scientific method) or whether each science is continuing to differentiate from the others as it develops. My only observations are that this is an important question to which the answer is unobvious (see also chapter 2, note 15). The second issue concerns the challenge to the classical post-Copernican conception of objectivity potentially posed by quantum mechanics. If we cannot theorize the objectively real in terms of what is invariant such that we may represent ourselves as among its objects, then it may be that significant parts of the metaphysics, methodology, and regulatory modeling of cognitive process will have to be modified. At present, neither the import of quantum theory, nor what consequences (if any) would follow from that import, are clear (cf. Hooker 1989a, 1991b, 1992).

6.II.8. Conclusion

Let us assume that the tasks of the previous sections can be successfully completed. Then, I contend, we would have developed a naturalistic theory of reason. I can see lots of unknowns and complications, but no principled reason why this program should not succeed. It is, I think, a more promising program than its traditional rivals and a more interesting one than is offered by those who attack reason. If we ever have to make sense of aliens, I would start with these regulatory notions rather than look first for predicate logic or merely assemble behavioral data. And this remains true for understanding those scarcely understood aliens that are ourselves.

NOTES

Chapter 1

1. *Homeorhesis* is Waddington's term (Waddington 1957) and refers to a teleonomic capacity for continuing a developmental process to a specified end. I shall use the term in a slightly more general sense to refer to any process stabilized against perturbations, for example, to a process homeostasis. Provided that it is supported internally and is not sustained only because of the coherence of environmental inputs, a homeorhesis can be regarded as a drive for homeostasis one regulatory order higher up. Order n: preserve some property H invariant; order n+1: preserve invariant the latter process. A stabilized set of regulations might utilize homeorhesis to ensure that should some homeostasis fail, then the process of reestablishing it is stable, and similarly should the reestablishing process fail . . .

The term *control system* could as easily have been used instead of *regulatory system*. Powers 1973 defines control as a capacity to reduce error or deviation from some reference condition (p. 47). But this term now has an ideosyncratic technical life in engineering systems theory, and it is preferable to use a term trailing fewer commitments. Regulatory systems may be realized, for example, in very different forms from the hierarchical controllers Powers envisaged (Hooker 1994e, cf. Hooker et al. 1992b, Stear 1987).

2. Since truth is by definition what results from using methodology M, it can't be a consistently formulated truth that M is inadequate to dis-

cover all the truth in some respect. What if M itself is discovered to be inadequate? Suppose, for example, we discover that the universe is too nomically symmetric and/or disaggregated for us to discover exactly all that happened over the first two billion years following the Big Bang—or, if you prefer, over the previous cycle of the universe before *our* big bang, or beyond our "light horizon." Or suppose that advances in decision theory turn up some flaw in M's risk-utility trade-off structure. How can one consistently express the need to learn an improvement in these circumstances?

3. The false path to coherence truth is made more attractive and less avoidable if one adopts the currently fashionable assumptions that truth is to be restricted to semantic theory and to the notion of reference within that. Rather, truth is a fundamental theoretical posit of cognitive theory in general, having manifold roles to play. See Hooker 1987, 8.3.1, and chapter 6 below.

4. The depth of evolutionary ignorance is emphasized by Campbell 1974, while the distinction between individual and species epistemic status echoes that of Lorenz 1977.

Reflexive consistency requires that for every order L of language there is a metalanguage in which it is asserted that the claims made at order L are fallible. It is, in short, fallibilism "all the way up."Only those, I think, who indefensibly demand logical closure in this context can charge systematic fallibilism with inconsistency.

5. The pervasiveness of theory in assessment—even the very theory nominally under test—is typically underestimated by philosophers, though it is a commonplace to scientists themselves. A version of the issues based on my experience as a working physicist is sketched in Hooker 1987, chapter 4. Now there are laboratory studies appearing that spell out the issues in more detail (e.g., Galison 1987 and Hacking 1983), and I have rejoined the issue in Hooker 1989b.

6. The account given there will be for truth and rationality only; nothing will be said about goodness and beauty. The difficulties in searching out what I think is an acceptably naturalist theory of the truth and rationality ideals were formidable (see e.g. the examination just of finitude in Cherniak 1986 and Hooker 1994c and the discussion of ideals in chapter 6). I believe that the difficulties with stripping out the ingrained antinaturalist assumptions surrounding goodness and beauty will prove more formidable. Any attempt at the task must await another occasion.

7. At least logic lies at the central focus of such systems, which may be held more or less extended, for example, to mathematics or to other synthetic-a priori truths supported by demonstration. Rationalists hold that these broadly logical truths are informative but necessary and hence contain no errors. Empiricists hold that logical truths are necessary but vacuous and hence contain no errors.

8. The literature here, which charts the fall of positivism, empiricism, and Popperianism, is immense, see chapter 3, part 3.I, for argument and some references.

9. See especially Popper 1979 for the reification of this confused dichotomy. For a discussion of Popper's Platonist schizophrenia, see Hooker 1981a, which has its roots in Popper's residual empiricism, see Hooker 1987, chapter 3. For a summary and further discussion of Popper's evolutionary epistemology, see chapter 3 below. On cause/logic, see also Hooker 1981b and Hahlweg/Hooker 1989a, part IV. These consequences of a formalist, logic-based rationality have of course been known for a long time. The classic attacks on simple-minded foundationalism and universal method date back to Kuhn's classic 1962 and Feyerabend's delightful 1961, and like comments can be pointed out in many other places. They are already clear in Brown's earlier work of 1979 and further developed in his 1988 and in Brown/Hooker 1994.

10. One should, at this point, discuss the introduction of logical probability measures on logical structures as ways of introducing risk to inductive methodologies. (Every classical probability is a map from a classical logical structure, a Boolean algebra, to the [0, 1] interval, see Kingman and Taylor 1966; Hooker 1975/79.) Certainly there will be particular methods among those appearing in this tradition that will prove useful in science, for example, application of various types of statistical inference (which are a paradigm of one kind of risk taking in hypothesis acceptance). But still the fundamental principles themselves are given a risk-free, because logically necessary, status. And the problems about the risky status of observational data remain. (A possible exception here may be Harper's revision system—see Harper 1977—but this is still a highly formalized logical system. Cf. the discussion of Holland et al. 1986 below.) The whole drift of the present analysis and other supporting analyses (e.g., Brown 1988; Hooker 1991a, 1994c) is that formal methods of this kind can only form a partial, context-independent, nondynamic fragment of rational procedure. But this is not the place to argue the case in detail; for that, see Brown/Hooker 1994. The reader is invited instead to reflect on the divergence between the logical and decision theoretic approach to rational methods and to the arguments presented in favor of the latter.

11. For other arguments to this end, see below and chapter 2, and Hooker 1987, chapter 7, and 1989b, 1991a.

12. There are various variants of this position depending on whether one does one's sociology from a structuralist, Marxist, etc., point of view. For a review and references, see Trigg 1980; cf. Hooker 1987, chapter 8.

13. On unification and the dispute about the nature of rational representation, see chapter 2, section 2.II.1, and Hooker 1992, 1994a, b.

14. See further chapter 2, section 2.II and Piaget on autonomy, chapter 3, plus Shapere 1984.

15. This negative claim can never be proven; rather, individual cases against embedding need to be considered on their merits. A couple of these are briefly considered at various places herein (e.g., Kuhnian discontinuities in chapter 2, section 2.II.2, Thagard in chapter 3, and Piaget's structure of science in chapter 5). On the matter of materialism, naturalism is committed to unity with the natural world, since it is through a general strategy of this kind that cognitive progress has been made over the past centuries; but that world may turn out to be the face of the divine, as Newton once thought and several prominent physicists currently think, and not least because of the introduction of the new systems ideas.

16. What follows is a partial and modified version of Hahlweg/Hooker 1989a, section IV.1. The original conception was a joint construction of Dr. Jane Azevedo and I, but I am responsible for the version here. In Hahlweg/Hooker 1989a, I noted a "base case," which, following Bradie 1986, I labeled evolutionary epistemology of mechanisms (EEM), namely, the view that the brain evolved under Darwinian selection. This is clearly also a very weak position, having in itself no logical implications whatever for cognitive function. There is one strengthening of essentially this position that is worth singling out, namely, what Hahlweg has called the bioepistemology of Lorenz and his Continental followers (see Hahlweg/Hooker 1989a, section I.3). The Continental bioepistemologists have aimed to underpin the notion of a quasi-Kantian categorical conceptual structure to mind on the basis of its being selected through the evolutionary process. This is possible if one conceives of adaptation as fit between evolving mind and some supra-species structure but is otherwise dubious. The Piagetian notion of dynamically constructed categories is, I believe, preferable (see chapter 5 below). I note this alternative in passing here. Its present interest for me is that it is one example where Bradie's distinction between EEM and EET (evolutionary epistemology of theories) breaks down. Once one develops a naturalist unified account of brain-mind, this breakdown becomes completely general (see Hahlweg/Hooker 1989a, section IV.2).

Chapter 2

1. The first idea derives from Hooker 1981b. In chronological order the remaining ideas have their roots in Hooker 1987 chapter 3 section 3.3 and chapter 5, 1978, 1980, 1981a and 1987 chapter 7, 1985, 1989a and 1991a. In various places in this chapter the content has been adapted from one or more of these sources. My idea of these ideas has benefited from discussion with many people, among which special mention is due to Drs. David Lane and Robert van Hulst, 1970-73, who taught me ecological systems energetics and dynamics; to several graduate students, in particular Scott Carley, John Collier, Kai Hahlweg and David Naor during 1975-80 at the

University of Western Ontario, where we all studied systems dynamics in biology and more widely; and to Hahlweg again during 1985-7 as research assistant; the ideas are given partial and often highly restricted expression in Hahlweg/Hooker 1989a.

2. In this case, a sentence such as "X has a gene for characteristic A, which is dominant" would not be perspicuously analyzable as "There is a y such that y is a gene, y is a component of X, y causes X to be A and y has the property of dominance," but instead as "There is some causal process P within X such that P causes X to be A under conditions C and X has P because of C," where C includes the specification of a range of input (fertilization) conditions. The former analyses set one off looking for some molecular property with which Mendelian dominance can be identified. But such searchers are in vain. In Hooker 1981b, Part III (partially repeated in Hahlweg/Hooker 1989a, Part IV.2), I provided a brief analysis of the reduction of transmission or Mendelian to molecular genetics, summarized the (very complex) constraints that apply, and argued that careful attention to the distinction between functional and causal theories resolved the differences between the major philosophical positions in the field.

I inherited the idea for this treatment of Mendelian genetics from Armstrong's schematic use of the alleged gene-DNA identity as a model for mind/brain identities; see Armstrong 1968. But Armstrong presses on with treating the reduction as a uniform microreduction (see Causey 1977), that is, as if gene and amino acids stood in a part/whole relation. This is mistaken. At the core of the disagreement between Armstrong and myself is the importation of the mistaken model of reduction into the mind/brain field. One aspect of this is the defective treatment of perception and the secondary qualities because they are not treated as dynamic systems (see Hooker 1978).

3. Note that (1) this reducive analysis is not committed to materialism, that depends on the further characterization of neurones (cf. Hooker 1978, 1981b, 1993d, and (2) this analysis is able quite generally to include Collier's important notion of relative causal independence between macro- and microstates (see Collier 1988b).

4. The term *information* is being used here in the sense of Information Theory and not in the distinct though related sense of semantic information. What the precise relation is between these two senses is a difficult question to which there are evidently no good answers at the present time. For a promising beginning, and a critique of the assumption that information incorporation is some kind of encoding, see Bickhard 1980a,b, 1993; cf. Ward 1989. Similarly, the task of relating information to thermodynamic order is a complex and unresolved one that cannot be pursued here. For an introduction to the area, see the asterisked bibliographical entries.

5. Which particular subset of incorporated information specifies the cognitive ones? This is a difficult question to which there is no good answer at the present time. Surely the incorporated states have to be systematical-

ly organized and at least partially hierarchial, but what else? How weak a connection to output is admissible? How weak an internal protological ordering? Fodor would go so far as to demand that they constitute symbolic representations over which computations are made and that these latter be sufficiently rich (both internally and in connection to output) as to represent a structure for which a notion of mistaken belief makes sense. I believe that this is likely an excessive demand, that at bottom intelligence will prove to be much richer than this narrow sense of computation; see note 4 above, chapter 1 and Brown 1988; Dreyfus and Dreyfus 1986; Hooker 1988, 1994e.

6. This framework provides the basis for an analysis of those claims, by Lorenz 1977 originally and more recently in Plotkin 1982 and others, that all of evolution is a knowledge process. In assessing these claims, it is essential to distinguish thermodynamic order accumulation from the particular case of information incorporation and information incorporation from the still more particular case of cognition. How this is to be done, precisely, is no easy thing to say, and no attempt will be made here; cf. notes 4, 5. With respect to the first transition, see Collier 1988a and Ward 1989.

7. See Bonner 1982; Goodwin et al. 1983, 1989; Raff/Kaufman 1983; Waddington 1957, 1966, 1975.

8. Indeed, because phenotypes are complex regulatory systems that exhibit resilience and adaptability, there is no 1:1 map between genotype and phenotype; the same phenotypic characteristics may be produced by disparate genotypes, and the same genotype may give rise to disparate phenotypes in disparate environments. This has especially been emphasized by Waddington (1957, 1966, 1975); see also Piaget 1970/72, 57, and his phenocopy process, 1974/80, and Hooker 1994d. There is a functional similarity cognitively. Because scientists are complex cognitive regulatory systems that also exhibit resilience and adaptability, there is no 1:1 map between their cognitive commitments and their public scientific activities; the same public scientific activities may be produced by scientists holding disparate cognitive commitments (e.g., because additional countervailing evidence accepted still leaves a particular line of investigation the best one to pursue, or because theory disagreement can leave agreement on laboratory methods, etc.), and the same cognitive commitments may give rise to disparate public scientific activities in disparate environments (because there is a new range of phenomena to investigate, or new instrumental methods are acquired, etc.).

9. In most cases, these transitions were made aeons ago but in others—notably slime molds—we can watch them occurring still. Slime molds normally behave as collections of individual unicellular organisms, but under starvation conditions, they assemble and specialize to form a new single multicellular animal, which moves out in search of food and produces a reproductive organ. These processes have only recently begun to be understood. (For a discussion that relates them to human social processes, see Garfinkel 1987.) And beyond the current populations—what? Social insect populations show superorganism features and many have speculated simi-

larly about humans. Are computers the currently forming ganglia in a planetary parallel-distributed-processing nervous system?

10. Despite the emphasis on the systems structure of populations, there is no particular commitment to group selection implied. Sober has argued that Wright's shifting balance process, which demonstrates how population structure can make a difference in evolution, can occur without group selection taking place. Group selection requires that membership in a group would have to be a positive *causal* factor in an organism's survival and reproduction. (See Sober 1984, 318–23.) This may apply in some cases (cf. Wade 1978). How frequent it is, is still an open empirical question. Similarly, it is an intriguing question whether or not scientific groups can be units of cognitive selection. It is certainly tempting to accept that membership in a scientific group can be a causal factor in a scientist's failure or success. On the other hand, Hull has made a strong case for individual selection (see Hull 1988b). Only detailed sociological and other empirical investigation will clarify the issue.

11. Of course, we can always embed systems descriptions in the formal language of sets, members, and maps, so both models can be expressed in these formal terms, but just for this reason, it is in itself uniformative.

I shall take individuals generically to be systems whose dynamics show some suitable set of invariances, where this includes at least a basis for uniquely identifying them. Thus rigid object dynamics leaves the mass, volume, rotational inertia, etc. of the set of constituting molecules invariant, and the nonintersecting joint space-time trajectories of the constituents provide a sufficient basis for individuating and counting these objects. Individuals can constitute a suitable level of a system. Individuals must have relevant structure to exhibit the dynamical invariances that define them. But structure per se does not suffice. Waves on a pond are not individuals, though they are structured, because their capacity for superposition destroys too many invariances; neither are photons. While the much more complex dynamics of living individuals does not exhibit all the invariances of a rigid object, it leaves at least a range of functions invariant, together with whatever structure that requires (e.g., approximate organ location and constitution), and in consequence, living systems show similar unique space-time trajectories while they persist. Because of their structures and processes, individuals interact in structured ways. Often the notion of the limit of zero interaction makes sense compatibly with preservation of individuality (e.g., by sufficiently spatially separating individuals), but it may not. This very basic conception does not exhaust the subtlety of the notion of individuality in biology—see Buss 1987; Gilbert 1992; Tauber 1991; cf. the diversity of realization of Hull's replicator/interactor scheme (Hull 1988a, b)—but the discussion must be left here.

12. Properly speaking, we should go on to include statisticians, laboratory technicians, computer programmers, and other functionally essential persons, such as secretaries and administrators, but here I shall preserve a

traditional artificial simplicity and confine attention to the core scientific populations. The omission of the others is not trivial. Ultimately, we shall need to provide a principled cognitive sociology based on characterizing cognitive activities in terms of information extraction and organization. From this perspective, learning will occur in complex patterns across all human institutions (e.g., partially in businesses in the course of pursuing profits) wherever its conditions are met, and conversely, all institutions will have a basic description that is prior to any principled division into cognitive and other activities. This applies equally to scientific institutions. See also notes 20, 21.

13. See especially his 1978a and later his 1978b and 1987. For an assessment of Feyerabend in the present perspective, see Hooker 1991a. He has certainly not been alone in voicing these criticisms (cf. Churchland/Hooker 1985), and it is now commonplace in the philosophy of science to agree on the failings of the classic formal methodologies, those of the empiricists, of Popper and of Lakatos (see chapter 3) and of later descendants, which include Newton- Smith 1981, Suppe 1974, and van Fraassen 1980. Each of these contains, from its own point of view, a critique of the classic accounts, yet from the present perspective, each continues essentially the same formal idealization.

There are several aspects to this rejection of formal philosophy of science as more than approximative idealization. First, there is the rejection of logic as the paradigm of reason and its replacement with a wider decision theoretic concept (chapter 1); logic emerges as a special case under appropriate circumstances. Second, there is the rejection of formalism as an adequate framework for a theory of rationality and its replacement with a theory of nonformal, systematically constrained judgment (Brown 1988; Hooker 1991a, 1992, 1994c); formal representation emerges as a degenerate special case. Third, there is the rejection of propositions as the paradigm units of rational cognitive structuring, replacing them with a wider notion of concepts and nonformal conceptual relations (Churchland 1989; Hooker 1975, 1987, 1993d); propositional structuring emerges as a special case under restricted circumstances. All of these points apply first to rational processes in individuals and by extension to science as a whole. Though the requirements of intersubjective communication push the design of scientific institutions toward realizing those circumstances where formal, logico-propositional representation applies, that representation will never, and can never, wholly apply.

14. See Cavalli-Sforza/Feldman 1981; Popper 1979; Toulmin 1972. Here Popper is typical in requiring selection ultimately to be understood in terms of formal logical relations (chapter 3).

15. The product of scientific process will be a proliferation of scientific disciplines, each with its distinctive domains, theories, and methods that are specialized to the conditions under which the many facets of reality are epistemically accessible to us. This does not look different from the biologi-

cal case. While lying behind this scientific diversity will be growing metaphysical and nomological unities (this is central to naturalism), these are only representable as maps among the various theoretical models. This circumstance does not look obviously different from the occurrence of maps across the genetic regulatory structures of species corresponding to the common exploitation of the properties of gravity, light, etc. Both structures will reflect reality only piecemeal. While science has stronger mapping resources for making connections than do genetic controllers, it is not obvious that the planetary collection of genetic regulatory systems is not a partial but unifiable operational representation of planetary reality within its constraints in much the same way as our sciences are. (Cf. the widespread idea of genes as constituting a "language" or code of regulation, see Campbell 1982 and Pattee in Jantsch 1981.) At least it requires more careful theoretical reflection before a cybernetic or regulatory clash is admitted. Science may still be a specialization of biological regulatory dynamics in this respect as well.

16. The constraints may be nomic (e.g., the constraint on velocity derived from relativity theory or that on thermodynamic efficiency derived from the second law of thermodynamics) or structural (e.g., the size of the Earth, or locations of the continents) or some combination of the two (e.g., the rate of continental drift).

17. So-called climax forests are examples of local equilibria. Recognizably stable states are a function of scale; if we look at short time scales over relatively small spatial areas, then we may be able to locate many local relatively stable ecologies, but as the spatial and temporal scales are enlarged, locatable stabilities, even of an oscillatory kind, decline. Over geological times, the planet as a whole shows little sign of stability, although specific conditions, such as adaptation to an aerobic atmosphere, may become increasingly stabilized (entrenched). Here lack of stability does not imply random change (though if may), but complex dynamics.

18. A reasonable, if rough, example is provided by the biologists Hull studies in his examination of controversy within systematics (Hull 1988b). But note that a high-order commitment to a research program is often more important here than agreement in detail with laboratory methodology, acceptable data, or even theoretical principle. The workable, and desirable, diversity in commitment is briefly explored at note 47 and text.

19. Recognizably stable states are again a function of scale. If we look at short time scales, and over relatively small spatial areas (or a narrow range of institutional affiliations), then we may be able to locate many local relatively stable cognitive ecologies,that is, groups of scientists working in a particular country or institution with relatively stable sets of scientific beliefs and methods. But as the spatial and temporal scales are enlarged, locatable stabilities decline. On a historical time scale of even a century, or a planetary cross-cultural spatial scale, there is no evidence of any stable scientific equilibrium being approached, although low order practical compo-

nents, such as that most stone scratches skin, may become increasingly stabilized (entrenched). Here lack of stability does not simply random change, but complex dynamics.

20. Because of their intimate interactions (see chapter 1 and below), a principled distinction cannot be made between science and technology; rather, they both belong to a single dynamic, information-concentrating system, which shows different features in different contexts. What is cognitively relevant is a function of the causal structure of our cognitive regulatory processes. The reader then is to understand that the term *science* refers to this entire complex system. And beware of assuming simple institutional boundaries. Surely social fashion is a cognitively irrelevant environmental feature? It ought to have no influence on scientific behavior. But some important chemistry has been done on women's facial make-up preparations. What demarcates this from medical research on plastic surgery or military work on chemical warfare? I do not conclude that there are no important distinctions to be made in these cases—to the contrary. But there is no principled way to declare one a *purely* cognitive activity and the others not. See notes 12, 21.

21. This makes space for a principled cognitive sociology of science, in place of the mis-conceived logic/nonrational-social-cause dichotomy of traditional formal theory of science. (Cf. Kitchener 1989, but here more radically.) In principle, one would like to treat the collection of all populations of scientists in the larger social and inanimate environment as a single dynamic system whose time scales for change may range from the very short to the very long. Once formalist presumptions have been set aside, it is only the practical difficulties imposed by ignorance, finite resources, and constraints on the management of complexity that should force us to resort to the simplified treatment of an isolated, communally uniform science or one lumped into simple Kuhnian camps. See also notes 12, 20.

22. Here there is an enormous variety of 'programmes' possible. A crude analysis organizes them in a two-dimensional space ordered by the conditionalization on ordered environmental input sequences of either or both of development events and developmental programs.

23. This is only roughly expressed here, but the intended meaning should be clear enough. For the notion of stability, see chapter 1. With respect to the notion of process versus static state, here is a first step in analytic refinement. Dynamic process parameters are ones corresponding to state-characterizing predicates whose analysis essentially involves a temporal relation and static-state predicates are those involving no essential temporal relations. The temporal relations of dynamic-process predicates ultimately hold among system conditions described by static state predicates and their time of occurrence. It is for this reason that I think of the process properties as second order and the static properties as first order. But this is a crude distinction. There is much more complexity involved in regulato-

ry systems (see. Salthe 1985, 1993, Ulanowicz 1986, Yates 1987, and many other asterisked references) and there is no space to enter upon it here.

24. Exactly how this process works is more controversial. Catching a ball is certainly not done, as current models might suggest, by calculating the ball's trajectory; that is determined by a differential equation of high order that not even a fast computer can easily approximate. For an alternative, nonformal approach that supports the conception of reason developed in this book, that is, of reason as a nonformal self-organizing systems property of some sort, see Hooker et al. 1992a, b; cf. Hooker 1994e.

25. For some notions of viable systems, see Beer 1979a, b, and Espejo/Harnden 1989; cf. Naor 1979; Iberall 1972; Rosen 1985, 1991; Yates 1987. See also Piagetian autonomy, chapter 5.

26. For a current discussion of these old ideas, see Hooker 1992 and references. The present discussion was drawn from that paper.

27. See Hooker 1991a. A coherent simplification is one that reduces the number of in- dependent laws or parameters needed to explain a domain of phenomena. The two dimensions to explanatory depth, coherent simplicity and ontological depth, are intimately interrelated. Improvements along both dimensions go via ontological identifications that achieve unifications: identifications within a given ontology that allow coherent simplification and identifications across ontologies that allow a new, more systematic underlying ontology to be introduced. Achieving ontological depth typically provides the basis for unification across domains as well. This latter happens when the underlying ontology is already, or can then be, related to the laws of other domains; for example, the microbiological theory of disease relates medicine to various biological domains such as molecular genetics. When this happens, partial coherent simplifications are then made possible. Thus the relation between the two dimensions may be summarized in this way: Ultimately the achievement of a deeper ontology is more important than the achievement of coherent simplification; however, it is desirable to increase the systematicity and widen the scope of any particular ontology as far as possible, and this is done through increases in coherent simplification.

28. For the parallel case of the electron microscope, see the discussion in Bechtel 1989 and references, and see Hacking 1983. Incidentally, what is to count as "reasonably accessible" and what reasonably to do when stability fails in the various ways it can are important methodological issues whose resolution is a function of theory, resources, etc., not simply Popperian "decisions." These issues cannot be pursued here.

29. To return to a recurrent theme: Of course science is a social construction, as is its objectivity. And it is obviously an imperfect one. But it, and its working, is not *merely* a social construction. That is the essential point, one to which "strong" sociologists and others seem oblivious (ironically playing out the other side of empiricist metaphilosophy). And one

should distinguish moral, political, etc. criticism of its current working from its working being merely a social construction. (If it didn't really work, the military weapons, etc. would not be so very frightening.) See notes 12, 48.

30. The ceteris paribus clause is inserted here because there are also countervailing factors to consider. Increased adaptability may require "expensive" genetic or other resources to support it, for example, and it may come at some cost to refinement in existing adaptations. Less interestingly here, but certainly no less possible, the whole process may be subverted by unintended consequences of selection, by interference from random genetic drift and other factors arising from the limitations of selection, from an 'unlucky' choice of environment and from unanticipatable exogenous shocks, such as collisions with meteors. (Gould likes to emphasize these considerations; see Gould 1989 and Nitecki 1988.) It is sufficient for my purposes that the general processes descibed here do occur often in the relevant conditions—to which, I shall take it, the whole mammalian line bears witness.

31. But not conversely since increases in regulatory height and complexity may be used to achieve behavioral rigidity (witness many bureaucracies); it is the dynamical design which matters. Where these are populational processes, there will be corresponding genotypic change. But note that increasing genotypic complexity cannot be inferred from increasing phenotypic complexity, because simple construction rules applied to appropriate materials may still lead to complex products (see Cunningham 1972; Kauffman 1993; Sperry 1983 on feline cortex regrowth) and because environmental complexity may be exploited in the construction process. There is no doubt though that genotypes are complex regulatory systems and that increasing phenotypic complexity (adaptability) requires increasing genotypic complexity of some kind.

32. On the distinction between homogeneous and heterogeneous environments and its importance, see Levins 1979; Margalef 1968; Hahlweg/Hooker 1989a, Part II; Hahlweg 1991. I am indebted to Hahlweg for discussion of these topics.

The qualified *may* of three sentences back is there especially to acknowledge that if there is sufficient genetically based capacity for behavioral adaptability, then it may reduce the need for further genetic adaptability. This becomes especially prominant with humans, where our capacity to survive in polar regions or space, for example, is derived nearly entirely from behavioral adaptability. To take a parallel case, consider a machine M for building computing machines; as the required computational routines become more complex, so ceteris paribus must M. But this need only be so until M can construct a general computer like our current von Neumann architecture machines; after that, the software can go on developing much more independently of M's capacities. That is the point made here. And it is part of Piaget's point about achieving completed or closed groups of operations (see chapter 5). Of course, there may still be reason to continue to modify M, for example, to improve computational speed, or change architecture (cf. current massively parallel architectures); but from here on, the relation of M to our required computational capacities will be more subtle and indirect. See also note 31.

33. These simple feedback/feedforward loops scarcely capture the full complexity of the dynamics involved; many more such loops could be added. Nor has there been any attempt made to discuss in any detail such subtle and sometimes controversial matters as adaptability of genetic regulatory structures, or of interrelations between developmental sensitivity to environmental information and developmental adaptability, and so on. Nor has there been any attempt to quantify the dynamical relations in order to study them mathematically and empirically. First, these details are not needed to develop the general characterization of cognition sought in this book. Second, at this historical juncture attempts at quantitative dynamics are in general not warranted by our meager knowledge (by mine anyway). Nonetheless, we thus arrive at a conception of the evolutionary process that is more complex than a simple interplay of chance and necessity (the Darwinian slogan of Monod 1972). Here the systems dynamics drives the build up of genetic and phenotypic complexity expressed as phenotypic adaptability, its most common and probably most complex form is that of purposive behavioral adaptability supported by a complex nervous system. None of this need be inconsistent with Darwinism, though it certainly gives the phenotype a distinctively important role that traditional Darwinism tended to ignore. It is also open, perhaps even inviting, to embed the process in a wider theory of self-organizing systems (chapter 1). Be that as it may, it is the process that leads us to human cognition and science as its current manifestation.

34. The qualifier "amenable to behavioral adaptation for that species" is necessary because some problems posed for a species by heterogeneity can only be solved by nonbehavioral physiological adaptability. There is, for example, much more to hibernation—a solution to a large temporal variability—than just the behavioral maintaining of a motionless, relaxed posture; it includes the reduction in metabolic rate, reabsorption of urine, and so on. The metamorphosis of many plants through an annual seed production, death, and seedling regrowth cycle is a still more dramatic case. This is the converse of the point made at note 32. Actual species instance the detailed interplay of the two considerations.

35. The reader is reminded that this is metaphorical functional talk; abstract contents do not literally interact. Rather, the causal description, where interaction talk actually belongs, is concerned with the neural and behavioral causal consequences of the neural states that instantiate the corresponding cognitive commitments. It is a convenient shorthand to speak in the metaphorical manner, and does no harm where the functional outcomes of the real interactions are characterizable purely logically; but as we have seen, this simplification is only available in suitably confined contexts.

36. The explosion of new, alternative mathematical forms beginning with non-Euclidean geometries a century ago, and similarly for logic since 1900, bears witness that in these areas also we learn through experience.

(Cf. Brown 1988; Holdsworth/Hooker 1983; Hooker 1975/79, 1979, 1987.) On change in scientific method, see Blake et al. 1960 and Oldroyd 1989. Diagram 2.3 is taken from Hahlweg/Hooker 1989a, and see Brown 1988 on which it draws.

37. See Feyerabend 1978a. For recent work on Newton, especially the continuing value of his methdological heritage, see Harper 1989, 1993, Hooker 1994b, Stein 1990 and references.

38. For discussion of the difficulties, see Nitecki 1988. Part of the difficulty has to do with such basic terms as complexity, information, intelligence, and so on themselves being controversial and not well understood; for discussion, see notes 30–3 inclusive and references, and chapter 1.

39. In the first case, I refer respectively to the second Stone representation theorem and to the CTP symmetry theorem; see Bub 1974 or Hooker 1973 and Streater and Wightman 1964. With respect to the physical constants, one is referring to the discovery of e, h, and c — respectively, the electron charge, the quantum constant, and the velocity of light; their values are critical to the structure and stability of the universe as we know it. Not only do we not have a plethora of mathematical metaphysics that satisfy all these restraints, despite frequent naive assumptions to the contrary, it is still not obvious that we have even one; see Hooker 1992.

40. Just to keep track of the multidimensional character of progress, note that for each regulatory order N there can be (1) increased adaptive refinement (horizontal N progress) and (2) increased adaptability (vertical N progress), the conditionalizing of order N on order N+1, but that the latter can comprise either the creation of order N+1 and/or its increased refinement (horizontal N+1 progress).

41. Gould 1977 links these characteristics with r and K selection. He notes that K amelioration, which characterizes the mammalian evolutionary line, also takes the system dynamics beyond the original predator-prey models from which the notions of r and K selection spring. See also Prigogine and Stengers 1984. This is a good example of the kinds of system dynamics shifts with which this chapter (nay, book) is concerned.

42. See Hooker 1987, section 7.8; cf. sections 8.8.7, 8.3.9. The notion of an external nervous system is most obvious in the case of the computer, which at present is like a simple ganglion cluster in a distributed nervous system. Once we have self-programming, self-redesigning computers, they will start to look more like us, local quasi-autonomous intelligent information processing centres.

The discussion of the last few paragraphs has linked increasing phenotypic regulatory complexity and increasing social complexity, but the relationship itself is complex. On the one hand, social complexity may simply reinforce phenotypic differentiation rather than higher-order problem solving, and on the other hand the entire population may move in the direction of the formation of a superorganism (cf. slime molds, note 9). What interaction there may be between these two processes—in the human case there is clearly some, and could be much more—and how superorganism formation is to be understood are matters that cannot be explored here, but are crucial to our future. We could choose the social equivalent of the colony insects, developing powerful institutions coordinating rigid specialized social roles, or choose the opposite free market extreme—but surely both are unattractive and represent poor epistemic and social design; cf. Hooker 1994.

The characterization of regulatory complexity is further complicated by the recognition of two different structural dimensions along which it may develop; hierarchical-serial and parallel-distributed processing. The latter clearly also involves levels, if not strict hierarchy, but information processing capacity may be increased, and on occasion the order of regulation reduced, by increases in the parallel capacity. Human brains evidently exhibit substantial parallel and hierarchical architectures, as do human social institutions. Economic and scientific autonomy and specialization achieves both a parallel distributed mode of operation across goods or disciplines even while specialization provides for "chains of command" or equivalent, which permit hierarchially organized coordination. How these structural features interact, and how they interact with the development of phenotypic and societal complexities, are largely beyond human ken at the moment and certainly beyond detailed discussion here. For some discussion, see Dyke 1988. There has been some discussion of science as a parallel- distributed processing system beginning to emerge, but while it obviously is a feature of science as an institutionalized process, (1) it is clearly not the whole story, as discussion herein shows, and it is as yet unclear exactly how to describe its significance, so (2) in and of itself it is a relatively uninformative idea unless embedded in a regulatory systems theory of science, since it is the dynamic processes characterizing science that are important.

43. Detailed studies of experimentation are now being carried out, but typically in retreat from the simple logic-based conception of method; these studies display more (Franklin 1986; Galison 1987) and less (Latour 1987) overt openness to a dynamic process replacement. For a regulatory representation of experimentation, see diagram 4.2 and text to Hooker 1987, chapter 4. A working through of the many theories involved in one well-known experimental measurement, the cloud chamber, is presented in Hooker 1987, chapter 4, appendix 2. (N.b. Interchange C_4 and C_5 in the diagram.) See also the discussion of plasma probes in Hooker 1989a, summarized below, and of method in a systems setting (Hooker 1994a, b).

44. Artifact development/evolution has an independent interest, though it seems little worked on—Popper notes its importance (see chapter 3) but does no more; but see Cragg 1989 on artifact evolution and Gray 1992 for their roles in developmental systems.

45. See Hooker 1987, 1994b. It would be good to have to hand at this point a theory of the limitations of institutions, but I know of no theory on offer; for remarks on limits, see the conditions for superfoliation at section

2.II.2 above and chapter 5, notes 26, 30, and for other analyses of constraints on decision processes, see Brams 1976, Brennan 1985, Etzioni 1961, 1967, Merton 1972, and Simon 1969; cf. Hooker 1987, 287, and sections 8.3.7, 8.3.9, and their references. I believe that construction of this kind of analysis is an urgent necessity.

46. Or some suitable complication thereof. The centrality of risk forces choice among different rationality principles (e.g., between maximum expected utility and minimizing maximum expected loss), with the option of various additional constraints (e.g., eliminating possible losses above some ceiling). See the discussions generally in Brams 1976, Elster 1983, 1986, Hooker et al. 1978 and McClennen 1990, and specifically Hooker 1994c at note 8 and text. An issue that arises is the nature of the utilities to be attributed to scientists qua scientists; are they purely epistemic, irretrievably mixed epistemic and pragmatic, or purely pragmatic? (Here epistemic utilities are those directly related to the pursuit of truth, and pragmatic utilities are those related to other goals, for example, prestige, salary, institutional power, etc.) Levi 1967 adopts the first position, Nicholas 1984 the third, and I the second (Hooker 1987, chapter 5). I pursue the issue no further here except to comment that, insofar as pragmatic utilities play an essential role, the explanation of the epistemic character of science must focus on the capacity of its institutional design to so entrain individual behavior as to collectively pursue truth. So I return to institutional design.

47. A schematic example will serve to present some of the possibilities, possibilities whose relevance can be appreciated if taken in the light of the earlier discussion of plasma physics methodology (section 2.II.1). Consider a group of three scientists A, B, and C. Scientist A believes that the present theory T is fundamentally correct but agrees that it is worthwhile testing this belief further by attempting to construct alternative theories and testing them. Scientists B and C, on the other hand, disagree with A, and believe that T is defective and that there is an urgent need to construct more adequate alternatives. As it happens, A doubts that the experimental data on which B and C partly base their disbelief in T has the epistemic force which B and C take it to have, because A has some important doubts about the theories of the experimental arrangements involved in collecting that data, doubts that B and C do not share. Moreover, scientist B also believes that the specific experimental defects in the existing theory stem from its failure to satisfy certain very high-order theoretical principles, principles of symmetry. Scientist C, though, is sceptical of reliance on any such principles and believes that the defects in the existing theory are to be traced to its inadequacy to cope with certain quite specific features of the discomfirming experimental situations.

Now, as it happens, A agrees with B that the most fruitful alternative theories to pursue are a class of theories very similar to T except that they are reformulated so as to satisfy the preferred symmetry requirements. Scientists A and B agree that A is to attempt to develop experimental designs that will be sensitive to the satisfaction of such symmetries and to examine the experimental results, while B is to attempt the formal mathematical development of a theory of the required sort. Scientist C, on the other hand, believes it to be more epistemically valuable to examine experimentally a class of theories that are known to be more accurate in experimental situations of the sort that provided the original disconfirming evidence for T, even though the best formulated theories in this group happen not to have any deep theoretical relationship to the theories being investigated by A and B. All three are agreed, however, that the two proposed lines of research jointly present the greatest likelihood of epistemic payoff.

48. See Munevar 1989 and Hooker 1987, chapters 7, 8. The discussion of the preceding paragraph was drawn from Bjerring/Hooker 1979; now see also Ackermann 1986. On the mutual shaping of individual and institutional roles of relevance here, see Vickers 1968, 1983. Ultimately, we can bring the sociological, political, and economic literature on institutional design and functioning to bear on the design of epistemic institutions—consider the application of the note 45 references—and also the operations research and cognate literature, especially the viable and inquiring systems literature; see respectively note 25 and Churchman 1972.

49. Often these authors take themselves to be rejecting scientific reason, but like Feyerabend before them, it is really only the old formalist notion of method as simple logical rules that is rejected; the rest of their studies serve as grist for the mill of the present regulatory systems conception/decision theoretic account. On Feyerabend specifically, see Hooker 1991a; on the recovery of the cognitive in sociology, see notes 12, 21, 48.

Chapter 3

1. The text to follow is a joint product, co-written as equals. It is the product of a specific historical context. Barry E. Hodges completed Philosophy Honors in 1990 (1st class, university medal). C. A. Hooker's studies formed the framework for B.E.H.'s studies and this chapter. As part of his Honors work, B.E.H. provided studies of the structure of Popperian decisions (see Hodges 1990) and of Popper's evolutionary epistemology. The latter essay formed the initial basis of Part 3.II of this chapter. Both B.E.H.'s essay and the present, somewhat different version of it emerged from discussions between us. Both of us have reworked the text and contributed substantive content to Parts 3.II and 3.III. If detailed attribution is to be made, C.A.H. wishes to acknowledge B.E.H.'s contribution, not only to text analysis, but to developing the general conception of reason as the control or regulation of decisions. On balance, however, it seems least inaccurate to simply insist on joint authorship.

2. Consider, for example, the plethora of "conventional decision" introduced in the 1959 Preface and opening sections of Popper 1980, first published 1934, see Hodges 1990. Cf. the shift in Popper's treatment of convention over the course of his work (see Part 3.III below).

Readers are reminded that, as the term is used here, *cognitive* is synonymous with *epistemic* when the latter is construed in a naturalistic, fallible way (see chapter 1). Those more at ease with the latter terminology, which seems all too appropriate when discussing empiricists and rationalists, losing its sense of fallible conjecture, may mentally substitute it as they read.

3. See, in chronological order, Hooker 1987, chapter 3, 1981a, 1985, and 1991a.

4. Logical empiricism derives its metaphilosophical impulse from the Kantianism in which many of its practitioners were first trained. The principles of L constitute an a priori framework for intelligibility, just as Kant intended for his transcendental principles (and which he called a new transcendental logic). This complex relationship between Kantianism and the successor logical empiricism, in contradiction philosophically but in significant agreement meta-philosophically, is repeated in Popper's own relationship to the logical empiricism in which he was first trained; see below.

5. The simplest version of L is just the calculus of finite truth functions with observation sentences as recursive base. The metatheorems characterizing empiricism for this language all fail as soon as one extends the logic to the predicate calculus. Formal inductive logic was an attempt to restore them, but it has never succeeded. The failure is evidently principled, resting on the undecidability of the predicate calculus (and richer logics) and on the impossibility of characterizing inductive inference in purely formal terms; see note 8 and Brown 1988 and works cited at notes 3 and 6 and their references in turn.

6. See Achinstein/Barker 1969, Bjerring/Hooker 1981, Brown 1979, Easlea 1973, Feyerabend 1965a,b, 1969, Hooker 1987, and Suppe 1974 for these criticisms, and references to the empiricist literature. In perusing this literature, it will become clear to the reader that no one historical figure, let alone any collection of historical figures (e.g., those original members of the Vienna Circle), instantiates exactly the foregoing doctrines. Nonetheless, it is reasonable to isolate common presuppositions in an artificially sharp form in order to understand the doctrine, and that of Popper, which grew out of it.

7. This point, like many others, is found in Popper 1980, but now see Feyerabend 1978a, Hooker 1987, and below. For the general move away from a formalist conception of reason, which forms a larger framework for this book, see, chronologically, Hooker 1987, chapter 5, Brown 1979, 1988, Hooker 1991a, and Brown/Hooker 1994.

8. The essential point is nicely illustrated by an extension of Bertrand Russell's chicken story (Russell 1959, 35). Consider a young chick born in January and fed each day until December 22 of that year. This feeding evidence has no exceptions and thus, when combined with Reichenbach's straight rule of induction, yields the conclusion that the fowl will be fed on the morning of December 23. Russell took the failure of this conclusion, not only to note the limits of animal (and human) pattern extraction, but to

query the extent of lawful pattern in reality. But this last is not at issue here, for insert the very same evidence into a wider theory of cultural practices, and we may deduce with at least as high a probability that on the morning of December 23 this bird will be slaughtered for the dinner table. Something else of importance is at issue, however, for note that the very evidence that one may take in one context to support continued feeding is in itself evidence, in another context, that feeding will not continue. (The better the feeding, the fatter the chicken, the more attractive it is as dinner). Rationally chosen method then is a function not only of logic but of our representation of the substantive context in which formal logic is to be applied. Whether these representations are considered empirical generalizations or theories proper (in the present case, about cultural practices), on pain of anti-empiricist circularity, the choice of context representation cannot itself be dictated by inductive logic. Choice, and hence judgment, remain fundamental. (And if the context is specified theoretically, as it plausibly is here and certainly is in most science, then the difficulties for formal inductive logic are only increased.)

Carnap tried to circumvent this general problem by laying down a requirement that the total evidence is always to be used. But first, this principle is itself not a theorem of logic, so its adoption requires a judgment, and second, its use requires judgments concerning both the criteria for total evidence in a context and whether the criteria have been met. The decisions constituting these judgments cannot be forced by logic.

There is also the difficulty of understanding how theories can be inductively supported. Since theoretical descriptors designate unobservable properties and events, no single instance of them appears among the evidence; and so there seems a logical barrier to extending inductive inference to theories. One way out of this would be to claim that there are special principles of inference from empirical generalizations to theoretical entities focused, for example, on inferences to unobservable causes. There are at least two difficulties with this approach: first, to provide any plausible principle for the introduction of new theoretical concepts—after all, the history of science is replete with the introduction of surprising and counterintuitive theoretical notions—and second, to argue that these principles are purely formal. An alternative way out of the difficulty is to eschew theoretical terms as having any substantive content and to regard them instead as place holders in a formal logical structure. Rationality is confined to the organization of that logical structure, and the only novelty called for is formal. In one form or another, this has been the classic empiricist response (see Suppe 1974). A major problem with this alternative, besides that of induction itself, is to give a plausible account of the semantics of theory (cf. Hooker 1987, chapter 2 and references). There are further problems, but these must suffice here. For problems of induction, see also note 6 references.

9. This is so relative to any finite body of evidence. Briefly, the most secure truths are the analytic truths, but these are also vacuous. The most insecure truths are the analytic falsehoods, and these contain maximal content (since from an analytic falsehood all conclusions follow). In between

these extremes lie the range of synthetic statements and the more informative they are, the less likely they are, a priori, to be true. (This is so because the more informative a statement, the more possibilities it rules out and hence the more opportunities there are for it to turn out false.) In particular, the probability of a conjunction is less than or equal to either of the probabilities of its conjuncts $p(x \wedge y) < p(x), p(y)$; see Popper 1980, appendix *vii.

10. There is much more detail to add, for example, concerning severity of testing in terms of boldness of prediction. For further description and analysis, see Ackermann 1976, Burke 1983, Grünbaum 1976a, b, c, Hodges 1990, Lakatos 1970, Newton-Smith 1981, O'Hear 1980, and Schilpp 1974.

11. The only exception is the use of the concept of probability in corroboration for which Popper provides a measure in probabilistic terms, see Popper 1980, appendix *ix. This is still a formal concept, and since classical probabilities are maps from classical deductive logic into the unit real interval (a conception Popper helped pioneer), one not far removed from a purely logical one.

12. Cf. Hooker 1981a. Hooker 1987, chapter 3, concludes that Popperian metaphilosophy is essentially identical to empiricist metaphilosophy. Some of the details and emphases differ (e.g., there is less emphasis on translation into some privileged formal language), but overall Popper's metaphilosophy shows a striking parallel to that of empiricism. The most significant departure is Popper's abandoning the requirement of epistemological foundations, as he must given his account of observation terms, substituting instead the requirement that epistemology explain how progress in understanding is possible. This is in general a trivial issue for a foundational epistemology; one simply explains how the foundations can be improved, and improved knowledge follows. For empiricism, for example, the foundations are observations, and these are improved by adding more observations; induction then ensures that knowledge improves. But for Popper, this becomes a major, nontrivial issue (as we shall see). It is a particularly important one, since much of the force of Popper's position derives from the argument that it was through highly counterintuitive, counterinductive theories that science has made its most striking progress historically.

13. The general ideas here are well known (though perhaps not the multiple roles for the theory officially under test); see Duhem 1962 and Campbell 1957. An example of all these complications is worked out in moderate detail in Hooker 1987, chapter 4, and another in Hooker 1989b, chapter 2, section 2.II.1. See also Feyerabend 1978a on Galileo. These ideas are constantly being reinvoked in new studies; see Galison 1987.

14. Neither of course can inductive formalist method. Contradiction containment is not a trivial issue. A well-known case of anomaly containment is the motions of the heavenly bodies, particularly Uranus, the moon, and mercury vis-a-vis Newtonian mechanics (see Wilson 1970 and Stein 1990). All these anomalies need explanations, but some of them were a long

time coming and our current understanding of each of them is quite different. Most planetary wobbles, including those of Uranus, are explained by small, linearly additive interplanetary perturbations (in the case of Uranus, by Pluto, which was discovered in this way). The moon's complex behavior is explained through the complexities of nonlinear n-body dynamics, which we now know can also produce chaos, a fundamental departure from *expectations* of classical dynamical behavior. The behavior of Mercury is explained through relativistic theory, which represents a fundamental departure from classical *principles*. For methodological discussion and references, see above and Hooker 1994c.

15. See especially the results of Miller 1974, 1975. For recent discuss,ion see Niiniluoto 1984, 1987, and Oddie 1986.

16. Recent work in social-historical science studies has illustrated this point well (see Galison 1987 and Latour 1987) although there is a tendency to also deny that there is any rational pattern to comprehend. (On these latter issues, see chapter 2 and Trigg 1980.) And Feyerabend 1978a has particularly argued that, in context, it can be reasonable to violate any of the formal methodological rules that have been proposed.

17. What, for example, determines the weight that ought to be given to a problem, its intrinsic content or the answer that is (currently!) accepted? Consider a genetic theory G entailing that all bird species were single colored. The problem is: Why are some swans black and not white?, and the alternative answers are: (1) a dye in the food causes the olour change, (2) an extra, hitherto unnoticed gene mechanism M causes it but M reinforces the general principles of G; and (3) the color is caused by a new gene mechanism M′ incompatible with G. For further discussion, see Newton-Smith 1981 on Laudan 1977, a problems formulation. (But if problems are taken as indirect measures of cognitive utility in a decision theoretic epistemology, then much that Laudan says makes sense.)

18. "Construction of theories about the environment, in the form of physical organs or other changes of the anatomy. . . ." (Popper 1984, 244). This is a claim that is both widely held and hotly contested. Here we need only opt for the same kind of accumulation of internal regulatory orders, often expressed as levels, as the neo-Darwinian Campbell 1974 is happy to introduce.

19. Popper 1972, 47, says "we are born with expectations; with 'knowledge' which, although not *valid a priori*, is *psychologically* or *genetically* a priori, i.e., prior to all observational experience"; cf. Popper 1979, 72: "there is no sense organ in which anticipatory theories are not genetically incorporated."

20. Note the similarity of Popper's account here to Piaget's basic principle of an increasing shift from exogenous reaction to endogenous construction (chapter 5) and with Campbell's account of the "nested hierarchy of inductive achievements," which act to narrow the possible "search space"

for the emission of random trials; cf. Campbell 1974, 421 ff., or Campbell 1977, 502: "presumptive procedures." which "reduce the waste" involved in random trials.

21. Compare the role of abstractions here to that of reflective abstraction in Piaget, but contrast the attention Piaget gives to construction with its neglect in Popper (it is reduced to randomness, or "psychology"). In Popper, this is essentially because it cannot be reduced to formal (logical) operations. Both Popper and Piaget agree about the confinement of reason to formal systems, hence Piaget's attempts to characterize stages formally. But despite the internal tensions it caused, Piaget took the underlying naturalist regulatory model seriously, more seriously than does Popper.

22. It is not necessary to accept Popper's identification of this process with language alone to allow that the appearance of some such capacity to develop the reflexive nature of a symbolic consciousness, sufficiently rich to allow for the evaluation and criticism of abstractions, marks a crucial increase in the power of open programs and of the specifically human capacities of reflexive consciousness. Nor should we allow Popper to mislead us about the differences across levels. Any genetic or behavioral change that is to be directly subject to selection must also result in a public trait or traits, something which is there for intra- and interspecies detection and response. The peculiarities of World 3 variations lie in the detachment of their fate from that of their makers and the operation of rational evaluation as opposed to causal evaluation (cf. both Rescher, discussed in chapter 4, and Piaget, discussed in chapter 5). In particular, the mere fact of linguistic formulation per se does not seem to convey any greater publicity for variants. But in this essay, we do not intend to examine Popper's notions of consciousness, language, and mind, which run naturalist and quasi-Platonist anti-naturalist themes together.

23. This is a fairly standard treatment of this line of criticism, and Thagard applies it explicitly to Popper among others. See also O'Hear in Holland/O'Hear 1984, 196 and Ruse 1986, speaking in relation to Toulmin's evolutionary epistemology (49) but later extending it to Popper (62).

24. Remarkably, Popper often claims that these changes are not arbitrary (see Popper 1974, 110–1, the jury analogy). It is hard to see a principled Popperian basis for this (undoubtedly true) claim; we conjecture that Popper is again trading on the ambiguity in his use of the notion of convention (see Part 3.III).

25. This latter requirement is not trivial. Levi's decision structure, for example, however well it represents local decisions, is well known to be characterized by "global myopia," by lack of guaranteed global coherence. See Levi 1967, Bogdan 1976, and then Levi 1980. The construction of the sequence of environments (= problems) is a matter for decision: for Popper, conventional decision. The problem will, again, be the problem of how to achieve a global constraint on this artificial construction of the succession of problems such that it itself is progressive.

26. Popper's intellectual background includes German anarchism (e.g., Nietzsche, Stirner) and Vienna Circle positivist empiricism, and we can helpfully see his philosophy (especially of science) as attempting to marry the two, combining critical method and an open society with accepting empiricist metaphilosophy, in particular the assumption of the formalist program for reason and of commitment to the rational life as a non-cognitive, ethical act. The early period of Popper's life can be seen as the initial statement of this position (our early stage 1). The middle period can be seen as trying to work out the formal theory that would support this general line of research (late stage 1). The later period can be seen as an effort to salvage the formal program from its manifold problems by creating a metaphysics within which that program seems guaranteed to succeed (stage 2).

It is from within this framework that we can perhaps understand the thrust of Feyerabend's criticisms of Popper. Feyerabend also shares Popper's background. But his rejection of empiricism runs deeper than Popper's, for it includes a rejection of the formalist program for reason (and perhaps more; see Hooker 1991a). Thus we see the early part of Feyerabend's life as affirming the anarchist tradition but at the same time separately absorbed in the analytical exposition of the complex ideas of contemporary science. However, the anarchist critique of traditions is ambivalent toward science; on the one hand, science is praised as one of our foremost tradition-busting procedures, and on the other hand, it is itself to be condemned when it becomes another authoritarian tradition. (This ambivalence is beautifully illustrated already in Feyerabend's 1961 paper, "Knowledge without Foundations".) Feyerabend's middle period can be seen as being devoted to extending and deepening Popper's criticism of empiricism and of traditional philosophy of science insofar as it accepts an authoritarian tradition of knowledge (but with empiricist elements retained; for example, his incommensurability theory of meaning derives from an ostensive/implicit definition dichotomy in turn based on the empiricist observation/logic dichotomy.) His later period can be seen as an abandonment of any attempt at an authoritarian theory of knowledge (and of science in particular) in favor of an increasingly energetic reassertion of the anarchist attitude. (See his progression: What's so great about science? What's so great about truth? What's so great about reason?) The break is clear by 1974, when Feyerabend records his evaluation of Popper's third phase as degenerating (Feyerabend 1974).

But just as Feyerabend's anarchism ultimately led him to focus on the theory of traditions and their institutionalization, so Popper's anarchism drew him into the theory of society (Popper 1966), and like Feyerabend, this led to a very different conception of reason, viz., one focused around context-sensitive strategic problem solving decisions (cf. Stokes 1989). It is largely this sub-theme we take up in Part 3.III.

27. Clearly, the situation is far more complex than this: Popper offers us a sketch of "a procedure goverened by rules, " which governs the acceptance and rejection of basic statements—the jury analogy. This is an attempt to theorize our perceptual systems and to *institutionalize* means of error-correction; for example, this is Rconvention, sense (3), a theorized

practice. Ultimately for Popper, all such theories must eventually come down to judgments constituting decisions as to which basic statements to accept, with the group maintaining coherence among these judgments by consensus processes.

28. In some situations (e.g., initially in a research project), the criticism of the current pattern of accepted "background knowledge" will be on trial and error, ad hoc sort of basis, rather than as a formally theorized intervention. In this case, what we have is a higher-order practice governing conduct; cf. Popper's discussion of science as a "second order tradition" (Popper 1972, 127).

29. For the basic idea, see Bjerring and Hooker 1979, 1980; cf. Hooker 1987, chapter 7.

30. Popper reduces the difference between the amoeba and Einstein to just the presence and absence of the critical attitude toward error. In doing so, he misses all the intermediate complexity of nonrigid control. (This is the important functional counterpart to the dichotomy that his Worlds 1/2, World 3 reification creates.) Indeed, Popper offers his account of deliberative decision as a solution to what he calls Compton's problem—how is it possible that ["the world" of] abstract meanings can regulate human behavior—and Compton's postulate: "The solution must explain freedom; and it must also explain how freedom is not just chance but, rather, the result of a subtle interplay between *something almost random or haphazard, and something like a restrictive or selective control*—such as an aim or standard—though certainly not a cast iron control" (Popper 1979, 231-2). The solution lies in understanding the special role of language, in particular the critical function that abstract meaning makes possible. Popper's solution is presented as follows:

> Now in developing its higher functions, our language has also grown abstract meanings and contents; that is to say, we have learned how to abstract . . . and how to pay attention to its *invariant content or meaning* . . .
>
> What I have called "Compton's problem" . . . is now no longer a problem. Their [abstract meanings'] power of influencing us is part and parcel of these contents and meanings; for part of the function of contents and meanings is to control. (240)

But all this merely *asserts* that the problem is solved, no insight into the how of the solution is offered. It is an explanation of the same form as Voltaire's famous parody: opium causes sleep because it has a sleep-inducing capacity. Once again, PEE processes in themselves prove inadequate to provide a solution. The beginnings of a solution is sketched, we suggest, by combining naturalist accounts of teleology (see chapter 4, section 4.VIII) and of meaning as deriving from operational concept formation (see chapter 5 and cf. Hooker et al. 1992b) with the account of information and invariance offered in chapter 2. But this is well beyond where Popper would willingly go (and beyond the scope of this book).

31. The quotation is from Churchman 1968, 43; cf. Bjerring/Hooker 1979, 1980, 1981; Naor 1979. Our exposition below in indebted to Naor 1980, which he prepared while studying with C.A.H. and which drew attention to the importance of Popper's remarks about institutions. We thank him for its use. Both Naor and C.A.H. ultimately have James Leach to thank for drawing attention to the implications of Churchman's work for rethinking scientific activity.

32. This opens up a rich field of largely untouched research, by transferring strategic studies of the emergence of sociopolitical norms to the case of cognitive norms; Axelrod 1984 and Ullman-Margalit 1977; cf. Mitroff 1974 and Mitroff's mentor Churchman 1968, 1972.

Chapter 4

1. See Rescher 1977, 15-17. Rescher offers this translation from Montaign: "To adjudicate [between the true and the false] among the appearances of things we need to have a distinguishing method (un instrument judicatois); to validate this method we need to have a justifying argument; but to validate this justifying argument we need the very method at issue. And there we are, going round on the wheel" (*Essays*,book II, chapter 12). This line of argument is closely related to Fries's trilemma and to what I have called the God's Eye argument, for discussion see Hooker 1987.

2. It is now, unhappily, common practice to follow the empiricists in their deployment of a cognitive/pragmatic dichotomy together with confining the cognitive to truth considerations and hence to speak of these latter regress-stopping considerations as pragmatic as opposed to cognitive. I cannot find a principled foundation for such a dichotomy within a naturalist approach to intelligence for which praxis/practice is central (cf. Hooker 1987, especially chapters 3, 7, 8, and Hooker et al. 1992a). Similarly, it is clear for Rescher as well as myself that there is no principled distinction to be made between science and technology; rather the two are to be regarded as components of a single complex dynamical system (see chapter 1). Hereafter, I shall simply use the term *science* to name this entire system.

3. Substituting "fallibly rationally warranted" for claims to truth, the Wheel Argument becomes: Let C represent any epistemically accessible, hence practically usable, criterion of fallible rationally warranted scientific acceptance. What validates the employment of C? Only an argument that can independently show that C is fallibly rationally warranted. But any such argument must have fallibly rationally warranted premises and hence itself be in need of validation by the criterion C. But here Rescher's own analysis of interconnected regulatory warranting processes shows that neither the requirement of independence nor the conclusion are rationally warranted. What is left then is the requirement that all regulatory components of the system be fallibly rationally warranted, and this, as the following analysis will again show, is a requirement naturalists can meet.

4. As for the three pragmatist acceptance criteria, at one point there is the contrast between accepting p and rejecting (not accepting) p, at another, that between accepting p and accepting not-p, and yet elsewhere that between accepting p and accepting one of a number of alternatives inconsistent with p.

The vagueness derives from lack of discussion of exactly what kinds of utilities may be involved. If such clearly noncognitive utilities as sexual pleasure and social prestige are alone involved, then the remarks are true but trivially so. If on the other hand the only utilities involved are narrowly cognitive ones, say relief from ignorance and absence of error (Levi 1967), then the reading of this passage becomes problematic. Of course, one could always restrict utilities to the extreme cognitive utility of being right, as Rescher at one point wishes to do, but that would be to trivialize the problem in the opposite direction to that represented by the "flippant" non-cognitive utilities above. Rescher simply assumes that utilities can be neatly segregated according to a cognitive/pragmatic dichotomy, while I have offered arguments as to why they cannot; see note 2.

5. These simple claims are somewhat obscured by convoluted and controversial arguments; see, e.g., Collier 1979.

6. In RQ9, Rescher seems to advocate something like Reichenbach's straight rule of induction, but this is precisely not a rule we use without theory-governed context-dependent qualification (see chapter 3, Part 3.I, especially note 8). Here we have another illustration of the delicate interplay of theory and method.

7. Tellingly, Rescher remarks: "The *acceptance* of a thesis is, to be sure, a decisive act. But like other decisive acts (marriage, for example) one can take tentative and indecisive steps in its direction. Taken initially on some slight provisional and probatively insufficient basis, a thesis can build up increasing trust. A fundamentally economic analogy holds good here: a thesis, like a person, can only acquire a solid credit rating by being *given* credit (i.e., *some* credit) in the first place—provisionally and without any very solid basis" (Rescher 1977, 115). Confusingly, Rescher says in the very next paragraph that a presumption is a prima facie truth in exactly the sense in which one speaks of prima facie duty in ethics (ibid., 115). Notoriously, the deontological approaches to ethics clash with consequentialist decision- theoretic approaches. Presumably, talk of duties here should be translated in some way into a utilitarian context.

8. Newton's treatment of gravitational motion provides a beautiful and telling example; see Friedman 1986 or Hooker 1994b.

9. For an elaboration of this general point, see chapter 2, section 2.II.1, and Hooker 1989b, 1992a.

10. But there cannot be as few as the two loops he suggests, because of the difficulties with that idea already uncovered. On the idea of the machine analogy, cf. RQ20. The neutral and generic character of this basic

model suggests again a link to Piaget's unified account of regulatory systems. However, it is a relatively weak conception of self-organization in relation to even Piaget's model, since while the *content* of Rescher's system changes over time and to that extent we may consider the system not only self-regulating but self-organizing, the *regulatory structure* itself remains fixed in all other respects.

11. Following his earlier discussion of pragmatism (40), Rescher takes these needs, goals, or whatever to be segregable from cognitive utilities. I do not believe that this segregation is possible; see note 2. Moreover, I also doubt that the dubious distinction between needs and wants employed can be made good, for essentially similar reasons. I would also wish to take issue with the implied view that truth has only instrumental value, not intrinsic value, though I agree that its instrumental value powers the science-society machine; cf. chapter 1. But I set these issues aside here.

12. This process is illustrated dramatically in our growing capacity for genetic engineering generally but in human genetics in particular. We shall soon be capable of genetically selecting for cognitive abilities in a modestly strong way; unavoidably however, any attempt to do so will itself both be a cultural experiment in the consequences of designing that kind of political society and will set a new artifactual environment for the development of cognitive science. Indeed, the very history of science itself is an experiment biologically, socially, and cognitively (cf. chapter 2). It is as yet an open question whether science is a culture-bound experiment and, if so, in what respects.

13. This is an appropriate place to say that Rescher's philosophical thought has an impressively wide sweep to it and he is prolific in its expression. It is possible that the difficulties identified here and others identified earlier find their resolution within his many other books. While I have deliberately confined my attention to *Methodological Pragmatism* here and there is not the space to expand the analysis beyond it, I have formed the impression that doing so would not change the conclusions drawn above in any fundamental way. This is, for example, true of including Rescher 1990.

14. Nor should "neutral mutations"—ideas producing no immediately testable difference—be ignored, for they may prove important under scientific change, either because they then become functionally relevant or because their conceptual and/or methodological resources may be transformed into a new scientific regulatory setting that does produce important changes in commitment, for example, the long prehistory of atomic speculation or the relevance of tensor algebra, whose development predated its use in relativity theory.

15. One could attempt to strengthen the sense of teleonomy, and weaken Darwinism, by opting for some principle of the increase of order or complexity through the adoption of a non-Darwinian systems dynamics model of evolution, in the spirit of some of those now emerging (see Depew

and Weber 1985; Kauffman 1993; Weber et al. 1988; Wesson 1993), but the present conception of teleonomy ought, and can afford, to remain open-minded on this issue at this time.

Chapter 5

1. On the criticism of narrow neo-Darwinism, see Piaget 1971/77 and 1976/78; cf. the discussion of the role of phenotypic capacities at chapter 1, section 1.II, and of blindness at chapter 4, section 4.IX. While Piaget's enthusiasm for phenotypic capacities has led him to the edge of Lamarckism (or over), this can be excised, leaving an important and rich systems model compatible with Darwinism; see Hooker 1994d.

Piaget's ideas have had an important influence on the development of my naturalism over the past 20 years, but it was always "in a general way," because I never felt that I had the right tool to provide coherent insight into his diverse and eclectic works. I now believe that the regulatory systems approach that animates this book provides the missing tool. Amongst a quasi-infinite body of Piagetian literature, the highly selective list of references I provide are those I have found particularly pertinent. These references will effectively lead the reader into the remainder. I have found Haroutunian 1983 and Kitchener 1986, 1987 particularly useful to consult; each in their very different ways changed the shape of this essay. I shall add here a brief word on each.

Haroutunian 1983 provides an extended examination of Piaget's use of Waddingtonian-like biological models, providing many useful insights. While conceding much of value in Piaget, she is also highly critical. I am sympathetic to some of these criticisms (e.g., that Piaget does not have a full-blown theory of learning and that troubles arise from his peculiar approach; see section 5.I.2.2), but not to others, which seem to me to hinge on her assumption of a formalist or programming theory of mind; see herein. (The last chapter of Boden 1979 is afflicted with the same assumption.)

Kitchener's scholarly comprehensiveness is a comfort. Kitchener has read more extensively in Piaget's *oeuvre français* (and certainly more fluently) than have I. (Fortunately, most of the important works are now translated into English.) I have found his *reportage* of Piaget's doctrines fair and thorough and his reemphasizing the philosophical dimension of Piaget's thought especially valuable. If there is a criticism to make of Kitchener's work, it is that he is evidently so close to Piaget that his own writing reflects too much the confusing tendencies that Piaget himself shows, namely to eclectic collage and synoptic compression. By contrast, I have a simple analytic tool I bring to Piaget, the regulatory systems perspective, and I am prepared to slice through his texts to see if there is any underlying order. Happily, I believe there is (within limits). And this leads to the correction of what I take to be a serious *lacuna* in Kitchener, the absence of proper consideration and integration of Piaget's biology, and to a serious criticism of Kitchener's *interpretation* of Piaget's position; see sections 5.II.4, 5 below.

2. See Baldwin 1894/1968, 1896. For discussion by Piaget, see Piaget 1974/80, chapter 4; 1976/78, chapter 2. For the relevance of Baldwin's work, see Campbell 1974; Broughton/Freeman-Moir 1981.

3. It is precisely a general conception of this sort that also lies behind the approach taken to evolutionary epistemology in Hahlweg/Hooker 1989a and is taken throughout this book. Reading Piaget has motivated my approach in both these works and my naturalism generally. In Hahlweg/Hooker 1989a, Piaget's conception of genetic epistemology was dealt with extremely briefly and hence simplistically; his general regulatory systems approach was merely hinted at. Here and in Hooker 1994d I offer a more precise account of his position, accompanied by an extended critique. Along the way, I will reject two unqualified claims made in Hahlweg/Hooker 1989a, viz., that genetic epistemology is focused solely on individual development and that evolutionary and genetic epistemology are distinct. If one fails to adequately grasp the encompassing regulatory setting in Piaget, then one will arrive at limited and distorted notions of genetic epistemology. Many have done so, myself included. Here I shall try to have the regulatory conception dictate the analysis.

4. Piaget's own intellectual development can be traced through the sections of Gruber and Voneche 1977, especially the early sections, and Boden 1979, chapter 1, and cf. Flavell 1963. See also Kitchener 1986, chapter 1, for much more philosophically pertinant detail, but beware the absence of a biological, regulatory perspective, cf. Vuyk 1981. Chapman 1988 is an excellent biographical account, with a fine feel for the regulatory perspective, but too inclined to accept vague and generalized appeals to "analogies" between biology and cognition (which, alas, often enough reflect Piaget himself) to be helpful for systematic naturalist theory.

5. Piaget refers to equilibration as adaptation and sometimes to the features of the equilibria as adaptations (see also Chapman 1988). There is potential for confusion here. Successive equilibria represent successive operational adequacies across successively wider environmental ranges, an increasingly flexible capacity to respond successfully to environmental inputs. In the terminology used in this book, they represent increasing adaptability, not (or not merely) increasing adaptation. And the directionality to Piaget's sequences of developmental stages corresponds to that accorded adaptability in the evolutionary epistemology developed herein, whereas following Darwinism, sequences of adaptations per se are denied this property. (See also the discussion in section 5.II.3.) To avoid confusion, I shall avoid the terms adaptation and adaptability in what follows. On the other hand, because Piaget finds a common regulatory process underlying phylogenesis and ontogenesis, scientific change and individual psychogenesis (see section 5.I.2), I shall use the term *development* throughout, speaking even of evolutionary development.

6. Piaget and Garcia distinguish assimilation sharply from mere association (1983/89, 268–9), since the latter accepts stimuli as presented and

merely searches for correlations, while in the former the organism is active in imposing a set of salient features or categories that both actively reject some stimulus components as "noise" and also tend to idealize the remaining components as perfect feature exemplars (i.e., operate as binary filters). Cognitively, agents accommodate in order to better assimilate, and this is evidently true too in most kinds of neural net learning models.

7. Piaget says, for example, 'The 'reflecting abstraction' includes two inseparable aspects: a 'reflecting' in the sense of projecting on an upper level what is happening on a lower level, and a 'reflection' in the sense of a cognitive reconstruction or reorganisation (more or less conscious) of what has thus been transferred" (Piaget 1975/77, 35.) The "higher level" is to be one capable of widening the scope of the preceding one and integrating the "lower level" operations into it (Piaget 1967/71, 320); cf. text above. Why levels in this sense, including the group theoretic version propounded by Piaget, should require some hierarchy of operations on operations is not clear, nor whether we are dealing with an order or metahierarchy. Nor is the role of the "reconstruction or reorganisation" vis-à-vis this shift, since some such activities will be part of normal updating activity even within equilibria (see section 5.II.3), nor whether consciousness is a by-product in all this or has some distinctive role to play (cf. Piaget 1974/76). But these are only initial questions, and Piaget's texts would not permit clear answers to most of them. In what follows, I have tried to formulate doctrine with an eye to the questions that would arise if one wanted to develop naturalistic theoretical models of the processes at some later time, but no doubt I have also contributed my own obscurities.

8. The self-reproducing organization of living systems has been emphasized by Varela 1979. Recently, Dooley 1993 has also drawn attention to the regulatory systems or cybernetic terms in which Piaget conceptualizes self-reproduction and development. Dooley wants to relate Piaget's framework to that of Varela 1979 and to Maturana and Varela 1980. However, Piaget's naturalistic regulatory conceptions do not sit very well with the rather obscure and neo-rationalist treatment in Maturana and Varela, and I do not pursue this line.

9. The correspondence in the text is only intended as a rough one; it needs to be worked out in detail. But this is not easy to do, since Piaget has never done much of the detailed work on inter-theory relations and methodological structure in science, for example, to compare space- time structures or state space structures (cf. Hooker 1992 for some review and references) or place methods in a systems framework (Hooker 1989b, 1994a, b, and chapter 2); and see the Aristotle-Galileo shift schematized in diagram 2-3, chapter 2. While Piaget does attempt to relate some history of science, especially mathematics and physics, to the psychogenetic history of humans (see 1950 and 1970/72, chapter 3, section 3), this attempt is at best fragmentary, though often stimulating. Not until the very recent and welcome Piaget and Garcia 1983/89 is there any systematic study, and this book still has relatively little

to say about detailed intertheory relations and methodology. And I plead equally guilty vis-à-vis Piaget's cognitive categories. It would certainly be valuable to develop this analysis, but it is beyond the scope of this chapter.

10. This reading of the passage fits with the influence on Piaget of the "historico-critical" group of French philosophers who were all centrally concerned with the analysis of the conceptual development of scientific ideas and theories and emphasized the logical progression of these through time. (See Kitchener 1986 for discussion.) At one point, Piaget says that genetic epistemology "constitutes a simple extension of the historico-critical method" to individual psychological development (drawn from Piaget 1967 and quoted at Kitchener 1986, 12).

11. See e.g. Piaget and Garcia, 1983/89, 64. The only failure of process universality they accept is that, in biological evolution (and early ontogenesis?), the structures of earlier stages are not reflected and integrated into successor stages in the detailed way claimed for cognitive development proper (see 275). But the structural reconstruction involved in many changes of fundamental belief or epistemic framework (those that "eliminate pseudo-necessity," or involve regulatory ascent; see, for example, the schematization of the Aristotle-to-Galileo shift in chapter 2) can be quite radical. So this contrast seems a dubious proposition for cognitive development as well. This returns us to a search for common regulatory processes across the board (modulo complexity-dependent shifts; see section 5.I.2.3 below).

12. If one were convinced by Piaget's argument concerning the universality of the phenocopy process (see PQ14), then one could add this to the common (auto) regulatory process. Similarly, if one were convinced by the argument in Piaget and Garcia 1983/89, then one could add to this common process the intra-inter-trans succession (see PQ6). Of course universality can perhaps already be achieved cheaply by resorting to sufficiently general and vague process specifications, for example, the vaguer claim of increasing endogenous competence (as opposed to phenocopy) already included in the text formulation; this may provide an initially helpful orientation but must ultimately be given detailed instantiation. In my view, it is at this time good research strategy to attempt to maximize the specific content to common (auto)regulatory processes (it provides a bold conjecture to test), but we should expect modification in detail as the system exfoliates (section 5.I.2.3). Overall, it is premature to attempt to come to judgment on these matters.

13. This map is actually induced by the others if the notion of an inverse map, E^{-1}, is well- defined, formally $H_{po} = E_{pc}*H_{cp}*E^{-1}_{op}$, where * denotes some mapping composition. These may not be all the maps worth considering. The inverse of all four maps seem clearly constructible. I read Piaget's text as primarily supporting the maps given in diagram 5.1 below.

14. While this is more clearly so for E_{op}, an argument can be given for E_{pc} as well that derives from reducing the distinction between individual and community or population to a matter of regulatory degree and design along the lines indicated in chapter 2.

15. Before that question can be clearly addressed, there is a prior problem of clarifying what is involved. In PQ14, and it is typical, Piaget uses language that skates across both the distinction between individual and population levels of specification and the distinction between causal and information representations. By failing to clearly respect these distinctions, Piaget both makes it hard to decipher his claims and leaves himself unnecessarily vulnerable to important criticism; see Haroutunian 1983.

16. This will serve as advance warning against a Haekelian recapitulationist reading of diagram 5.1, a mistake Piaget encourages in his structuralist-formalist-stage mode; but Piaget himself rejects this reading of diagram 5.1 in terms of products rather than processes (see section 5.I.2.2 below).

17. That is, we should understand Piaget's method here in the same way as Piaget himself explains why he didn't research human prehistory (see PQ12). First, the contents of stages are studied because they are accessible; this focuses attention on how to characterize them. It is natural to choose formal characterization, because that is the dominant tool on offer. (Only recently, for example, have information theoretic characterizations of states been relevantly developed, and the tools are still weak by comparison with those of formal logic and group theory.) Then logical interstage relations again employ the dominant tool and moreover offer hope of characterizing sequence order without knowing the regulatory process details. As it turned out, this method failed; the interstage relations are functional (regulatory) but not formal. Considering how ignorant we still are of regulatory dynamics and how urgent it is that a working scientist find a usable methodology, formal characterization must have looked an attractive approach. For further remarks, see section 5.II.3 below.

18. And this means, I believe, that the debate on innateness returns to focus on the order of the regulatory structures required to provide an adequate model of cognitive development, exactly where Piaget located it. Haroutunian 1983, 15, and Kitchener 1987, note 84, by contrast, claim that Piaget is essentially a Chomskyan structuralist rationalist. But this conclusion turns out to presuppose a highly formalistic modeling of Piaget's biological and cognitive models, one at odds with Piaget's basic regulatory systems approach.

19. But too-quick or too-sweeping regulatory analyses may lead to difficulty. At one point, e.g., Piaget finds himself arguing that "it seems plausible to hold that those concepts which are the most resistent [to change under a scientific revolution] are also the most deeply-rooted from the psycho- and perhaps even bio-genetic point of view" (Piaget 1970/72, 76). Yet given a sufficiently sophisticated conceptual structure, empirical information is capable of forcing profound changes to our conceptual schemes, as the passage from classical to quantum theory demonstrates. We may expect that earlier regulatory structures will be embedded more deeply and so more "costly" to change (Wimsatt and Schank 1987, 1988), but it is far from obvi-

ous that the kind of psychological factors to which Piaget refers play any more than an indirect and indecisive role here. Rather, it is the regulatory coherence of science itself that dominates the outcome; see chapter 2. (The capacity for profound change, though, is the dual of that for cutting down possibilities discussed at the close of chapter 2; there are in this subtle issues concerning the resilience of regulatory systems that cannot be discussed here.) The complex kinds of interaction science exhibits also show that one should not rush into simple hierarchical models of regulatory order, as Piaget is inclined to do (cf. Dyke 1988 and chapter 2).

20. This seems to be Kitchener's view (see Kitchener 1986, 155), though earlier Kitchener had argued for the inclusion of cognitive development in the primates and in human prehistory as well (149). Kitchener offers a similar explanation to mine (note 17) to account for Piaget's choice of research field (ibid.).

21. See note 9. On the other hand, Piaget's remarks about the process of increasing objectivity and the separation of ego-centered projection from external attribution (1970/72, 82) can be made to correspond to a basic structural development in physics; see chapter 2, section 2.II.1.

22. A good example of the difficulties can, I suggest, be seen in Piaget's earlier search for scientific dynamics among the logical structural relations within the sciences. The result is a curiously hybrid structure, his "circle of the sciences." This notion, I believe, cannot be consistently embedded in a regulatory systems structure and indeed ultimately lacks a compellingly coherent rationale; it also fits ill with a naturalist constructivist position. (In an earlier version of this chapter, I had devoted an appendix to arguing this case, but space constraints dictated dropping it here.) This pessimistic reading is by no means universally held; Gillièron 1986 holds that the formal constructions of the circle of the sciences captures the essence of Piaget's genetic epistemology, and she concludes that genetic and evolutionary epistemologies are therefore very different from one another. Yet I am also not alone in finding it difficult to provide a coherent reading of this material; Kitchener, in his careful review of Piaget's epistemology (1986), completely omits this part of Piaget's thought and for precisely the same reason (private communication).

23. See Kitchener's discussion in 1986; cf. 1987, note 53 and Bickhard 1980b. The formulation in the text is crude because homeostatic subsystems will be complexly causally interrelated and needs recognized by the system will often represent amalgams of homeostases, and conversely, a given homeostatic system might be involved in the expression of many needs. Iron deficiency, for example, is involved in many human body malfunctions and hence recognized needs, but there is no direct felt need for iron per se. In what follows, I ignore such complications.

24. Kitchener/Kitchener 1981 have argued that formalism cannot do justice to what Piaget requires of reason. I support this view—see note 28

below and chapters 1, part 1.I and 6, section 6.I.2.2. There I offer reasons not to treat reason as fundamentally formal; see also Hooker 1991a, 1994c. The text makes obvious my use of Kitchener's helpful studies, particularly Kitchener 1987, in this section.

25. Piaget speaks of an equilibrium *between* assimilation and accommodation. This can't be quite right; confusion can easily arise here from use of ambiguous or vague terms. The basic equilibrium is between system and environment. It is also true that at equilibrium, there is no feedback from (failure of) microaccommodation to macroaccommodation, and this represents an internal equilibrium.

26. There are three kinds of significant qualifications to place on these claims, three kinds of limitation that provide significant clauses to the ceteris paribus condition. The first concerns limits imposed by our finitude; see Cherniak 1986. The second concerns limits imposed by our ignorance, especially ignorance of epistemic institutional design, which may lead to crippled inquiry processes; see Churchman 1972 and chapter 2. The third concerns limits imposed by our social/moral imperfections; see Mitroff 1974 and Munz 1989. Limits of some kind there are undoubtedly are, but developing a coherent account of them is a subtle business indeed (cf. Brown 1988 and Hooker 1994c on Cherniak). Without the aid of intense analytical effort by many researchers, for example, who would have thought that many different very simple devices could all be universal Turing machines? I shall not pursue any specific position on limits here.

27. Roughly, that is because it is unclear whether there could be sufficiently large microaccommodations to count as Piagetian development while not strictly adding any further regulatory orders (i.e., while not amounting to regulatory ascent). Micro grades into macroaccommodation whereas the refinement/ascent distinction is sharp.

On the other hand, Piaget does not directly discuss the feedback relations between macro and microaccommodation which prove central to the scientific case; see chapter 2, section 2.II.2 and text. Two of these latter correspond to the two routes by which equilibria are destabilized, but the others have no Piagetian counterpart. Even in Piaget and Garcia, where it is acknowledged that the process is more complex (and where detailed studies of selected developments in mathematics and physics are provided), there is only brief treatment (1983/89, 204–8, 239–40).

For these reasons, it must remain for another occasion to investigate a Piagetian treatment of "rational generalisation" in science.

28. See note 24. The requirement that reason as characterized be a formal structure runs into some thorny problems. If, for example, it is asserted that for every developmental change, there was already in existence a higher regulatory level that rationalized that change (say by representing it as an optimization, that is, as a certain kind of homeostasis), then either regulatory ascent is not open ended or we are literally infinite systems. The only alternative, as far as I can see, is to have environmental input play a

role in creating each higher regulatory level, but then it can scarcely come about by explicitly represented formal reason alone.

Note that these difficulties do not require relinquishing the representation of homeorheses as higher-order homeostases in general but do require rejecting the idea that regulatory developments are represented in the system in advance. This is also a thoroughly Piagetian idea (see PQ2, PQ18), but it raises the question of exactly what capacities the autoregulatory system must have in advance. The relevant homeorheses will be given in the causal representation as some causally active features of biological systems; they may even be formulable in the information representation as some very complex processes (but remember the system regulatory structure is itself changing) but will not in general appear as separable regulations, especially not at some imagined topmost heirarchial level (cf. Hooker et al. 1992b). This conclusion is reinforced by recalling that all regulatory levels interact, there is no distinguished hierarchy in that sense (chapter 2). Our understanding of self-organizing systems is still extremely limited, so perhaps there are other, subtler alternatives (cf. the discussion in chapters 2 and 6). In any event, I leave the difficult issue of the embodiment of reason for a time when developing scientific insights may permit a better conception of the problem and its solution.

29. And despite a variety of passages that can be read in a Kantian realist, or even more strongly idealist manner; see Kitchener 1986, 101–111. Kitchener remarks that when Piaget attacks realism, he is attacking only a naive realism that defends a "copy" or "mirror" theory of knowledge and not any more critical realism. See also note 31. Naturalist realism is of the critical kind (see the attack on the copy theory in Hooker 1978 and Hooker 1987, chapter 8).

30. This formulation goes beyond the few remarks Piaget offers; see Piaget 1950, 2; 250; 1967, 1, 244. Disturbance in itself need not preclude obtaining disturbance-free knowledge— witness classical mechanics where the disturbances themselves are theorized and calculable— though it might do so. But quantum theory raises quite profound problems of principle, in particular the issue of the form of objective representation; see Hooker 1992, 1994a. Piaget's conception of knowledge as a set of transformations is curiously adaptable to both Bohr's and Einstein's opposing positions and needs some deepened analysis at this point to be brought to bear on this fundamental issue. The limits discussed here are in addition to those discussed at notes 26, 28.

31. If it were argued that the base notion of truth here is a pragmatist one, a theory is true to the extent that it succeeds, I would counter that this ignores the self/other separation central to cognitive development. Even so, because of the way the ideal is modally formulated (. . . all *possible* explorations . . ., etc.), it still forms a realist correspondence concept (Devitt 1984; cf. Hooker 1987). Note that the formulation of the ideal in PQ24 is also modal, despite its surface grammar; see section 5.II.5 below.

32. And even though he recognizes that it does not fit with his Popperian reading of Piaget (see section 5.II.5 below) because of Popper's sharp—even ontologically reified—dichotomy between normative and descriptive.

33. Kitchener 1986 argues persuasively that in this respect Piaget has borrowed much of the Kantian program of constructing a transcendental psychology, as opposed to a transcendental philosophy which Piaget rejects. But precisely to this extent, these claims of necessity rub against Piaget's naturalist biological regulatory schema; cf. PQ3, PQ13, PQ14 (while placing him, ironically, in the same tradition as Lorenz).

34. There is a subtle but important conflation in Kitchener 1986, 26–7. There the ideal epistemic individual is equated with the normal or healthy individual and with a set of formal ideals. Though Kitchener recognizes that this can't be quite right because of limitations common to all individuals (e.g., occurrence of unavoidable mistakes), he does not see that for a naturalist a (fallible) theory of the former is the central way to characterize a (fallible) theory of epistemic and rational ideals, whereas for a non-naturalist, the latter may assume an independent form. See chapter 6 and Hooker 1991c.

35. At one point, Kitchener argues that Piaget's regulatory model of reason "invites the objection, however, that one can know a priori what the future will be, namely it will be 'more equilibrated.' How this avoids 'a priorism' is therefore not clear . . ." (Kitchener 1987, note 31.) But first, all the content to the concept of an equilibrium is given a posteriori; even what reason is, let alone good science, is only disclosed en route, see PQ19, PQ23, PQ32. And second, the characterization of the whole process in terms of equilibria is strictly conjectural, surely, since even for Piaget it is derived from scientific observation. If we find intermittant rationalist elements in Piaget here, then let them be set aside.

36. At one point, Kitchener describes Piaget as an "enlightenment rationalist," meaning that he believed in science, progress, etc. with the driving force being reason. This may be a useful way to understand much of the seemingly rationalist tone in Piaget's writing. But so long as reason is taken naturalistically, it does not require commitment to rationalism.

With respect to Piaget and naturalism, it was unfortunate that I did not read Bickhard 1988, 1989 until this book was in press. Bickhard has developed a naturalist approach to cognition similar in concept to my own, but often exploring complementary aspects to those I develop (see Bickhard 1980a, b, Bickhard and Ritchie 1983), and he illuminates Piaget accordingly. He has, for example, developed a powerful critique of encoding or formal correspondence theories of cognitive representation and an alternative interactionist account of the causal origins of representation which he then applies to a critique of current formal linguistics (Bickhard and Campbell 1992) and artificial intelligence (Bickhard 1991a, 1993). Unsurprisingly, Bickhard and I largely agree about the outlines of a viable Piagetian naturalism. Bickhard contributes a rather more elaborate analysis of Piaget's formalist difficulties than I have offered, focusing on the structuralist assumption of truth as requiring exact formal correspondence between internal and external (cognitive and cosmic) structures. Before rescuing Piaget for naturalism, he paints Piaget after the rationalist manner of Kitchener here. Bickhard also provides important insights into Piaget, for example, concerning the importance of phenocopy and the inadequacy of '1 level' (un-nested) models of variation, selection and retention (VSR) processes, which complement well the analysis I offer. But ultimately I would contend that Bickhard's presentation of structuralist Piaget does not do justice to Piaget, and for the same kinds of reasons that have just been provided in criticism of Kitchener. Though Bickhard 1988, 1989 surely do not aim at a complete, or even a balanced, presentation of Piaget's whole position, it is perhaps indicative that he discusses neither Piaget's conception of reason as a constructed ideal (even while elsewhere he sketches a first rough idea of reason as self-organization through nested VSR processes—Bickhard 1991b), nor the status of logic itself as a fallible construction, nor the interplay between normative and factual. In short, while I have no objection to Bickhard's critique of structuralism and welcome his alternative interactionist account, I credit Piaget with far more naturalist interactionism than does Bickhard and I would argue that Bickhard's interactionism needs further development in certain key respects (indeed in something like the manner of this and the next chapter).

37. I evidently occupy, then, exactly the position Piaget describes as tempting but not ultimately defensible; see Piaget 1967/71, 342. However, perhaps he only means there that, even if mathematics is mutable, still it has its own developmental dynamics and does not develop simply as an extension of science. But various versions of this latter are certainly naturalistically allowable.

Chapter 6

1. Throughout this chapter, *reason* and *rationality* are used interchangeably; the choice of one of these terms in the text is dictated solely by considerations of current grammatical practice. Neither term should be construed narrowly (e.g., as confined to formal argument).

2. See Hooker 1987 chapter 2, note 5; similar remarks came from many other philosophers as well at the time. Though I did learn more from Piaget than from Quine, Quine's article did serve to draw my attention to the idea that in a naturalized conception, "the criteria of cognitively justified action merge much more conspicuously into the general criterion of rational action" (Ibid.), and this has proven a liberating insight that culminates here and in Brown/Hooker 1994.

3. Material in this section is adapted from Hooker 1991a, but now modified in the light of subsequent discussion, especially with Hal Brown.

4. Involuntariness is not enough. Not every aspect of what is pushed involuntarily on us conveys truth. Most perceptual illusions cannot be voluntarily corrected, and we must involuntarily listen on occasion to advertisements, political speeches, and white lies. It is indeed an example par excellence of the exercise of reason to learn to distinguish what respects of what involuntary signals reliably convey information. ("Messages" heard by schizophrenics are involuntary yet we conclude that no external signal at all is present.)

5. For phenomenalist empiricism, "external" is read strictly as "external to consciousnes" while for naturalist or materialist empiricism it can be read as "external to the embodied person." But this is not the place to spell out the niceties of philosophical distinction, which are of small moment in this context.

6. Again, the distinction of note 5 applies. The transcendence is closer to the battle for clarity and efficiency within nature than to any transcendence of nature, though there is also ambiguity of the same kind on this point.

7. See Hooker 1994c, 1987, chapter 8, cf. especially Brown 1988; Brown/Hooker 1994.

8. See Putnam 1981; Devitt 1984. This last assumption derives its force from Tarski's satisfaction theory of truth. That this theory is a semantic one has no doubt encouraged the first assumption. For discussion, see Devitt 1984. For Tarski's original work, see Tarski 1956. I have long held that Tarski's semantic theory of truth is a theory of how to use the predicate "is true" consistently in a linguistic framework and that it is no more than this. (See Hooker 1987, chapter 2, section 2.6.1.) In particular, my view is that it does not express any significant correspondence theory of truth and does not reduce truth to satisfaction or reference.

9. For some discussion, see Hooker 1975, 1987, section 8.4.3.10.

10. For discussion, see Hooker 1987, sections 5.4, 7.11, 8.3.11.

11. Judgments are insufficiently precise, for example, when reality permits truthful judgments with quantitative ranges smaller than those made. Partially true judgments are those that could be conjoined to other true judgments and the whole conjunct remain true. Approximately true judgments are those that employ concepts that are applicable only at some level of systems approximation to reality and in contrast to the concepts of a more unified theory yielding deeper understanding.

This is a relatively specific use of the notion of approximate truth, general philosophic usage often intends to refer to all three categories above. Among other things, however, it has been the tendency of philosophers to ignore these distinctions, which have made it so difficult to articulate a plausible notion of approximate truth. See Hooker 1987, section 8.4.3.9.

12. Which inferences these latter are itself depends on the possibility structure of the world. See Hooker 1987, sections 8.2.7 and 8.4.3.1, 2. In particular, I take up no detailed position here on inductive logics, Bayesian confirmation theories, etc., except to insist that truth must play a key role in their structuring if they in turn are to play a key role in rational confirmation.

The concept of truth plays a key role in structuring rational procedures generally (not just in logics, which are the minor algorithmic part of rationality), but it cannot be the sole founding concept. Even for deductive logics, for example, we need to start with true premises if we are to reach only true conclusions, and this is not a matter logic itself can decide. And even within epistemology, whose ideal is (rationally held) truth, there are all the strategic compromises among valuable cognitive goals to be decided (see the discussion of truth proxies in section 6.II.2).

13. The quoted portions here and in the next few paragraphs refer to Putnam 1982. Putnam's arguments against a correspondence notion of truth are developed in Putnam 1981.

14. See chapter 3, sction 3.I and earlier Hooker 1987, chapter 2, and 1981a. And compare this last passage of Putnam's with the last paragraph of Hooker 1987, chapter 7, written independently but at about the same time.

15. The capacities to speak a language, enter into a culture, and a host of related capacities are at least relative capacities. The degree to which we actually possess these capacities is relative to the environments in which we were raised. (Note that, like the capacity of a pipe to carry any given fluid, the actual capacities possessed in a specific context are not themselves relational.) Further, all cognitive capacities have fundamentally to do with capacities to receive environmental inputs and generate selected environmental outputs and are thus relational in form. Their intrinsic bases in phenotypes will be whatever supports those systems capacities. These will be very complex regulatory capacities and could scarcely be captured in any simple formula.

16. I owe this and several other nice examples and distinctions in this book to the careful but supportive commentary of Norton Jacobi.

17. Note that reasonedness or reasonableness is to be sharply distinguished from rationalization, both in the sense of confabulation in the service of other goals and in the sense of an overly rigid structure to life that prevents exploration of the other life qualities (those that are entered as proxies for beauty and goodness) as well as spontaneity and creativity. It is itself part of the character of reasonableness to maintain a balance among these pursuits and that too is part of the self-reflexive character of reason.

18. I want to emphatically reject the view that there is no reasoning entering into either being cunning or playing tennis or a violin, etc. But I also recognize the distinction between having reasons to adopt a given strategy and that strategy being a reasoning one. This does not destroy the idea

of judgment as an extension of skill organized by reason, for there is a capacity to shape and then rely directly on judgment in a way that may (but also may not be) directed at aiming for truth, beauty, and goodness (and reason), and this capacity is more clearly visible in the exercise of skills. Moreover, aiming for these regulatory ideals are themselves generalized skills. But this is no place for laying out a theory of skills; see Brown 1988 and Brown/Hooker 1994.

19. Of course, what an organism possesses is a constructed internal representation of the ideals, not the ideals themselves. Not all intelligence is thus rationally organized. Moreover, not all of mind or spirit is intelligence in the narrower sense. Norton Jacobi, pursuing a Buddhist point, put it this way:

> The existence of pure, ecstatic, non-contingent joy (i.e., of joy independent of needs and wants, expectations, goals and achievements) also demands an explanation in terms of the ontology of mind and the metaphysics of ultimate reality—and it too must be naturalised. As you know, I think that joy is the fundamental regulatory principle of *Homo sapiens*—and that its experience almost certainly lies outside of rationality as either of us have chosen to conceive it. (1992; Private communication)

I largely accept these claims, though I would set love alongside joy and understand them both as more intimately rooted in intelligence. It is part of what especially complicates an understanding of the goodness and beauty ideals. I only remind that naturalism does not entail any narrow materialism, only the rejection of division by human fiat instead of empirical experience. But extending a naturalist account to a full-blown theory of mind is well beyond the confines of this book.

20. It might, for example, begin with a construction of proposition along the lines of Stalnaker 1984, as a division in a space of represented possibilities and elaborate cognitive and other states as operators on these. Such constructions promise to grade back suitably to simpler nervous structures. But I am non-ommittal about these matters at this point.

21. I have elsewhere consistently expressed the view that language is unlikely to provide the most fundamental insight into the nature of rational intelligence and its evolution. See Hooker 1987, chapter 8, note 188 and text. Work by people like Churchland 1986 and Churchland 1989 provide a refreshing alternative to the typical assumption by philosophers that language is central to intelligence. The very intimacy of their contacts with serious efforts to develop a neurobiology of mind suggests the fruitfulness of a nonlanguage-based approach. And more recently, I have become further entrenched in this view through new work in the development of control theory, initially laid out in Hooker et al. 1992a, b.

22. For further discussion of these relatively recent, but potentially very important and fascinating ideas, see Bechtel and Abrahamson 1991, Churchland 1989, Hooker 1994e, Hooker et al. 1992b, Smolensky 1988 and references. Generating a fitting solution efficiently will have to do with economy in the use of neural modeling resources and the like no doubt, but it evidently need not require reference to any system of formal logic.

23. It would certainly boost the plausibility of the naturalist position to be able do so! Meanwhile, one could say such vague things as that neurally organized intelligence involves the development of complex sensory and motor control systems and the memory and association structures that link them, and likely be correct. But these vague notions in themselves take us almost no distance toward understanding intelligence. Today's computers, for example, can be equipped with all of these devices in large quantity, and consequently they can be capable of making very large numbers of judgments, indeed of making them much faster and in much more detail than we can; yet today's computers stand among the lowest ranks of intelligent things. Although opinions by experts about the organizational secret of both neural and mental structures abound in the published literature, we are at too early a stage in our knowledge, there is too much ignorance and too much speculative diversity, to form any firm commitment on the matter. Consider the recent rise of connectionist models of intelligent organization, a real alternative to the hitherto dominant digital processing models, and yet they still have deep difficulties attending them; in particular they are only just beginning to be provided with their necessary regulatory systems framework (see note 22 references).

24. Cf. the discussion of tensions in Hooker 1987, chapter 8 and 1991c. The complex range of activities listed above encompasses the much narrower definitions of reason usually offered, e.g., that it is the capacity to bring evidence to bear on making judgments (included in the relevance and method components) or that it is the capacity to design solutions to problems (included in the method and self-reflexive components). See Brown 1988. Yet Brown's clear and helpful book shows how complex even these component activities are. See further Brown/Hooker 1994.

25. This analysis is argued for in detail in Hooker 1994c, building from a critical appraisal of Cherniak's critique of the traditional theory of reason (Cherniak 1986) and will be further discussed in Brown/Hooker 1994. See also note 27.

26. The general formulation of the material on ideals to follow is a product of joint discussions between Hal Brown and myself. He contributed as much as did I to its creation, and (so) he is as responsible as I am for its general drift. But the precise version given here, the particular detail of its formulation, is my own current account of the matter, and criticisms that turn on detail must accrue to me. Our joint efforts will appear in Brown/Hooker 1994 in due course.

27. Cherniak proposes a weakened version of A, A″, to replace it—roughly, the requirement that *all* beliefs be consistent, *all* inferences made, etc. is replaced by the requirement that *at least some* beliefs be checked for consistency, etc.—and a rejection of I, R as an ideal. In Hooker 1994c, I have argued that, although the "direction" is right, Cherniak does not provide a fully adequate analysis because of the neglect of degeneracy. But Cherniak's arguments, extended, ultimately reveal, not merely the complications of finitude, but current analytic theory of rationality as degenerate. An adequate or nondegenerate theory of reason for finite human agents will not be recapturable as simply a complication of, or correction to, idealized rationality, though the latter may be recaptured as a limit degeneracy of the former. Establishing this claim clarifies and extends Cherniak's analysis. More importantly, it opens the way to re-casting the philosophical theory of reason in a more adequate form, one supporting a new rapprochement with the sciences.

28. See the discussion of regulatory ideal formation in regulatory systems below. I should also be prepared to consider giving up naturalism if it turned out that there was no suitable account of regulatory ascent available that provided a persuasive naturalist embedding of rational processes. (This might be the case if, for example, powerful support emerged for *certain forms of* Chomskyan rationalism; cf. Piattelli-Palmarini 1980.)

Until we reach some understanding of the key processes in relation to the fundamental properties of matter, we must remain unsure of their implications. On the other side, our rational capacities are lost in mystery, mystery circumscribed by the Gödel, Lowenheim- Skolem, and other theorems, and which in any case we do not understand in any clear sense; I refer to our capacities for definite description, self-reference, and the like. Naturalism has no a priori commitment to materialism, only to a unified account. It may well be that, even in the domain of reason, there lie grounds for supposing that there is more to the natural world than meets the eye of natural science, as that eye has developed historically so far. Other dimensions to life may of course equally provide such reasons. We are left with a field of open questions, whose pursuit will further help to clarify our natural understanding of life processes in this universe.

29. Leaving the matter here won't do; Stich has recently argued that truth is not intrinsically valuable and should not play a fundamental role in a theory of reason (see Stich 1989). Similar arguments would apply to the other regulatory ideals. I shall return briefly to these arguments in section 6.II.6 and illustrate in just the one case the manner of response. But the basic argument is already present here: without this internal structure to reason we cannot adequately account for our actual regulatory capacities. This is my view of reason. It would not disturb the naturalist program, of course, were the regulatory ideals reduced or dispensed with in any of the ways mentioned in the text.

30. We will have embedded in brain function from a long prehuman evolution the projective structure of space and time. This projective struc-

ture already encodes logical structure geometrically. See Hooker 1973. Many simple operational structures based on space-time structure exhibit even more complex logical structure; see Giles in Hooker 1979.

31. For the interested reader, some basic ideas about efficiency in economic theory are rehearsed briefly here that may serve to clarify what is said and will anyway give it some relation to economic models. (Here I am indebted to discussion with Norton Jacobi.) The general question of efficiency arises only in relation to action under some constraints (i.e., limits on what actions are possible) and in respect of the use of at least qualitatively identifiable, and (typically) quantitatively measurable, resources (whatever these be: capital, equipment, time, labor, materials, knowledge, goodwill, etc.). Given a constrained resource context of this kind, we can study how to reduce resources to achieve a given goal within the constraints or how to increase the output value obtainable from the use of fixed resources, etc. Changes of these kinds represent increases in efficiencies.

To transform these general qualitative and comparative (ordinal) efficiency problems into an optimization problem (i.e., one yielding maximization of utility through minimisation of resource waste) requires that further conditions be met so that an optimization problem is well defined. Economic theorists standardly make life algorithmically easy for themselves by assuming that the constraints are linear (and any other independently given variables likewise), that there exists an objective function that sums up all operative values in an appropriate way and yet can be expressed in terms of a single dependent variable (utility, typically money), and that this function is "well behaved" (i.e., is continuous and differentiable, convex to the constraint hyperplane). Under these conditions, one has a linear optimization problem for which a unique solution exists (though there is no algorithm for finding it in any finite time). As these austere constraints are relaxed toward reality the well-definedness of the optimization problem becomes increasingly problematic; for example, there may be several optima rather than a unique "best" one, or their existence becomes practically unspecifiable or even uncomputable (which may be different from their not existing; physics, for example, is filled with uncomputable functions). Only recently have economists begun to take nonlinear dynamic models seriously, although they have fundamental consequences for its principles; see asterisked bibliography items.

Economists distinguish between static and dynamic efficiency. Static efficiency is efficiency determined within fixed available resources and constraints. In dynamic efficiency one examines the possibilities for applying available resources to alter the constraints and the available resources, each such alteration then defining a new context with its static efficiency problem. But since these alterations may occur continuously, it is really necessary to evaluate outcomes (i.e., the objective function, if there is one) across the whole time path of the solution space. In this case the criteria for selecting among time paths are no longer univalued; for example, one could choose the time path with the maximum value after time T, or that with the maximum time-discounted value overall to time T, or the most stable time path with outcome value at time T greater than some floor, etc. Thus dynamic efficiency is always a matter of judgment, no matter how formally accessible the problem domain and its static efficiencies are. The distinction between static and dynamic efficiency is essentiallly that made below in the text between NCE and WCE, except that WCE incorporates a still wider search problem to allow for value alteration, proxy learning, etc. (i.e., with a changing objective function and efficiency criterion allowed). A fortiori it will not be formally determined, which is the essential point made in the opening discussion.

BIBLIOGRAPHY

Note: In those references [to Piaget] where two dates are indicated, the earlier one refers to a first appearance (in Piaget's case, of the French edition) and the later one to that of the relevant current edition (in Piaget's case, the English edition actually used).

* Abraham, R., and C. Shaw [1982], *Dynamics—The geometry of behavior*. 4 vols. Vismath: The Visual Mathematics Library. Santa Cruz: Ariel Press.

Achinstein, P. and S. Barker (eds.) [1969], *The legacy of logical positivism*. Baltimore: Johns Hopkins University Press.

Ackermann, R. J. [1976], *The philosophy of Karl Popper*. Amherst: University of Massachusetts Press.

Ackermann, R. J. [1986], "Consensus and dissensus in science" in Fine, A., and Machamer, P. (eds.), PSA 1986. East Lansing, Mich.: Philosophy of Science Association.

* Allen, P. M. [1983], "Self-organization and evolution in urban systems" in Crosby, R. (ed.), *Cities and regions as nonlinear decision systems*. AAAS Selected Symposia Series. Boulder: Westview Press.

* Anderson, P. W., K. J. Arrow and D. Pines (eds.) [1988], *The economy as an evolving complex system*. Redwood City, Cal.: Addison-Wesley.

Apostol, L. [1982], "The future of Piagetian logic." *Revue Internationale de Philosophie* 142–3: 567–611.

Armstrong, D. M. [1968], *A materialist theory of mind*. New York: Humanities Press.

Ashby, W. R. [1970], *Design for a brain*. London: Chapman and Hall Science Paperbacks.

Axelrod, R. M. [1984], *The evolution of co-operation*. New York: Basic.

* Bak, P., and K. Chen [1991], "Self-organized criticality." *Scientific American* 264, no.1: 46–53.

Baldwin, J.M. [1894/1968], M*ental development in the child and the race*. New York: Augustus M. Kelly. (The 1897 French edition has additional passages, see Piaget 1976/78.)

Baldwin, J.M. [1896], "A new factor in evolution." American Naturalist.

Bechtel, W. [1989], "An evolutionary perspective on the re-emergence of cell biology" in Hahlweg/Hooker 1989b.

Bechtel, W., and A. Abrahamson [1991], *Connectionism and the mind: an introduction to parallel processing in networks*. Oxford: Blackwell.

Beer, S. [1979a], *The heart of enterprise*. Wiley: New York.

Beer, S. [1979b], *Platform for change*. Chichester: Wiley.

* Berge, P., Y. Pomeau, and C. Vidal [1984], *Order within chaos* (trans. by Tuckerman). New York: Wiley.

Bickhard, M. H. [1980a], *Cognition, convention and communication*. New York: Praeger.

Bickhard, M. H. [1980b], "A model of development and psychological processes." *Genetic Psychology Monographs* 102: 61–116.

Bickhard, M H. [1988], "Piaget on variation and selection models: Structuralism, logical necessity, and interactivism." *Human Development* 31: 274–312. Reprinted in L. Smith [1992], *Jean Piaget: Critical Assessments*. London: Routledge.

Bickhard, M. H. [1989], "Interactivism and Genetic Psychology." *Archives de Psychologie* 57: 99–121.

Bickhard, M. H. [1991a], "The import of Fodor's anti-constructivist argument" in Steffe, L. P. (ed.), *Epistemological foundations of mathematical experience*. New York: Springer.

Bickhard, M. H. [1991b], "A pre-logical model of rationality" in Steffe, L. P. (ed.), *Epistemolgical foundations of mathematical experience*. New York: Springer.

Bickhard, M. H. [1993], "Representational content in humans and machines" *Journal of Experimental and Theoretical Artificial Intelligence* 5: 285–333.

Bickhard, M. H., and R. L. Campbell [1993], "Some foundational questions concerning language studies: with a focus on categorical grammars and model-theoretic semantics" *Journal of Pragmatics* 17: 401–433. (See also replies to criticism at pp. 557–602.)

Bickhard, M. H., and D. M. Ritchie [1983], *On the nature of representation: A case study of James J. Gibson's theory of perception*. New York: Praeger.

Bjerring, A. K., and C. A. Hooker [1979], "Process and progress: the nature of systematic inquiry" in Barmark, J. (ed.), *Perspectives in metascience*. Lund: Berlings.

Bjerring, A. K., and C. A. Hooker [1980], "The implications of philosophy of science for science policy" in Hooker, C. A., and T. Schrecker (eds.), *The human context for science and technology*. 2 vols. Ottawa: Social Sciences and Humanities Research Council.

Bjerring, A. K., and C. A. Hooker [1981], "Lehrer, consensus and science: the empiricist watershed" in Bogdan, R. J. (ed.), *Profiles: Keith Lehrer*. Dordrecht: Reidel.

Blake, R. M., C. J. Ducasse, and E. H. Madden [1960], *Theories of scientific method*. Seattle: University of Washington Press.

* Blatt, J. M. [1983], *Dynamic economic systems*. Brighton, Sussex: Wheatsheaf.

Boden, M. [1979], *Piaget*. Brighton: The Harvester Press.

Bogdan, R. J. (ed.) [1976], *Local induction*. Dordrecht: Reidel.

Bohm, D. [1965], *The special theory of relativity*. New York: W. A. Benjamin.

Bonner, J. T. [1982], *Evolution and development*. New York: Springer-Verlag.

Boon, L. [1987], "Variation and selection: scientific progress without rationality" in Callebaut/Pinxten 1987.

* Boulding, K. E. [1981], *Evolutionary economics*. London: Sage.

Bradie, M. [1986], "Assessing evolutionary epistemology." *Biology and Philosophy* 1: 401–59.

Brainerd, C.J. [1978], "The stage question in cognitive-developmental theory." *The Behavioural and Brain Sciences* 2; 173–213.

Brams, S. J. [1976], *Paradoxes in politics*. New York: Free Press.

Brennan, G. [1985], "Economics at the margin: natural and institutional constraints on the acquisition of knowledge." *Search* 16: 17–22.

* Brooks, D. R., and E. O. Wiley [1986], *Evolution as entropy. Toward a unified theory of biology*. Chicago: University of Chicago Press.

Brooks, R. A. [1991], "Intelligence Without Representation." *Artificial Intelligence* 47: 139–159.

Broughton, J., and D.J. Freeman-Moir (eds.) [1981], *The cognitive-developmental psychology of James Mark Baldwin*. Norwood, N.J.: Ablex.

Brown, H. I. [1979], *Perception, theory and commitment: the new philosophy of science*. Chicago: University of Chicago Press.

Brown, H. I. [1987], *Observation and objectivity*. Oxford: Oxford University Press.

Brown, H. I. [1988], *Rationality*. London: Routledge.

Brown, H. I., and C. A. Hooker (with D. Naor) [1994], *Beyond formalism: a naturalist theory of reason*. In preparation.

Bub, J. [1974], *The interpretation of quantum mechanics*. Dordrecht: Reidel.

Burke, T. E. [1983], *The philosophy of Karl Popper*. Manchester: Manchester University Press.

Buss, L. W. [1987], *The evolution of individuality*. Princeton, N.J.: Princeton University Press.

Butts, R.E. [1984], *Kant and the double government methodology*. Dordrecht: Reidel.

Butts, R.E. (ed.) [1986], *Kant's philosophy of physical science*. Dordrecht: Reidel.

* Caglioti, G., H. Haken, and L. Lugiato [1988], *Synergetics and dynamical instabilities*. Amsterdam: North-Holland Physics Publishing.

Cairns, J., J. Overbaugh, and S. Miller [1988], "The origin of mutants." *Nature* 335: 142–5.

Callebaut, W., and R. Pinxton (eds.) [1987], *Evolutionary epistemology: a multiparadigm program*. Dordrecht: Reidel.

Campbell, D. T. [1969], "Ethnocentrism of disciplines and the fish-scale model of omniscience" in Sherif, M., and C.W. Sherif (eds.), *Interdisciplinary relationships in the social sciences*. Chicago: Aldine.

Campbell, D. T. [1974], "Evolutionary epistemology" in Schilpp 1974.

Campbell, D. T. [1977], "Comment on 'The natural selection model of conceptual evolution'." *Philosophy of Science* 44: 502–507.

Campbell, D. T. [1986], "Science's social system of validity-enhancing collective belief change and the problems of the social sciences" in Fiske, D. W., and R. A. Shwader (eds.), *Pluralisms and subjectivities*. Chicago: University of Chicago Press.

Campbell, D. T. [1987], "Selection theory and the sociology of scientific validity" in Callebaut/Pinxton 1987.

* Campbell, J. [1982], *Grammatical man*. New York: Simon and Schuster.

Campbell, N. R. [1957], *Foundations of science*. New York: Dover.

* Caplan, S. R., and A. Essig [1983], *Bioenergetics and linear nonequilibrium thermodynamics*. Cambridge, Mass.: Harvard University Press.

Causey, R. [1977], *Unity of science*. Dordrecht: Reidel.

Cavalli-Sforza, L. L., and M. W. Feldman [1981], *Cultural transmission and evolution*. Princeton: Princeton University Press.

Chapman, M. [1988], *Constructive Evolution: Origins and Development of Piaget's Thought*. Cambridge: Cambridge University Press.

Cherniak, C. [1986], *Minimal rationality*. Cambridge, Mass.: Bradford/MIT Press.

Christensen, W. [1992], "Teleology." Honours thesis, University of Newcastle. Unpublished.

Churchland, P. M. [1979], *Scientific realism and the plasticity of mind*. Cambridge: Cambridge University Press.

Churchland, P.M. [1989], *A neurocomputational perspective*. Cambridge, Mass.: Bradford/MIT Press.

Churchland, P. M., and C. A. Hooker [1985], *Images of science: essays on realism and empiricism*. Chicago: University of Chicago Press.

Churchland, P.S. [1983], "Consciousness: the transmutation of a concept." *Pacific Philosophical Quarterly* 64: 80–95.

Churchland, P.S. [1986], *Neuro-philosophy: toward a unified understanding of the mind-brain*. Cambridge, Mass.: MIT Press.

Churchman, C.W. [1948], *Theory of experimental inference*. New York: MacMillan.

Churchman, C. W. [1968], *Challenge to reason*. New York: McGraw-Hill,

Churchman, C. W. [1972], *The design of enquiring systems*. New York: Basic Books.

Collier, J. [1979], "Information content as a measure of epistemic utility." *Methodology and Science* 12: 206–12.

* Collier, J. [1986], "Entropy in evolution." *Biology and Philosophy* 1: 5–24.

* Collier, J. [1988a], "The dynamics of biological order" in Weber/Depew/Smith 1988.

Collier, J. [1988b], "Supervenience and reduction in biological hierarchies." *Canadian Journal of Philosophy*. Supp. vol.: Biology and Philosophy. Calgary: University of Calgary Press.

* Collier, J. [1992], "Out of equilibrium: new approaches to biological and social change." *Biology and Philosophy* 8, 445–56.

Conant, R. G., and W. R. Ashby [1970], "Every good regulator of a system must be a model of that system." *International Journal of Systems Science* 1: 89–97.

Cragg, C. B. [1989], "Evolution of the steam engine" in Hahlweg/Hooker 1989b.

Cummins, R. [1983], *The nature of psychological explanation*. Boston: Bradford/MIT Press.

Cunningham, M. [1972], *Intelligence: its origin and development*. New York: Academic Press.

* Cvitanovic, P. (ed.) [1984], *Universality in chaos*. Bristol: Hilger.

* Depew, D., and B. Weber [1985], *Evolution at a crossroads*. Cambridge, Mass.: MIT Press.

Devitt, M. [1984], *Realism and truth*. London: Blackwell.

Dewan, E. M. [1976], "Consciousness as an emergent causal agent in the concept of control system theory" in Globus, G, G. Maxwell, and I. Savodnik (eds.), *Consciousness and the brain*. New York: Plenum.

* Diner, S., D. Fargue and G. Lochak [1986], *Dynamical systems: a renewal of mechanism*. Singapore: World Scientific.

Dooley, J. [1993], "Piaget, self-organising knowledge, and critical systems practice." *Systems Practice* 6: 359–81.

Dreyfus, H. L., and S. E. Dreyfus [1986], *Mind over machine*. Oxford: Blackwell.

Duhem, P. [1962], *The aim and structure of physical theory*. New York: Atheneum.

* Dyke, C. [1988], *The evolutionary dynamics of complex systems*. Oxford: University Press.

Easlea, B. [1973], *Liberation and the aims of science*. London: Chatto and Windus for Sussex University Press.

Edelman, G. M. [1989], *Neural Darwinism*, Oxford: Oxford University Press.

* Eigen, M., and R. Winkler [1981], *The laws of the game: how the principles of nature govern chance* (trans. by R. Kimber and R. Kimber). New York: Knopf.

* Ekeland, I. [1988], *Mathematics and the unexpected*. Chicago: University of Chicago Press.

Ellis, B. [1985], "What science aims to do" in Churchland/Hooker 1985.

Elster, J. [1983], *Sour grapes: studies in the subversion of rationality*. Cambridge: Cambridge University Press.

Elster, J. (ed.) [1986], *Rational choice*. Oxford: Blackwell.

Espejo, R., and R. Harnden (eds.) [1989], *The viable system model: interpretations and applications of Stafford Beer's VSM*. Chichester: Wiley.

Etzioni, A. [1961], *A comparative analysis of complex organizations: on power, involvement, and their correlates*. New York: Free Press.

Etzioni, A. [1967], "Mixed scanning: a third approach to decision making." *Public Administration Review* 24: 389.

Feldman, R. [1988], "Rationality, reliability and natural selection." *Philosophy of Science* 55: 218-27.

Feyerabend, P. K. [1961], "Knowledge without foundations." Oberlin College (mimeographed).

Feyerabend, P. K. [1965a], "Problems of empiricism" in Colodny, R. (ed.), *Pittsburgh Studies in the Philosophy of Science*. Vol. 2. New Jersey: Prentice-Hall.

Feyerabend, P. K. [1965b], "Reply to criticism" in Cohen, R. S., and M. W. Wartofsky (eds.), *Boston Studies in the Philosophy of Science*. Vol. 2. New York: Humanities Press.

Feyerabend, P. K. [1969], "Problems of empiricism II" in Colodny, R. (ed.), *Pittsburgh Studies in the Philosophy of Science*. Vol. 4. Pittsburgh: University of Pittsburgh Press.

Feyerabend, P. K. [1974], "Popper's objective knowledge." *Inquiry* 17: 475–507.

Feyerabend, P.K. [1978a], *Against method*. London: Verso.

Feyerabend, P. K. [1978b], *Science in a free society*. London: New Left Books.

Feyerabend, P. K. [1987], *Farewell to reason*. London: Verso.

* Firrao, S. [1983], *The theory of self-organizing systems in physics, biology and psychology*. Milano: CENS.

Flavell, J.H. [1963], *The developmental psychology of Jean Piaget*. New York: Van Nostrand.

Fodor, J.A. [1980], "On the impossibility of acquiring 'more powerful' structures" in Piatelli-Palmarini 1980.

* Foster, J. [1987], *Evolutionary macroeconomics*. London: Allen and Unwin.

Franklin, A. [1986], *The neglect of experiment*. Cambridge: Cambridge University Press.

* Frehland, E. (ed.) [1984], *Synergetics—from microscopic to macroscopic order*. New York: Springer-Verlag.

Friedman, M. [1986], "The metaphysical foundations of Newtonian science" in Butts 1986.

Galison, P. [1987], *How experiments end*. Chicago: University of Chicago Press.

Garfinkel, A. [1987], "The slime mould *Dictyostelium* as a model of self-organisation in social systems" in Yates 1987.

Giere, R. [1979], *Understanding scientific reasoning*. New York: Holt, Rinehart and Winston.

Giere, R. [1988], *Explaining science: a cognitive approach*. Chicago: University of Chicago Press.

Gilbert, S. F. [1992], "Cells in search of community: critiques of Weismannism and selectable units in ontogeny." *Biology and philosophy* 7; 473–87.

Gillièron, C. [1986], "Is Piaget's 'genetic epistemology' evolutionary?" in Callebaut/Pinxton 1986.

* Glass, L., and M. Mackey [1988], *From clocks to chaos*. Princeton: University Press.

Glastone, S., and R. H. Lovberg [1960], *Controlled thermonuclear reactions*. New York: Van Nostrand.

Goldman, A. [1986], *Epistemology and cognition*. Cambridge, Mass.: Harvard University Press.

Goodwin, B. C., N. Holder, and C. C. Wylie (eds.) [1983], *Development and evolution*. Cambridge: Cambridge University Press.

Goodwin, B., and P. Saunders [1989], *Theoretical biology: epigenetic and evolutionary order from complex systems*. Edinburgh: Edinburgh University Press.

Gould, S.J. [1977], *Ontogeny and philogeny*. Cambridge, Mass.: Harvard/Belknap.

Gould, S.J. [1989], *Wonderful life: the Burgess shale and the nature of history*. New York: W. W. Norton.

Gray, R. D. [1992], "Death of the gene: developmental systems strike back," in Griffiths, P. E. (ed.), *Trees of life: essays in philosophy of biology*. Dordrecht: Kluwer.

Gruber, G. E., and J. J. Voneche [1977], *The essential Piaget*. New York: Basic Books.

Grünbaum, A. [1976a], "Ad hoc auxiliary hypotheses and falsificationism." British Journal for the Philosophy of Science 27: 329-62.

Grünbaum, A. [1976b], "Can a theory answer more questions than one of its rivals?" *British Journal for the Philosophy of Science* 27; 1–23.

Grünbaum, A. [1976c], "Is the method of bold conjectures and attempted refutations justifiably the method of science?" *British Journal for the Philosophy of Science* 27: 105–6.

* Guckenheimer, J., and P. Holmes [1983], *Nonlinear oscillations, dynamical systems and bifurcations of vector fields*. New York: Springer-Verlag.

Hacking, I. [1983], *Representing and intervening*. Cambridge: Cambridge University Press.

Hahlweg, K. [1983], "The evolution of science: a systems approach." Ph.D. thesis. University of Western Ontario, Canada. Unpublished.

Hahlweg, K. [1991], "On the notion of evolutionary progress." *Philosophy of science* 58: 436–51.

Hahlweg, K., and C. A. Hooker [1988] "Evolutionary epistemology and relativism" in Nola, R. (ed.), *Epistemology and relativism*. Boston: Kluwer (Reidel).

Hahlweg, K., and C. A. Hooker [1989a], "Evolutionary epistemology and philosophy of science" in Hahlweg/Hooker 1989b.

Hahlweg, K., and C. A. Hooker [1989b], *Issues in evolutionary epistemology*. Albany N.Y.: State University of New York Press.

* Haken, H. [1973], *Synergetics: Cooperative phenomena in multi-component systems*. Stuttgart: B. G. Teubner.

* Haken, H. [1977a], *Synergetics: an introduction*. Berlin: Springer-Verlag.

* Haken, H. [1977b], *Synergetics: a workshop*. Berlin: Springer-Verlag.

* Hannan, M. T. and J. Freeman [1989], *Organisational ecology*. Cambridge, Mass.: Harvard University Press.

* Hao B.-L. [1984], *Chaos*. Singapore: World Scientific.

Haroutunian, S. [1983], *Equilibrium in the balance*. New York: Springer-Verlag.

Harper, W. [1977], "Rational belief change, Popper functions and counterfactuals" in Hooker et al. [1978].

Harper, W. [1989], "Consilience and natural kinds reasoning (in Newton's argument for universal gravitation)" in Brown, J. R., and O. Mittelstrass (eds.), *An intimate relation*. Boston: Kluwer.

Harper, W. [1993], "Reasoning from phenomena: Newton's argument for universal gravitation and the practice of science. in Theerman, P. and A. F. Seeff (eds), *Action and reaction: Proceedings of a symposium to commemorate the tercentenary of Newton's Principia*. Newark, Del.: University of Delaware Press.

Harper, W., and C. A Hooker (eds.) [1976], *Foundations of probability theory, statistical inference and statistical theories of* science. 3 vols. Dordrecht: Reidel.

Hodges, B. E. [1990], "Two problems for one Popper." Honors Thesis, University of Newcastle. Unpublished.

Holdsworth, D. G., and C. A. Hooker [1983], "A critical survey of quantum logic." *Scientia, Logic in the Twentieth Century:* 1–130.

Holland, A., and A. O'Hear [1984], "On what makes an epistemology evolutionary." *Proceedings of the Aristotelian Society*. Supp. vol. LVIII: 177–217.

* Holland, J. H. [1992], *Adaptation in natural and artificial systems*. Cambridge, Mass.: MIT/Bradford.

* Holland, J. H., K. L. Holyoak, R. E. Nisbett and P. R. Thagard [1986], *Induction*. Cambridge, Mass.: MIT Press.

Hooker, C. A. [1975], "The philosophical ramifications of the information-processing approach to the brain-mind." *Philosophy and Phenomenological Research* 36: 1–15.

Hooker, C. A. [1978], "An evolutionary naturalist realist doctrine of perception and a nihilist doctrine of the secondary qualities," in C.W. Savage (ed.), *Perception and cognition: issues in the foundations of psychology*. Minneapolis: University of Minnesota Press.

Hooker, C. A. [1980], "Explanation, generality and understanding." *Australasian Journal of Philosophy* 58: 284–290.

Hooker, C. A. [1981a], "Formalist rationality: the limitations of Popper's theory of reason." *Metaphilosophy* 12: 247–266.

Hooker, C. A. [1981b], "Towards a general theory of reduction." *Dialogue* XX: Part I, Historical framework, 38–59, Part II, Identity and reduction, 201–36, Part III, Cross-categorial reduction, 496–529.

Hooker, C. A. [1984], "Review of Prigogine, I. *Being and becoming.*" *Australasian Journal of Philosophy* 51: 355–7.

Hooker, C. A. [1985], "Surface dazzle, ghostly depths" in Churchland/Hooker 1985.

Hooker, C. A. [1987], *A realistic theory of science*. Albany, N.Y.: State University of New York Press.

Hooker, C. A. [1988], "Critical Notice: Churchland, P.S., *Neurophilosophy.*" *Australasian Journal of Philosophy* 66: 240–8.

Hooker, C. A. [1989a], "Bell, book and candle." in Kafatos, M. (ed.), *Bell's theorem, quantum theory and conceptions of the universe*. Boston: Kluwer.

Hooker, C. A. [1989b] "From logical formalism to control system" in Fine, A., and M. Forbes (eds.), *PSA 1988*. East Lansing: Philosophy of Science Association.

Hooker, C. A. [1989c], "Towards a philosophy and practice of energy policy making." *The McMaster Institute of Energy Studies Journal* 1: 130–43.

Hooker, C. A. [1991a], "Between formalism and anarchism: a reasonable middle way" in Munevar, G. (ed.), *Beyond reason: essays on the philosophy of Paul Feyerabend*. Boston: Kluwer..

Hooker, C. A. [1991b], "Discussion review of Brown, H. R. and Harré, R. (eds.), *Philosophical foundations of quantum field theory.*" *Philosophy of Science* 58: 324–9.

Hooker, C. A. [1991c], "Responsibility for the environment: a systematic perspective" in Cooper, D. E., and J. A. Palmer (eds.), *The environment in question: ethics and global issues*. London: Routledge.

Hooker, C. A. [1992] "Physical intelligibility, projection, objectivity and completeness: the divergent ideals of Bohr and Einstein." *British Journal for the Philosophy of Science* 42: 491–511.

Hooker, C. A. [1994a], "Bohr and the crisis of empirical intelligibility" in Faye, J., and H. Folse (eds.), *Niels Bohr and contemporary philosophy*. Boston: Kluwer.

Hooker, C. A. [1994b], "From Phenomena ro Metaphysics," in Prawitz, D., and D. Westerstáhl (eds.) *Logic and Philosophy of Science in Uppsala*. Dordrecht: Reidel.

Hooker, C. A. [1994c], "Idealism, naturalism and rationality: Some lessons from minimal rationality." *Synthese* 99: 181–231.

Hooker, C. A. [1994d], "Piaget's psychology, biology and evolutionary epistemology: a regulatory systems approach." *Biology and Philosophy* 9: 197–244.

Hooker, C. A. [1994e] "Toward a naturalised cognitive science" in Kitchener, R., and W. O'Donohue (eds), *Psychology and philosophy*. London: Allyn and Bacon.

Hooker, C. A. [1994f], "Value and System: Notes toward the definition of agri- culture", *Journal of Agricultural and Environmental Ethics*, 7 special Supplementary vol.

Hooker, C. A. (ed.) [1973], *Contemporary research in the foundations and philosophy of quantum theory*. Dordrecht: Reidel.

Hooker, C. A. (ed.) [1975/79], *The logico-algebraic approach to quantum mechanics*. 2 vols. Dordrecht: Reidel.

Hooker, C. A. (ed.) [1979], *Physical theory as logico-operational structure*. Dordrecht: Reidel.

Hooker, C. A., J. Leach, E. McCLennen (eds.) [1978], *Foundations and applications of decision theory*. 2 vols. Dordrecht: Reidel.

Hooker, C. A., H. B. Penfold, and R. J. Evans [1992a], "Cognition under a new control paradigm." *Topoi* 11: 71–88.

Hooker, C. A., H. B. Penfold, and R. J. Evans [1992b], "Control, connectionism and cognition: toward a new regulatory paradigm." *British Journal for the Philosophy of Science* 43: 517–36.

Hull, D. L. [1984], "Historical entities and historical narratives" in Hookway, C. (ed.), *Minds, machines and evolution*. Cambridge, Mass.: Cambridge University Press.

Hull, D. L. [1988a], "A mechanism and its metaphysics: an evolutionary account of the social and conceptual development of science." *Biology and Philosophy* 3: 123–55.

Hull, D. L. [1988b], *Science as a process*. Chicago: University of Chicago Press.

* Iberall, A. S. [1972], *Toward a general science of viable systems*. New York: McGraw-Hill.

* Jantsch, E. [1975], *Design for evolution*. New York: George Braziller.

* Jantsch, E. [1980], *The self-organizing universe*. New York: Pergamon Press.

* Jantsch, E. (ed.) [1981], *The evolutionary vision: toward a unifying paradigm of physical, biological and sociocultural evolution*. Boulder, Colo.: Westview.

* Jantsch, E., and C. H. and Waddington [1976], *Evolution and consciousness: human systems in transition*. London: Addison-Wesley.

* Kauffman, S. A. [1993], *Origins of order: self-organisation and selection in evolution*. New York: Oxford University Press.

Kingman, J. F. C., and S. J. Taylor [1966], *Introduction to measure and probability*. Cambridge: Cambridge University Press.

Kitchener, K. S., and R. F. Kitchener [1981], "The development of natural rationality: can formal operations account for it?" *Contributions to Human Development* 5: 160-81. (Reprinted in Meacham, J., and M. R. Santilli (eds.), *Social development in youth*. New York: Karger.)

Kitchener, R.F. [1986], *Piaget's theory of knowledge*. New Haven: Yale University Press.

Kitchener, R.F. [1987], "Genetic epistemology, equilibration and the rationality of scientific change." *Studies in History and Philosophy of Science* 18: 339–66 .

Kitchener, R.F. [1988], "Is genetic epistemology possible?" *British Journal for the Philosophy of Science* 38: 283–99.

Kitchener, R.F. [1989], "Genetic epistemology and the prospects for a cognitive sociology of science: a critical synthesis." *Social Epistemology* 3: 153–69.

Kornblith, H. [1993], *Inductive inference and its natural ground, an essay in naturalistic epistemology*. Boston, Bradford/MIT.

* Krcevinac, S. (ed.) [1981], *Global modeling*. Lecture notes in control and information science. Berlin: Springer-Verlag.

* Krohn, and G. Kuppers [1989], "Self-organisation: a new approach to evolutionary epistemology" in Hahlweg/Hooker 1989.

Kuhn, T. S. [1962], *The structure of scientific revolutions*. Chicago: University of Chicago Press. (Also in Neurath et al. [1970].)

* Kurzhanski, A. B., and K. Sigmund [1985], *Dynamical systems*. Lecture notes in economics and mathematical systems. Berlin: Springer-Verlag.

Lakatos, I. [1970], "Falsification and the methodology of scientific research programmes" in Lakatos, I. and A. Musgrave (eds.), *Criticism and the growth of knowledge*. Cambridge: Cambridge University Press.

Latour, B. [1987], *Science in action*. Cambridge, Mass.: Harvard University Press.

Latour, B., and S. Woolgar [1979], *Laboratory life: the social construction of scientific facts*. Beverley Hills: Sage.

Laudan, L. [1977], *Progress and its problems*. Berkeley: University of California Press.

Laudan, L. [1984], *Science and value*. Berkeley: University of California Press.

Levi, I. [1967], *Gambling with truth*. New York: A. A. Knopf.

Levi, I. [1980], *The enterprise of knowledge*. Cambridge: MIT Press.

Levins, R. [1979], "Co-existence in a variable environment." *American Naturalist* 114: 765–83.

Lewontin, R. C. [1982], "Organism and environment" in Plotkin 1982.

Lloyd, D. [1989], *Simple minds*. Cambridge, Mass.: Bradford/MIT Press.

Lorenz, K. [1977], *Behind the mirror: a search for a natural history of human knowledge* (translated by R. Taylor). London: Methuen.

Maes, P. [1990], *Desiging autonomous agents: theory and practice from biology to engineering and back*. Amsterdam: Elsevier.

Margalef, R. [1968], *Perspectives in ecological theory*. Chicago: University of Chicago Press.

Maturana, H., and F. Varela [1980], *Autopoiesis and cognition*. Dordrecht: Reidel.

McClennen, E. [1990], *Rationality and dynamic choice: foundational explorations*. New York: Cambridge University Press.

Merton, R. K. [1972], "The institutional imperatives of science," in Barnes, B. (ed.), *Sociology of science*. Harmondsworth, Middlesex: Penguin.

Miller, D. [1974], "Popper's qualitative theory of verisimilitude." *British Journal of the Philosophy of Science* 25 : 166–77.

Miller, D. [1975], "The accuracy of predictions." *Synthese* 30: 139–148.

Mitchell, M. [1993], *Analogy-making as perception: A computer model*. Cambridge, Mass.: Bradford/MIT Press.

Mitroff, I. I. [1974], *The subjective side of science*. Amsterdam: Elsevier.

Monod, J. [1972], *Chance and necessity*. New York: Vintage.

Morowitz, H. J. [1968], *Energy flow in biology: biological organization as a problem in thermal physics*. New York: Academic Press.

Morowitz, H. J. [1986], "Entropy and nonsense." *Biology and Philosophy* 1: 473–6.

Munevar, G. [1989], "Science as part of nature" in Hahlweg/Hooker 1989b.

Munz, P. [1989], "Taking Darwin even more seriously" in Hahlweg/Hooker 1989b.

Naor, D. [1979], "A viable system model of scientific rationality." Unpublished manuscript. The University of Western Ontario, London, Canada.

Naor, D. [1980], "What Sir Karl does not learn from Dr. Popper." Unpublished manuscript. The University of Western Ontario, London, Canada.

Newton-Smith, W. H. [1981], *The rationality of science*. London: Routledge and Kegan Paul.

Nicholas, J. M. [1984], "Scientific and other interests" in Brown, J. R. (ed.), *Scientific rationality: the sociological turn*. Dordrecht: Reidel.

Nickles, T. [1980a], *Scientific discovery: case studies*. Dordrecht: Reidel.

Nickles, T. [1980b], *Scientific discovery, logic and rationality*. Dordrecht: Reidel.

* Nicolis, J.S. [1986], *Dynamics of hierarchical systems: an evolutionary approach*. Berlin: Springer-Verlag.

Niiniluoto, I. [1984], *Is science progressive*? Dordrecht: Reidel.

Niiniluoto, I. [1987], *Truthlikeness*. Dordrecht: Reidel.

Nitecki, M. H. (ed.) [1988], *Evolutionary progress*. Chicago: University of Chicago Press.

Oddie, G. [1986], *Likeness to truth*. Dordrecht: Reidel.

Odum, H.T. [1971], *Environment, Power and Society*. New York: Wiley-Interscience.

O'Hear, A. [1980], *Karl Popper*. London: Routledge and Kegan Paul.

Oldroyd, D. [1989], *The arch of knowledge*. Kensington, N.S.W.: University of NSW Press.

* Pagels, H. [1988], *The dreams of reason*. New York: Simon and Schuster.

* Pattee, H. H. [1973], *Hierarchy theory: the challenge of complex systems*. New York: Geo. Braziller.

* Peacocke, A. R. [1983], *An introduction to the physical chemistry of biological organization*. Oxford: Clarendon Press.

Penrose, R. [1989], *The emperor's new mind*. Oxford: Oxford University Press.

Piaget, J. [1924], "L'experience humaine et la causalite physique de L. Brunschvicg," *Journal de Psychologie* 21: 586–607.

Piaget, J. [1933/77], "L'individualité en histoire: l'individu et la formation de la raison," in Piaget, J., *Etudes sociologiques*, 2nd ed. Geneva: Droz.

Piaget, J. [1950], *Introduction á l'épistémologie génétique*. 3 vols. Paris: Presses Universitaires de France.

Piaget, J. [1955], "Les lignes generales de l'épistémologie génétique." *Actes du IIe Congres de L'Union Internationale de Philosophie des Sciences*. Neuchatel: Editions du Griffon.

Piaget, J. [1963/68], "Explanation in psychology and psychophysiological parallelism" in Fraisse P. and Piaget J. (eds.), *Experimental psychology: its scope and method*. Vol. I (trans. J. Chambers). New York: Basic Books.

Piaget, J. [1965/71], *Insights and illusions of philosophy* (trans. W. Mays). New York: Meridian Books.

Piaget, J. [1967/71], *Biology and knowledge: an essay on the relations between organic regulations and cognitive processes* (trans. B. Walsh). Chicago: University of Chicago Press.

Piaget, J. [1970], *Genetic epistemology*. New York: Columbia University Press.

Piaget, J. [1970/71], *Psychology and epistemology* (trans. A. Rosin). New York: Viking.

Piaget, J. [1970/72], *The principles of genetic epistemology* (trans. W. Mays). London: Routledge & Kegan Paul.

Piaget, J. [1970/73], *Main trends in psychology*. New York: Harper.

Piaget, J. [1971], *Genetic epistemology*, New York: W.W. Norton.

Piaget, J. [1971/77], "Chance and dialectic in biological epistemology: a critical analysis of Jacques Monod's thesis" in Overton, W.F., and J. McCarthy Gallaher (eds.), *Knowledge and development*, Vol. 1. New York: Plenum.

Piaget, J. [1974/76], *The grasp of consciousness* (trans. S. Wedgwood). Cambridge, Mass.: Harvard University Press.

Piaget, J. [1974/78], *Success and understanding* (trans. A. J. Pomerans). Cambridge, Mass.: Harvard University Press.

Piaget, J. [1974/80], *Adaptation and intelligence: organic selection and phenocopy* (trans. S. Eames). Chicago: University of Chicago Press.

Piaget, J. [1975/77], *The development of thought: equilibration of cognitive structures* (trans. A. Rosin). New York: Viking.

Piaget, J. [1976], "The possible, the impossible and the necessary" (trans. G. Voyat). *The Genetic Epistemologist* 6: 1–12.

Piaget, J. [1976/78], *Behaviour and evolution* (trans. D. Nicholson-Smith). New York: Pantheon.

Piaget, J. [1978], "What is psychology?" *American Psychologist* 33: 648–52.

Piaget, J. [1980], "The psychogenesis of knowledge and its epistemological significance" in Piattelli-Palmarini 1980.

Piaget, J. and R. Garcia [1983/89], *Psychogenesis and the history of science* (trans. H. Feider). New York: Columbia University Press.

Piaget, J. et al. (eds.) [1967], *Logique et connaissance scientifique.* Paris: Gallimard.

Piatelli-Palmarini, M. (ed.) [1980], *Language and learning: the debate between Jean Piaget and Noam Chomsky.* Cambridge, Mass.: Harvard University Press.

* Pines, David [1988], *Emerging synthesis in science.* Redwood City, Calif.: Addison-Wesley.

Plotkin, H. C. (ed.) [1982], *Learning development and culture.* New York: Wiley.

Polanyi, K. [1968], *The great transformation.* Beacon Press: Boston.

Popper, K. R. [1957], *The poverty of historicism.* London: Routledge and Kegan Paul.

Popper, K. R. [1966], *The open society and its enemies.* 2 vols. London: Routledge and Kegan Paul.

Popper, K. R. [1967], "La rationalité et la statut du principe de rationalité," in Claassen, E. M. (ed.), *Les Fondements Philosophiques des Systemes Economiques.* Paris: Payot.

Popper, K. R. [1972], *Conjectures and refutations.* London: Routledge and Kegan Paul.

Popper, K. R. [1974], "Replies to my critics" in Schilpp 1974.

Popper, K. R. [1975], "The rationality of scientific revolutions," in R. Harré (ed), *Problems of scientific revolution.* Chichester: Wiley.

Popper, K. R. [1976], "The logic of the social sciences" in T. W. Adorno, H. Albert, R. Dahrendorf, J. Habermas, H. Pilot, and K. R. Popper, *The positivist dispute in German sociology* (translated by G. Adey and D. Frisby). London: Heinemann.

Popper, K. R. [1979], *Objective knowledge: an evolutionary approach* (rev. ed.). Oxford: Oxford University Press.

Popper, K. R. [1980], *The logic of scientific discovery.* London: Hutchinson. (First published as *Logik der Forschung*, Wien, 1934.)

Popper, K. R. [1984], "Evolutionary epistemology," in Pollard, J. W., *Evolutionary theory: paths into the future.* Chichester: Wiley.

Popper, K. R. [1987], "Natural selection and the emergence of mind," in Radnitzky, G., and W. W. Bartley III (eds.), *Evolutionary epistemology, theory of rationality and sociology of knowledge.* La Salle: Open Court.

Popper, K. R., and J. Eccles [1977], *The self and its brain.* New York: Springer- Verlag.

Powers, W. T. [1973], *Behaviour, the control of perception*. London: Wildwood House.

* Prigogine, I. [1980], *From being to becoming*. San Francisco: W. H. Freeman.

* Prigogine, I. and I. Stengers [1984], *Order out of chaos*. Boulder, Colorado: Shambhala.

Putnam, H. [1981], *Reason, truth and history*. Cambridge Mass.: Cambridge University Press.

Putnam, H. [1982], "Why reason can't be naturalised." *Synthese* 52: 3–23.

Quine, W. V. O. [1963], *From a logical point of view*. New York: Harper and Rowe.

Quine, W. V. O. [1969], "Epistemology naturalised" in Quine, W. V. O., *Ontological relativity and other essays*. New York: Columbia University Press.

Raff, R.A. and T. C. Kaufman [1983], *Embryos, genes, and evolution*. New York: Macmillan.

Rescher, N. [1977], *Methodological pragmatism*. London: Blackwell.

Rescher, N. [1990], *A useful inheritance*, Savage, Maryland: Rowan and Littlefield.

Richards, R. J. [1977], "The natural selection model of conceptual evolution," *Philosophy of Science* 44: 494–501.

Rips, L. [1990], "Reasoning," *Annual Review of Psychology*, 41: 321–53.

* Rosen, R. [1985], *Anticipatory systems: Philosophical, mathematical, and methodological foundations*. New York: Pergamon.

* Rosen, R. [1991], *Life itself: a comprehensive inquiry into the nature, origin and fabrication of life*. New York: Columbia University Press.

Rosen, R. (ed.) [1985], *Theoretical biology and complexity*. Orlando: Academic Press.

Ruse, M. [1982], *Darwinism defended: a guide to evolutionary controversies*. New York: Addison-Wesley.

Ruse, M. [1983], "Darwin and philosophy today," in Oldroyd, D., and Langham, I. (eds.), *The wider domain of evolutionary thought*. Dordrecht: Reidel.

Ruse, M. [1986], *Taking Darwin seriously*. Oxford: Blackwell.

Russell, B. [1959], *The problems of philosophy*. Oxford: Oxford University Press.

* Salthe, S. N. [1985], *Evolving hierarchical systems: their structure and representation*. New York: Columbia University Press.

* Salthe, S. N. [1989], "Self-organisation of/in hierarchically structured systems." Systems Research 6: 199–208.

Salthe, S. N. [1993], *Development and Evolution: Complexity and Change in Biology*. Cambridge, Mass.: Bradford/MIT Press.

Sarkar, S. [1990], "On the possibility of directed mutations in bacteria: statistical analyses and reductionist strategies," in Fine, A., Forbes, M., and Wessels, L. (eds.), *PSA 1990*. East Lansing, Mich.: Philosophy of Science Association.

* Saviotti, P. P. and S. Metcalfe (eds.) [1991], *Evolutionary theories of economic and technological change*. London: Hartwood Press.

Schilpp, P. A. (ed.) [1974], *The philosophy of Karl Popper*. La Salle: Open Court.

* Serra, R., G. Zanarini, M. Andretti and M. Compiani [1986]. *Introduction to the physics of complex systems*. Oxford: Pergamon Press.

Serra, R., and G. Zanarini [1990], *Complex Systems and Cognitive Processes*. New York: Springer.

Shapere, D. [1984], *Reason and the Search for Knowledge*. Dordrecht: Reidel.

Siegel, L. and Brainerd C. (eds) [1978], *Alternatives to Piaget: critical essays on the theory*. New York: Academic Press.

Simon, H. [1969], *The sciences of the artificial*. Cambridge, Mass.: MIT.

Skinner, B. F. [1953], *Science and human behaviour*. New York: MacMillan.

Slote, M. [1989], *Beyond Optimizing*. Cambridge, Mass.: Harvard University Press.

* Smale, S. [1980], *The mathematics of time*. New York: Springer-Verlag.

Smith, K. C. [1992], "Neo-rationalism versus neo-Darwinism: integrating development and evolution." *Biology and Philosophy* 7: 431–51.

Smolensky, P. [1988], "On the proper treatment of connectionism." *Behavioural and brain sciences* 11: 1–74.

Sober, E. [1981], "The evolution of rationality." *Synthese* 46: 95–120.

Sober, E. [1984], *The Nature of Selection*. Cambridge, Mass.: MIT Press.

Solomon, M. [1992], "Scientific rationality and human reasoning." *Philosophy of science* 59: 439–55.

Sommerhof, G. [1974], *The logic of the living brain*. London: Wiley.

Sperry, R. [1983], *Science and moral priority: merging mind, brain and human values*. New York: Columbia University Press.

Stalnaker, R. [1984], *Inquiry*. Cambridge, Mass.: MIT/Bradford Press.

* Stear, E. B. [1987], "Control paradigms and self-organisation in living systems" in Yates 1987.

* Steeb, W.-H., and J. A. Louw [1986], *Chaos and quantum chaos*. Singapore: World Scientific.

Stein, E., and P. Lipton [1989], "Where guesses come from: evolutionary epistemology and the anomaly of guided variation." *Biology and philosophy* 4: 33–56.

Stein, H. [1990], "From the phenomena of motions to the forces of nature: hypothesis or deduction?" in Fine A., M. Forbes and L. Wessels (eds.), *PSA 1990,* vol. 2.

* Stewart, I. [1989], *Does God play dice? The mathematics of chaos*. Oxford: Basil Blackwell.

Stich, S. [1985], "Could man be an irrational animal?" *Synthese* 64: 115–35.

Stich, S. [1989], *The fragmentation of reason*. Cambridge, Mass.: Bradford/MIT Press.

Stokes, G. [1989], "From physics to biology: rationality in Popper's conception of evolutionary epistemology" in Hahlweg/Hooker 1989b.

Streater, R. F. and A. S. Wightman [1964], *PCT, spin, statistics and all that*. New York: Benjamin.

Suppe, F. (ed.) [1974], *The structure of scientific theories*. Urbana, Ill.: University of Illinois Press.

Tarski, A. [1956], *Logic, semantics, meta-mathematics* (transl. J.H. Woodger). Oxford: Oxford University Press.

Tart, C. T. [1972], *Altered states of consciousness*. New York: Doubleday.

Tauber, A. I. (ed.) [1991], *Organism and the origin of self*. Dordrecht: Kluwer.

Tawney, R. H. [1962], *Religion and the rise of capitalism*. London: Penguin.

Thagard, P. [1988], *Computational philosophy of science*. Cambridge, Mass.: MIT Press.

* Thomas, R. (ed.) [1977], *Kinetic logic: A Boolean approach to the analysis of complex regulatory systems*. Lecture notes in biomathematics, no.29. Berlin: Springer-Verlag.

* Thompson, J. M., and H. B. Stewart [1986], *Nonlinear dynamics and chaos*. Chichester: Wiley.

Toulmin, S. [1972], *Human understanding*. Princeton, N.J.: Princeton University Press.

* Trappl, R. (ed.) [1986], *Power, autonomy, utopia: new approaches toward complex systems*. New York: Plenum.

Trigg, R. [1980], *Reality at risk: a defense of realism in philosophy and the sciences*. Brighton, Sussex: Harvester Press.

* Ulanowicz, R. [1986], *Growth and development: ecosystems phenomenology*. New York: Springer-Verlag.

Ullman-Margalit, E. [1977], *The emergence of norms*. Oxford: Clarendon Press.

Van Fraassen, B. C. [1980], *The scientific image*. Oxford: Clarendon Press.

Varela, F. [1979], *Principles of biological autonomy*. New York: Elsevier.

Vickers, G. [1968], *Value systems and social process*. London: Penguin.

Vickers, G. [1970], *Freedom in a rocking boat*. London: Penguin.

Vickers, G. [1980], *Responsibility: its sources and limits*. Seaside, Cal.: Intersystems Publications.

Vickers, G. [1983], *Human systems are different*. London: Harper and Row.

Vuyk, R. [1981], *Overview and critique of Piaget's genetic epistemology* 1965–80. New York: Academic Press.

Waddington, C. H. [1957], *The strategy of the genes*. London: Allen and Unwin.

Waddington, C. H. [1966], *Principles of development and differentiation*. New York: MacMillan.

Waddington, C. H. [1975], *The Evolution of an evolutionist*. Edinburgh: Edinburgh University Press.

Wade, M. [1978], "A critical review of the models of group selection." *Quarterly Review of Biology* 53: 101–14.

Ward, P. J. [1989], 'A model of end-directedness' in Hahlweg/Hooker 1989b.

* Weber, B., D. Depew and J. Smith (eds.) [1988], *Entropy, information, and evolution*. Cambridge, Mass.: MIT Press.

* Wesson, R. [1993], *Beyond natural selection*. Cambridge, Mass.: MIT Press.

Wilson, C.A. [1970], "From Kepler's laws, so-called, to universal gravitation: empirical factors." *Archive for the history of the exact sciences* 6: 89–170.

Wimsatt, W. C., and J. C. Schank [1987], "Generative entrenchment and evolution," in Fine, A., and Machamer, P. (eds.), *PSA 1986*. East Lansing, Mich.: Philosophy of Science Association.

Wimsatt, W. C., and J. C. Schank [1988], "Two constraints on the evolution of complex adaptations and the means of their avoidance," in Nitecki 1988.

*Winfree, A. [1987], *When time breaks down*. Princeton: Princeton University Press.

Wuketits, F. (ed.) [1983], *Concepts and approaches in evolutionary epistemology*. Dordrecht: Reidel.

*Yates, F. E. (ed.) [1987], *Self-organising systems: the emergence of order*. New York: Plenum.

*Zurek, W. H. (ed.) [1990], *Complexity, entropy and the physics of information*. Redwood City, Cal.: Addison-Wesley.

Name Index

Subject Index